Dahlem Workshop Reports
Physical, Chemical, and Earth Sciences Research Report 8
The Environmental Record in Glaciers and Ice Sheets

Goal of this Dahlem Workshop:
to assess and interpret the
environmental record in glaciers

Physical, Chemical, and Earth Sciences Research Reports

Series Editor: Silke Bernhard

Held and published on behalf of the

Stifterverband für die Deutsche Wissenschaft

Sponsored by:

Senat der Stadt Berlin
Stifterverband für die Deutsche Wissenschaft
Stiftung Winterling Marktleuthen
im Stifterverband für die Deutsche Wissenschaft

The Environmental Record in Glaciers and Ice Sheets

H. Oeschger and C. C. Langway, Jr., Editors

Report of the Dahlem Workshop on
The Environmental Record in Glaciers and Ice Sheets
Berlin 1988, March 13–18

Rapporteurs:
W. F. Budd, A. D. Hecht, G. I. Pearman,
J. W. C. White

Program Advisory Committee:
H. Oeschger and C. C. Langway, Jr., Chairpersons
P. J. Crutzen, W. Dansgaard, R. J. Delmas,
K. O. Münnich

A Wiley–Interscience Publication

John Wiley & Sons 1989
Chichester · New York · Brisbane · Toronto · Singapore

Copy Editors: J. Lupp, K. Klotzle

Photographs: E. P. Thonke

With 5 photographs, 92 figures and 29 tables

Library of Congress Cataloging-in-Publication Data

Dahlem Workshop on the Environmental Record in Glaciers and Ice Sheets
(1988 : Berlin, Germany)
The environmental record in glaciers and ice sheets : report of the Dahlem Workshop on the Environmental Record in Glaciers and Ice Sheets, Berlin, 1988 March 13–18 / H. Oeschger and C.C. Langway, Jr., editors; rapporteurs, W.F. Budd . . . [et al.].
p. cm. — (Dahlem workshop reports) (Physical, chemical, and earth sciences research report ; 8)
'A Wiley–Interscience publication.'
Includes bibliographies and indexes.
ISBN 0 471 92185 8
1. Paleoecology—Congresses. 2. Ice—Composition—Congresses. 3. Glaciers—Congresses. 4. Ice sheets—Congresses. I. Oeschger, H. (Hans) II. Langway, Chester C., 1929– . III. Budd, W.F. (William Francis) IV. Title. V. Series. VI. Series: Physical, chemical, and earth sciences research report ; 8.
QE720.D34 1989 88-8493
551.7'92—dc19 CIP

British Library Cataloguing in Publication Data

Dahlem Workshop on the Environmental Record in
Glaciers and Ice Sheets (*1988: Berlin, Germany*)
The environmental record in glaciers and ice sheets.
1. Glaciation
I. Title II. Oeschger, H. III. Langway, C.C.
551.3'1

ISBN 0 471 92185 8

Typeset by Photo·graphics, Honiton, Devon
Printed and bound in Great Britain
by The Bath Press Ltd, Bath, Avon

Table of Contents

Group Report

The Dahlem Konferenzen

Founders

Recognizing the need for more effective communication between scientists, the Stifterverband für die Deutsche Wissenschaft*, in cooperation with the Deutsche Forschungsgemeinschaft**, founded Dahlem Konferenzen in 1974. The project is financed by the founders and the Senate of the City of Berlin.

Name

Dahlem Konferenzen was named after the district of Berlin called *Dahlem*, which has a long-standing tradition and reputation in the sciences and arts.

Aim

The task of Dahlem Konferenzen is to promote international, interdisciplinary exchange of scientific information and ideas, to stimulate international cooperation in research, and to develop and test new models conducive to more effective communication between scientists.

The Concept

The increasing orientation towards interdisciplinary approaches in scientific research demands that specialists in one field understand the needs and problems of related fields. Therefore, Dahlem Konferenzen has organized workshops, mainly in the Life Sciences and the fields of Physical, Chemical, and Earth Sciences, of an interdisciplinary nature.

Dahlem Workshops provide a unique opportunity for posing the right questions to colleagues from different disciplines who are encouraged to

* The Donors Association for the Promotion of Sciences and Humanities, a foundation created in 1921 in Berlin and supported by German trade and industry to fund basic research in the sciences.

** German Science Foundation.

state what they do not know rather than what they do know. The aim is not to solve problems or to reach a consensus of opinion, the aim is to define and discuss priorities and to indicate directions for further research.

Topics

The topics are of contemporary international interest, timely, interdisciplinary in nature, and problem oriented. Dahlem Konferenzen approaches internationally recognized scientists to suggest topics fulfilling these criteria. Once a year, the topic suggestions are submitted to a scientific board for approval.

Program Advisory Committee

A special Program Advisory Committee is formed for each workshop. It is composed of 6–7 scientists representing the various scientific disciplines involved. They meet approximately one year before the workshop to decide on the scientific program and define the workshop goal, select topics for the discussion groups, formulate titles for background papers, select participants, and assign them their specific tasks. Participants are invited according to international scientific reputation alone. Exception is made for younger German scientists. Invitations are not transferable.

Dahlem Workshop Model

Since no type of scientific meeting proved effective enough, Dahlem Konferenzen had to create its own concept. This concept has been tested and varied over the years. It is internationally recognized as the *Dahlem Workshop Model*. Four workshops per year are organized according to this model. It provides the framework for the utmost possible interdisciplinary communication and cooperation between scientists in a period of $4\frac{1}{2}$ days.

At Dahlem Workshops 48 participants work in four interdisciplinary discussion groups. Lectures are not given. Instead, selected participants write background papers providing a review of the field rather than a report on individual work. These papers, reviewed by selected participants, serve as the basis for discussion and are circulated to all participants before the meeting with the request to formulate written questions and comments to them. During the workshop, each of the four groups prepares reports reflecting their insights gained through the discussion. They also provide suggestions for future research needs.

Publication

The group reports written during the workshop together with the revised background papers are published in book form as the Dahlem Workshop

Reports. They are edited by the editor(s) and the Dahlem Konferenzen staff. The reports are multidisciplinary surveys by the most internationally distinguished scientists and are based on discussions of advanced new concepts, techniques, and models. Each report also reviews areas of priority interest and indicates directions for future research on a given topic.

The Dahlem Workshop Reports are published in two series:

1) Life Sciences Research Reports (LS), and
2) Physical, Chemical, and Earth Sciences Research Reports (PC).

Director

Silke Bernhard, Dr med., Dr phil. h.c.

Address

Dahlem Konferenzen
Tiergartenstrasse 24–27
D-1000 Berlin (West) 30

Tel.: (030) 262 50 41

THE DAHLEM WORKSHOP MODEL

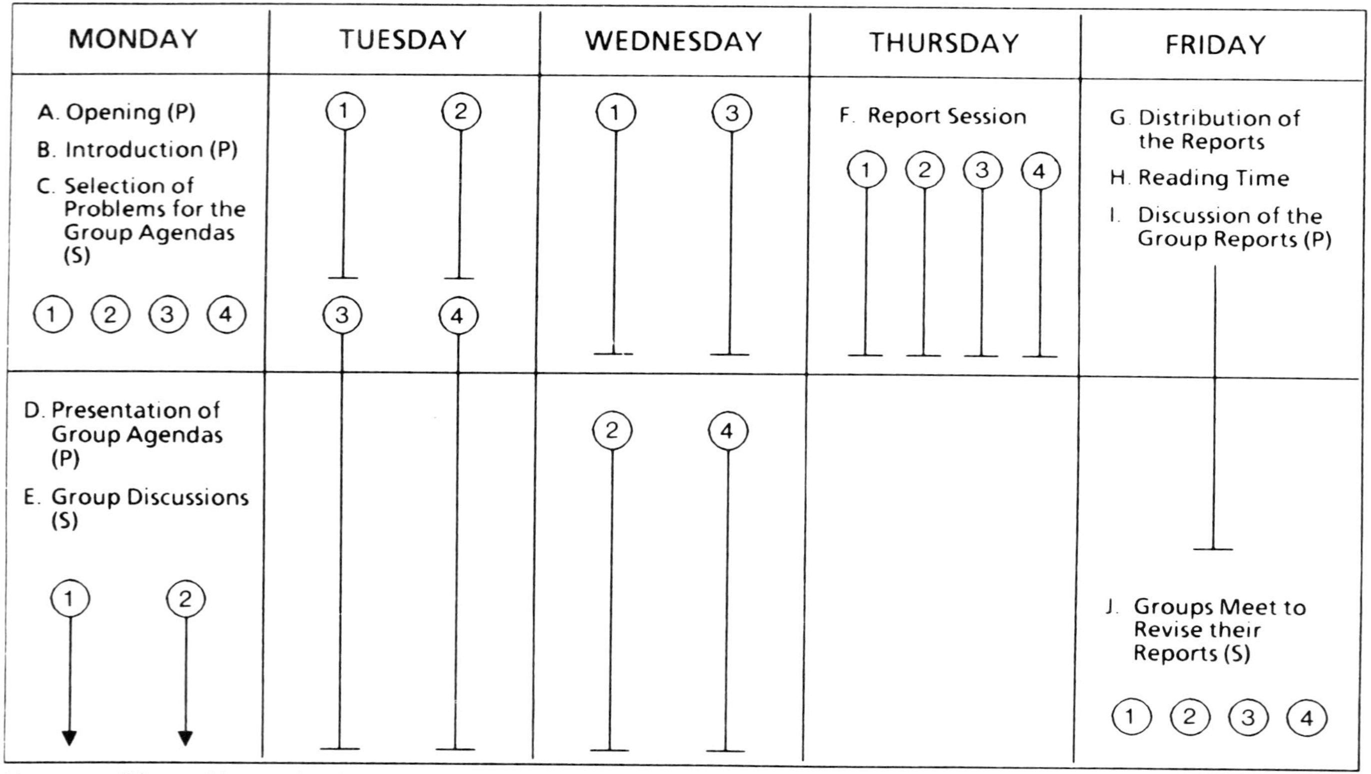

Key: (P) = Plenary Session;
(S) = Simultaneous Sessions;
○ = one discussion group

Explanation of the Dahlem Workshop Model

A. Opening
Background information is given about Dahlem Konferenzen and the Dahlem Workshop Model.

B. Introduction
The goal and the scientific aspects of the workshop are explained.

C. Selection of Problems for the Group Agenda
Each participant is requested to define priority problems of his choice to be discussed within the framework of the workshop goal and his discussion group topic. Each group discusses these suggestions and compiles an agenda of these problems for their discussions.

D. Presentation of the Group Agenda
The agenda for each group is presented by the moderator. A plenary discussion follows to finalize these agendas.

E. Group Discussions
Two groups start their discussions simultaneously. Participants not assigned to either of these two groups attend discussions on topics of their choice.
The groups then change roles as indicated on the chart.

F. Report Session
The rapporteurs discuss the contents of their reports with their group members and write their reports, which are then typed and duplicated.

G. Distribution of Group Reports
The four group reports are distributed to all participants.

H. Reading Time
Participants read these group reports and formulate written questions/comments.

I. Discussion of Group Reports
Each rapporteur summarizes the highlights, controversies, and open problems of his group. A plenary discussion follows.

J. Groups Meet to Revise their Reports
The groups meet to decide which of the comments and issues raised during the plenary discussion should be included in the final report.

This Dahlem Workshop Report is dedicated to the late A.P. Crary, visionary and dedicated Chief Scientist at the U.S. National Science Foundation (NSF), who encouraged and supported the early continuous deep core drilling efforts, and B. Lyle Hansen, inventive and industrious drilling expert who, while at USA CRREL, succeeded in accomplishing the first core drilling through a polar ice sheet at Camp Century, Greenland, in 1966.

The Environmental Record in Glaciers and Ice Sheets
eds. H. Oeschger and C.C. Langway, Jr., pp. 1–11
John Wiley & Sons Limited

Introduction

C.C. Langway, Jr.* and H. Oeschger**

**Ice Core Laboratory, Department of Geology
State University of New York at Buffalo
Amherst, N Y 14226, U.S.A.
**Physikalisches Institut, Universität Bern
3012 Bern, Switzerland*

GLACIERS AS A SOURCE OF EARTH HISTORY

The Earth's chronological evolution and development is a gigantic and complicated subject. The multiple physical, chemical, and biological systems acting at, on, or near the Earth's surface are in a state of constant interaction or change. It is an acknowledged axiom that nature abhors a vacuum. The interacting systems are always in various phases of near balance or unstable equilibrium or in dynamic disequilibrium, depending upon diversified controlling factors, their extent and magnitude. The constant changes or events occurring in the systems are often registered in various earth materials as deposits, layers, or growths in time-unit increments on either a local, regional, or global basis. The recognized global sources of paleoenvironmental information are somewhat limited. They include sea sediments, lake and bog deposits, tree-rings, periglacial features, and snow and ice deposits.

There is ample evidence that throughout much of the Earth's history, glacial epochs recurred at very distant intervals of time. These cold periods were relatively long-term intervals when continental-sized glaciers, i.e., ice sheets, waxed and waned over as much as two-fifths of the Earth's land surface. The most recent glacial epoch, the Pleistocene, is best known. It occurred during the last 2 to 3 million years and, in the Northern Hemisphere, made at least seven or eight major advances during the past 700 000 years. Less is certain about the details of the Southern Hemisphere's ice ages but recent studies on deep ice cores from both Greenland and Antarctica reinforce the view that the Pleistocene epoch terminated by rapid deglaciation in both the Northern and Southern Hemispheres at about the same time, some 11 000 years ago.

Why did the last global ice age, the Wisconsin, abruptly terminate some 10000 years ago? What is the relationship between ice ages and climatic change? What environmental conditions existed during the ice ages? When will the next ice age occur? These are a few of the perplexing questions still unanswered today some 150 years after Louis Agassiz presented the first of his ice-age theory papers at a meeting of the Swiss Society of Natural Sciences in Neuchâtel. The authors are among the growing number of scientists who believe that the principal answers to some of the preceding and related questions lie in the stratigraphic sequences existing in present-day glaciers and ice sheets. As remnants of the last ice age, it became self-evident that the Greenland and Antarctic ice sheets are the prime source areas for obtaining chronological information needed to reconstruct the climate and environmental history of the late Pleistocene epoch. The original goals were, of course, to develop a drilling method to obtain continuous polar ice cores over an extended depth profile and to develop the techniques of extracting the paleodata from the cores.

BACKGROUND

Continuous ice core drilling into polar glacier ice for scientific purposes was first proposed on a major scale by H. Bader in 1954 (Bader 1958). The initial pre-International Geophysical Year (IGY) ice core drilling operations were conducted at Site 2, N.W. Greenland in 1956 (Lange et al. 1959) and again in 1957 (Langway 1967). These ice cores reached depths of 305 m and 411 m, respectively. Surprisingly few ice cores have been obtained from various Greenland and Antarctic locations over the past thirty years. Merely five ice cores have been recovered from depths of over 1000 m (only three to bedrock), less than ten from depths of about 300 to 500 m, and only about two dozen from depths near 100 m, a very small sampling when considering the demonstrated potential and importance of the paleoinformation contained in the sequential deposits. These ice cores were studied by small numbers of individuals and international teams who mostly worked together in an exciting atmosphere of discovery and new ideas. The mutually collaborative field operations and laboratory studies cultivated scientific progress and opened up a new field in earth science which provided fresh information to modern thinking about earth processes.

The first deep ice core to completely penetrate the total thickness (1387 m) of an inland polar glacier was recovered at Camp Century, Greenland, in 1966 (Hansen and Langway 1966; Ueda and Garfield 1968). The Greenland ice sheet was core drilled to bedrock for the second time (2037 m) at Dye 3, Greenland, in 1981 (Gundestrup and Johnsen 1985; Langway et al. 1985). The only other polar ice core to reach bedrock (2164 m) was obtained at Byrd Station, Antarctica, in 1968 (Ueda and Garfield 1969; Gow 1970), although two other cores have been recovered to great depths in Antarctica—Dome C (1000 m, Lorius et al. 1979), and Vostok (2083 m, Lorius et al. 1985). Today, methods of shallow and intermediate depth ice core drillings have been perfected into nearly standard operations of relatively short field duration (Splettstoesser 1976; Holdsworth et al. 1984; Rand and Mellor 1985). A typical drilling to bedrock is considerably more complicated and costly (Langway 1976; Tillson and Kuivinen 1982).

Since the inception of the first intermediate ice core drilling project at Site 2, many new field methods have been developed for handling, processing, sampling, and study of ice cores. Concurrent with the advancement of core drilling technology, refinements have been made in ice core stratigraphy and logging routines as well as surface pit/core correlation methods which assure the accurate and consistent recording of stratigraphic features and, of great importance, provide a firm chronological datum for the cores. The basic field procedures, processing, sampling, and study techniques have been largely developed (Langway et al 1985).

The amount of information derived from an ice core investigation is dependent upon the depth and physical condition of the core, the continuity of the core column, and the extent and completeness of the core study program. The scientific results from the various shallow, intermediate, and deep ice cores recovered during the past three decades have been widely published (e.g., Oeschger et al. 1968; Dansgaard et al. 1969; Oeschger et al. 1972). These studies have demonstrated that ice sheets are unique natural depositional environments, and, like the bottom sediments of the world's oceans, they preserve in layered sequences an abundance of paleoenvironmental data (Dansgaard et al. 1975; Oeschger 1985). To date, the deep ice core records extend back about 150 000 years and more. Glacial ice of much greater age probably exists at the bottom in Central Greenland and certainly in East Antarctica. On the other hand, due to their restricted depths, shallow and intermediate cores furnish abbreviated but equally important environmental records. The data return from these shorter 100-year to 1000-year records is often greatly enhanced when the studies are organized and integrated within the broader data base of a deep ice core investigation. Conversely, a deep ice core investigation benefits greatly from the supplemental local and regional data derived from an array of shorter length cores spread out along a major flow line or extending out radially from the main drilling site. One should also recall that old ice exists at favorable locations along the marginal slopes of the ice sheets where it may be easily recovered by simple excavating techniques. The marginal ice zone is a relatively untapped major source of ice sheet history and ice flow dynamics that should be astutely factored into the analysis and interpretation of deep ice cores. There is also much to be learned about smaller scale earth and climate processes from Alpine glaciers.

OVERVIEW AND PREVIOUS RESULTS

Ice sheets and glaciers are essentially products of atmospheric processes and important components of the Earth's air/sea/land/climate system. Their waxing and waning during the geological, historical, and more recent past have had significant effects on global climate. Glaciers also serve as semipermanent reservoirs for atmospherically produced or transported substances. Within the frozen stratigraphic layers of glaciers are preserved a variety of aerosol particles; volcanic dusts and acids; other solid and soluble particulate matter; natural and artificial radioactive and stable isotopes; natural and artificial atmospheric gases; pesticides and industrial chemicals; and numerous trace substances. All these chemical constituents or components originally served as snow crystal nuclei or were incorporated as dry-fallout and became entrapped during the sedimentary/metamorphic processes of snowfall and glacier growth.

In addition to their diagnostic chemical components, glaciers and ice sheets have definite physical characteristics related to the depositional, diagenetic, and metamorphic processes involved in their formation and growth. A proper interpretation of the geochemical parameters measured in an ice core requires additional knowledge of the rheological conditions and physical properties occurring or existing at the surface, the bedrock interface, and within the glacier mass itself. Similar data are required from upstream and downstream directions. A variable composite of the chemical and physical measurements is necessary to accurately date past events, to determine the often changeable background values in the unpolluted atmosphere, and to develop reliable environmental and climate models essential for predicting future trends.

Although relatively new, the results of the multidisciplinary ice core studies made to date have provided exciting and fresh information on the history of the Earth and the planetary system. Among the important published results are the reconstructed records for the past 0.1 to 0.15 M years related to:

1. The climate of the Earth and the polar regions using $^{18}O/^{16}O$ and $^{2}H/^{1}H$ stable isotope ratio measurements. One significant result of these studies was the revelation that the major low-frequency ice-age climatic shifts were synchronously recorded in the deep ice core records from both poles and that multiple high-frequency climatic perturbations or cycles are superimposed on these records.
2. The evolution of the atmospheric dust load which has provided information on bipolar atmospheric circulation patterns, aerosol deposition and trajectory routes, global cycling paths, magnitude and composition of volcanic disturbances, and the fluxes of extraterrestrial matter on the earth's surface.
3. The evolution of the major and trace gaseous components and total

composition of the atmosphere, including the relationships between and concentration changes in CO_2, CH_4, and N_2O up to the time when precise and continuous direct atmospheric measurements began. One result of these studies was the astonishing revelation that the CO_2/air ratio varied contemporaneously with the climatic conditions experienced during glacial/interglacial shifts.

4. The modulation of cosmic radiation by changing solar and terrestrial magnetism. High ^{10}Be concentrations in ice reveal periods of the quiet sun, like the Maunder Minimum (A.D. 1640–1710) and are compatible with the short-term ^{14}C variations observed in tree-rings.

5. The chemical record of the Earth's changing atmosphere since the Industrial Revolution due to anthropogenic input. Of added significance was the establishment of preindustrial baseline concentration levels and the discovery of gradual and rapid coincidental shifts which periodically occurred in the levels of principal and trace atmospheric constituents such as SO_4^{-2}, NO_3^-, Cl^-, heavy metals, and others during the Wisconsin period as well as the dramatic and sharp change in these constituents at the Holocene/Wisconsin climatic boundary.

During the past several years, ice core research has profited significantly from the rapid progress made in the development and application of state-of-the-art laboratory analytical measuring devices (e.g., automatic stable isotope analyzer, tandem accelerator, ion chromatograph, solid conductivity gauge, rotational fabric analyzer, and various unique field measuring instruments). Even newer dimensions are possible which are being viewed on the horizon. The full potential and proper understanding of the complex natural processes which may be unfolded by ice core studies is in its infancy.

Many of the above-mentioned topics were the subjects of in-depth ice core investigations which are component parts of several papers contained in this volume.

THE WORKSHOP

The original impetus to conduct a Dahlem Workshop on ice cores was provided by Ed Goldberg. It seems that at a number of recent Dahlem Workshops in the Physical, Chemical, and Earth Science series, considerable discussion took place related to results from the rapidly emerging publications on ice core studies. Goldberg was an early contributor to the study of dating glacier ice and developed the ^{210}Pb measuring method. At his suggestion, Dr. Silke Bernhard invited us to prepare a workshop proposal which was subsequently approved. It was our viewpoint that a Dahlem Workshop would be an appropriate forum to incorporate and summarize past accomplishments in ice core research. The workshop would also present an opportunity to outline for the future the more prominent scientific study target areas and to formulate recommendations and technological directions to accomplish the task. Furthermore, a workshop would be particularly timely within the framework of the contemplated major international deep ice core drilling programs being planned for both Greenland and Antarctica.

A program advisory committee meeting was held in West Berlin in May, 1987, where the specific goal of the workshop was developed and the participants selected. As with all Dahlem Workshops, participation is limited; the final selection was based on professional expertise and achievement in ice core related research and international balance. The workshop on "The Environmental Record in Glaciers" was held at the Europa Center, West Berlin, Germany, between March 13 and 18, 1988.

The workshop participants worked in four discussion groups of twelve members each. Each group was responsible for a specific problem:

1. How do glaciers record environmental processes and preserve information?
2. What anthropogenic impacts are recorded in glaciers?
3. How can an ice core chronology be established? and
4. What does the long-term ice core record tell us about global changes in the environment?

A total of sixteen authors were invited to write background papers on specific topics. These background papers were distributed to all participants one month prior to the meeting. The papers were written to serve as a common information base for the workshop participants. During the week of the meeting each group engaged in debate and extensive exchange of ideas which led to a draft written report by the groups' respective rapporteurs. These draft reports were distributed to all workshop participants on the last day of the meeting. They were discussed then in a plenary session attended by all participants and were finally revised by the rapporteurs into the Group Reports as published in this volume.

In short, it was ascertained by Group 1 discussants that interpretation of data derived from ice core studies is neither straightforward nor simple. Unambiguous translation of these data involves knowledge of firn and ice processes, the transfer mechanism from atmosphere to snow and firn, and also information on atmospheric transport paths from the source region to the glaciers. It was concluded that the environmental record is well preserved in cold glaciers but that more research is needed to study the atmosphere and surface snow transfer functions and related processes. Some of the fundamental questions regarding interpretation of the records concern the sorting of variations in the concentrations of the multiple materials incorporated in the ice as related to changing snow deposition rates or variations in the incoming chemical fluxes.

Group 2 deduced that the determination of a deep ice core age-scale is currently a weak point in the research. All current methods of dating ice were discussed in detail. They ranged from numerical modeling and isotope dating to changes in the chronology brought about by variations in the impurity concentration levels and *in situ* temperature changes. Much attention was given to methods employing annual layer counting, solar and terrestrial radioactivity, comparison with sea sediment and dendrochronological records, and the Earth's orbital changes. The group agreed that there is much potential in new dating techniques based on AMS and RIMS. They also concluded that annual layers should be continuously measured further back in time, if possible to the last interglacial period (Sangamon/Eem). There is renewed hope that this objective will be accomplished by recent advances made in the establishment of well-defined global index horizons based on CH_4, CO_2, volcanic dust layers, ^{18}O in gas bubbles, and sharp chemistry boundaries. Further study is recommended to accurately date and interpret the observed low-frequency, high-frequency, and abrupt changes shown in the records and to cross-correlate the chronologies with other proxy data. It was further assessed that marginal zone polar ice and Alpine glaciers deserve additional attention.

Group 3 focused their considerations on the anthropogenic records contained in ice cores. Earlier research had established that the preindustrial concentration levels of the atmospheric gases CO_2, CH_4, and N_2O are now clearly being increased by Man's activity. In view of the large-scale exchange of gaseous fluxes between the atmosphere/ocean/biosphere CO_2 reservoirs, some concern was expressed regarding why the CO_2 system is not capable of buffering the relatively small anthropogenic input. In the reconstructed CO_2 record, an increase in concentration levels is evident starting around A.D. 1800, and substantial increase is shown since 1959. It was inferred that if it were not for the CO_2 emission due to deforestation and the increase in fossil fuel consumption, today's atmospheric CO_2 concentration would probably have remained in the 280 ± 10 ppm range. The group allowed that aerosol source areas, pathways, and deposition mechanisms are complex but important subjects of study in ice core research. It was observed that during the past 2000 years the reported lead concentrations in Greenland deposits shows a 200-fold increase and, as is usually the case for the less industrialized Southern Hemisphere, the Antarctic signal is much smaller (a 4- to 10-fold increase). Increasing trends in NO_3^- and SO_4^{2-} concentration levels are also evident during the past 200 years in Greenland. Ice sheets have been shown to be settling tanks for radioactive substances following nuclear weapon tests. These pollutant layers are chronologically identifiable horizons and reveal pertinent environmental information on atmospheric transport and snow deposit processes.

Group 4 evaluated the long-term ice core records and assessed the major and minor environmental changes as they are represented in ice sheets. They concluded that the physical and chemical information from the Greenland and Antarctica deep ice cores extends over the entire last glacial cycle and represents a most significant archive of paleodata. These data have already expanded our vision of the biogeochemical system in steady state for the last 1500 centuries. Measurements of CO_2 and CH_4 gases have shown general variations which are coherent with those documented by ^{18}O concentrations in the water molecules. The changes in the concentration levels of these gases might well have contributed, via the "greenhouse effect," to the climate coupling between the Northern and Southern Hemispheres.

All elements cycled through the atmosphere are recorded in polar ice strata and each parameter carries a message on an earth system process. Some parameters, like volcanic debris or cosmogenic isotopes, are indicators of possible climatic forcing. Others, like the O and H isotopes in water molecules, the aerosol and particle components of continental dust and sea spray, and the measurement of annual net snow accumulation rates, describe the physical state of the earth system. Parameters such as CO_2, CH_4, and their isotopes are derivatives of organic substances. Here measurements on the organic parameters return information on the terrestrial and marine biosphere as well as feedback or even primary forcing data used to understand the climate system. Since the Milankovitch phenomenon is in antiphase at opposite global poles, a comparison of time evolution parameters as revealed in ice cores from Greenland and Antarctica is of special interest. A key question concerns the synchronism of events at the interglacial/glacial transitions, which represent a potential means of disentangling the direct Milankovitch forcing influence from the coupling between the hemispheres.

If our knowledge of the earth system is to be significantly improved in the near future, the discussants of Group 4 feel that studies made on deep ice cores will help provide answers to many of the key questions.

CONCLUSIONS

The Dahlem Workshop on "The Environmental Record in Glaciers" clearly demonstrated that ice core data have revolutionized our present view on the evolution of the earth system and some of its important mechanisms. Ice core research has significantly contributed to the rethinking of the Earth as a highly interactive system wherein a wide spectrum of biogeochemical processes are interacting within the purely physical processes. This ensemble of processes is now often referred to as the organism Earth.

One of the most important scientific programs being developed for the next decades will be the "International Geosphere-Biosphere Program; A Study of Global Change," implemented by the International Council of Scientific Unions (ICSU). The vast scope and goal of the program is to understand the evolution of the earth system, the interaction of Man, and projections into the future. The discussions at the Dahlem Workshop underscored the importance of studying past systems processes in natural archives, especially the ice sheet records, for such a program. As a result of the new and in some cases completely unexpected information gained from ice core studies, new data sets are being included in the IGBP to evaluate global models.

In spite of the general international agreement by scientists involved in polar regions research and the positive endorsement in numerous study documents by various national scientific commissions, backbone national support and activity for ice core research is still inadequate. The current state-of-the-art research is still carried out by only a few laboratories with minimum to low funding levels. The last international deep drilling project was successfully completed over eight years ago. Conservative estimates show that with the experience, momentum, and equipment existing at the time at least two other drillings to bedrock could have been made by now. Funding for new deep drilling operations has been delayed for more than half a decade, during which time competent, experienced personnel and committed laboratories which made contributions to or lead in the early phase of this research have dispersed or are no longer supported.

In the atmosphere of the healthy spirit and enthusiasm which existed at the Dahlem Workshop it is our sincere hope that the conference represents an upward turning point in ice core research by signaling its recognition from a larger spectrum of scientific disciplines. We further trust that by the wider exposition of the treasury of data contained in ice cores that its base will be broadened and further support will be mustered to contribute to its future development.

ACKNOWLEDGEMENTS

We feel that this workshop served as an important and unprecedented coupling of active ice core researchers with a mixed group of interdisciplinary scientific participants who have deep interest in the variety of data being obtained from ice cores. The nature and format of the workshop allowed for free exchange of ideas and represents a very effective method of communicating problem-oriented ideas and topics across disciplinary boundaries. The group moderators encouraged and engaged in free-wheeling controversial panel discussions and kept the sound level of serious disputes to low uproars.

We also feel that the workshop was an overall success and fulfilled our expectations, goals, and objectives. It brought a sector of the scientific community interested in the results and application of ice core research up-to-date on the current status of glaciological studies and provided them with a fuller perspective on where their interests fit into this glaciological research. On the other side, it offered ice core investigators the opportunity to become intimately informed of where their research factors into the matrices of the various other disciplines with which their results interact or overlap.

We hope that this volume, resulting from the industry and expertise of the authors and participants, will contribute to advancing the general knowledge of the importance and potential of ice core research. We wish to deeply thank Silke Bernhard and her entire staff for their organizational planning and their excellent workshop arrangements. In conclusion, we sincerely thank the Senate of the city of Berlin, the Stifterverband für die Deutsche Wissenschaft, and the Stiftung Winterling Marktleuthen for providing the ways and means of making this workshop possible.

REFERENCES

Bader, H. 1958. United States polar ice and snow studies in the International Geophysical Year. *Geophys Monog*. **2**: 177–181. Washington, D.C.: Amer. Geophys. Union.

Dansgaard, W.; Johnsen, S.J.; Moller, J.; and Langway, C.C., Jr. 1969. One thousand centuries of climatic record from Camp Century on the Greenland ice sheet. *Science* **166**: 377–381.

Dansgaard, W.; Johnsen, S.J.; Reeh, N.; Gundestrup. N.; Clausen, H.B.; and Hammer, C.V. 1975. Climatic change, norsemen, and modern man. *Nature* **225**: 24–28.

Gow, A.J. 1970. Preliminary results of studies of ice cores from the 2164 m deep drill hole, Byrd Station, Antarctica. International Symposium on Antarctic Glaciological Exploration (ISAGE) ICSU–SCAR and IASH. *IASH Pub*. **86**: 78–90.

Gundestrup, N.S., and Johnsen, S.J. 1985. A battery powered, instrumented deep ice core drill for liquid filled holes. In: Greenland Ice Core: Geophysics, Geochemistry and the Environment, eds. C.C. Langway, Jr., H. Oeschger, and W. Dansgaard. *Geophys. Monog*. **33**: 19–22. Washington, D.C.: Amer. Geophys. Union.

Hansen, B.L., and Langway, C.C., Jr. 1966. Deep core drilling in ice and core analyses at Camp Century, Greenland, 1961–1966. *Antarc. J. US* Oct 1966: 207–208.

Holdsworth, G.; Kuivinen, K.C.; and Rand, J.H. 1984. Ice drilling technology. Special Report 84–34, U.S. Army Corps of Engineers. Hanover, NH: Cold Regions Research and Engineering Laboratory.

Lange, G.R.; Langway, C.C., Jr.; and Hansen, B.L. 1959. Deep core drilling in glaciers: Proc. U.S. Army Science Conf., 1959, West Point, NY., vol.2, pp.97–107.

Langway, C.C., Jr. 1967. Stratigraphic analysis of a deep ice core from Greenland. Research Report 77, U.S. Army Corps of Engineers. Hanover, N.H: Cold

Regions Research and Engineering Laboratory. Also Geol. Soc. Amer. Special Paper 125, 1970.

Langway, C.C., Jr., ed. 1976. GISP Science Plan (Committee Report). Lincoln, Nebraska: Univ. of Nebraska.

Langway, C.C., Jr.; Oeschger, H.; and Dansgaard, W., eds. 1985. Greenland Ice Core: Geophysics, Geochemistry and the Environment. *Geophys. Monog.* **33**. Washington, D.C.: Amer. Geophys. Union.

Lorius, C.; Jouzel, J.; Ritz, C.; Merlivat, L.; Barkov, N.I.; Korotkevich, Y.S.; and Kotlyakov, V.M. 1985. A 150000 year climatic record from Antarctic ice. *Nature* **316(6029)**: 591–596.

Lorius,C.; Merlivat, L.; Jouzel, J.; Pourchet, M. 1979. A 30000-year isotope climatic record from Antarctic ice. *Nature* **280(5724)**: 644–648.

Oeschger, H. 1985. The contribution of ice core studies to the understanding of environmental processes. In: Greenland Ice Cores: Geophysics, Geochemistry and the Environment, eds. C.C. Langway, Jr., H. Oeschger, and W. Dansgaard, pp. 9–17. Geophysical Monograph 33. Washington, D.C.: Am. Geophysical Union.

Oeschger, H.; Adler, B.; Loosli, H.; Langway, C.C., Jr.; and Renaud, A. 1968. Radiocarbon dating of ice. *Earth Plan. Sci. Lett.* **1**: 49–54.

Oeschger, H.; Stauffer, B.; Bucher, P.; Fromm, E.; Moll, M.; Langway, C.C., Jr.; Hansen, B.L.; and Clausen, H. 1972. C^{14} and oxygen isotope results on natural ice. Proc. of the 8th Intl. Conf. on Radiocarbon Dating, vol. 1. Wellington, New Zealand.

Rand, J., and Mellor, M. 1985. Ice-coring augers for shallow depth sampling. Research Report 85–21, U.S. Corps of Engineers. Hanover, NH: Cold Regions Research and Engineering Laboratory.

Splettstoesser, J.F., 1976. Ice-Core Drilling. Lincoln, Nebraska: Univ. Nebraska Press.

Tillson, R.A., and Kuivinen, K.C. 1982. An ice core science trench for use by glaciologists on the Greenland ice sheet. In: Antarctic Logistics Symposium. Leningrad: Scientific Committee on Antarctic Research.

Ueda, H.T., and Garfield, D.E. 1968. Drilling through the Greenland ice sheet. Special Report 126, U.S. Army Corps of Engineers. Hanover, NH: Cold Regions Research and Engineering Laboratory.

Ueda, H.T., and Garfield, D.E. 1969. Core drilling through the Antarctic Ice Sheet. Technical Report 231, U.S. Army Corps of Engineers. Hanover, NH: Cold Regions Research and Engineering Laboratory.

The Environmental Record in Glaciers and Ice Sheets
eds. H. Oeschger and C.C. Langway, Jr., pp. 13–27
John Wiley & Sons Limited

Aerosol Transport from Sources to Ice Sheets

G. E. Shaw

Geophysical Institute
University of Alaska
Fairbanks, AK 99775–0800, U.S.A.

Abstract. The presence of ultrafine particles and ionic species in meltwater throughout the polar ice cores demonstrates that the transport and nucleation of gases and the diffusive transport of aerosols is a regular occurrence, taking place on global scales of distance. A very minor fraction of sea salt and crustal eolian debris, generated by mechanical grinding and disruption of liquid films, ends up in submicron sizes and makes its way to the polar ice repositories; the fraction activated depends very critically on some kind of "storminess" index. In addition, a more vigorous atmospheric circulation would increase the severity and frequency of episodal "sodium storms" (rapidly transported sea salt) and "aluminium storms" (injection of eolian material) into the regions of the central ice sheets. However, though the above can be evoked as causal for "dusty ice ages," it fails to explain the rather invariant size distribution of insoluble particles, as reported by Ram and colleagues, for both shallow and deep ice cores. The atmospheric "machine" with its sundry complex mechanisms for removing particles has the tendency to build up a mode of removal-resistant particles around a fraction of a micron in diameter. We have to understand more about the physics, physical chemistry, and production of submicron-sized, mechanically generated particles if we are to be able to "read" the climate record from aerosols locked up in the polar ice sheets.

INTRODUCTION

Polar ice sheets are repositories of particles which were once suspended in the atmosphere. The presence of particles throughout the ice sheets demonstrates that the transport and nucleation of trace gases and the diffusive transport of aerosols themselves is a regular occurrence, taking place on global scales of distance. It is very interesting to note that there has never been an ice sample acquired, no matter how remote the location or deep the ice, that has not been found to contain insoluble particles and ionic species derived from dissolved particles; in a similar vein, there has never been a sample of atmospheric air taken, that I know of, that has been found to be free of suspended aerosols.

At this writing it is not at all certain what mechanisms are responsible for the observed fluctuations in the record of particle concentration with depth; for instance, there is the ambiguity of whether dustiness in the late glacial maximum age (De Angelis et al. 1984, 1987) was caused by greater aridity or was due to an increase in storminess. The dust could, in principle, also be volcanic or extraterrestrial, though the chemistry makes this improbable.

Intuitively, the ice sheet aerosol record tells us something about past climatic conditions, but the understanding necessary to bridge the gap from reading the aerosol ice record to making definitive statements about the condition of the planet at the time of deposition is so far sadly lacking.

In this document, I provide order of magnitude estimates of the major known or suspected source regions of atmospheric particles, provide some ways of treating the problem of atmospheric transport currents, and discuss the physics of the machinery of particle removal from the atmosphere. This "removal" machinery is operating at varying speeds for different kinds of aerosols: the engines and gear boxes of the machine processing small particles differ completely with those milling large particles. It turns out that no atmospheric machine is very efficient at removing an intermediate size range of particles, which is centered around a few tenths of a micron in diameter. This is an important point because it may be that this removal-resistant debris is about all that manages to get through to extremely remote places such as ice sheets. What once may have only been a minority species of the aerosol subgroup near mid-latitude source regions, may loom into dominance in polar ice.

MAJOR AEROSOL SOURCES

Soil-derived or Crustal Aerosol

The mechanism of wind erosion is responsible for injecting an estimated 100 to 1000 million tons of aerosol into the atmosphere each year, some extremely minor component of which is carried over a pathlength of many thousands of km. Well-known specific instances of desert dust transport are the periodic outfalls of red-colored dust in the Swiss Alps (Mörikofer 1941), yellow-colored "Kosa" sand fallout on southern Japan, widespread dust episodes in the middle east, Saharan dust falls in the Carribean Islands (e.g., Prospero and Carlson 1972), and many reports of fallout of gritty sands on ships at sea. Darwin deduced, for instance, that dust that fell on his ship must have originated from deserts thousands of miles away. Of course there is an absence of washout of aerosols over deserts themselves.

In April, 1979, I witnessed a spectacular Gobi desert dust veil over the Hawaiian Islands. From the Mauna Loa Observatory at 3500 m altitude, the dust veil could be seen edge-on as a dark band, but at least half of the dust was above the observatory (as deduced from optical depth measurements). The episode lasted for about ten days. Particles were in a mode centered around one micron diameter. One micron in diameter is considered "small" in the parlance of desert dust size distribution (as measured in deserts), but is "large" for atmospheric aerosols; for instance, less than one particle out of several thousand or tens of thousands is larger than 1 micron for background tropospheric aerosol.

Crustal aerosol is created by saltation, erosion, or other mechanical disturbance such as glacial movement or running water. It requires an enormous expenditure of energy to "grind down" material; for instance, it is virtually impossible to produce submicron aerosol by hammering rock. Crust-devised aerosols are large, typically possessing two, roughly log-normal, mass distributions: an "A" mode usually consisting of clay, around 1–10 μm, and a "B" mode around 10 to 100 μm, frequently consisting of quartz grains, whose surface is covered with clay dust (Kondratyev et al. 1985). Larger particles skid across aerodynamic flow lines and are rapidly lost from the atmosphere. They do not participate in long-range transport.

The residence time for particles larger than a micron or so is α dia^{-2}, so the remnants of dust from deserts on loess or soil deposits found over the central oceans or above ice caps are winnowed down in size very considerably from their original distribution. This fractionation by size with distance is likely to have a chemical component as well since, for instance, certain clay particles preferentially activate as ice nuclei, while plate-like minerals have different aerodynamic characteristics than spherical grains of the same mass. It might be too much to expect that a great deal of "memory of the original source" — say in its mineralogy — will remain after global-scale transport.

In dust storms, the mass concentration, γ, can go up to 10^{-1} to 10^{-2} g/m^3. In a region of powerful convection γ averages $2–3 \times 10^{-4}$ g/m^3 (Kondratyev et al. 1985), while in well vegetated regions with low levels of soil erosion γ is $2–5 \times 10^{-5}$ g/m^3. In the central Northern Hemisphere, Atlantic $\gamma \sim 0.5$ to 1×10^{-6} g/m^3, while in the Antarctic γ may be <0.1 μg/m^3 (Tuncel et al., submitted).

An interesting, relatively isolated source of desert aerosols is the small arid continental regions of the Southern Hemisphere, notably the Kalahari and Atacoma deserts and the deserts of central and western Australia; they are the prime candidates for providing crustal material to the southern ice sheet. Taken together, they have an area of 8×10^6 km^2 and lie at an average latitude of 32°S and probably inject 10–20 million tons yr^{-1} of micron-sized, transportable particles. A uniform hemispheric fallout would provide a downward flux of 3×10^{-14}g $cm^{-2}s^{-1}$. The unit 10^{-14} g $cm^{-2}s^{-1}$ is useful to express particle fluxes into ice sheets: the downflux of nonsoluble material at the South Pole is a "few" $\times 10^{-14}$ g $cm^{-2}s^{-1}$. The "ice flux" unit (IFU) is equivalent to about 10 cm of soil sedimentation per 100 my.

Oceanic Aerosols

Disruption of surface films in breaking waves on the ocean and jet drops from rising bubbles are a well-known source of sea salt aerosol, estimated to be 1000 million t yr^{-1} (1000 IFU). Particles produced by this mechanism fall into a range extending from a few tenths of a micron to >100 μm, depending upon the specific character of the bubble mechanism. Typically, one finds 0.5–2 sea salt particles cm^{-3} with $r > 1$ μm in the marine boundary layer and a mass loading of 10–20 μg m^{-3}, but the particles being "giant" in size fall off rapidly with height, going to <1% of their concentration at 2–3 km due to the cloud filter. Interestingly, along the coastline over land, the sea salt aerosols reach greater heights because of increased turbulence over land. The smaller of these coastline aerosols participate in long-range transport. This fraction is poorly measured and nearly unknown, unfortunately.

Near-surface sea salt concentrations fall off with distance inland from the ocean roughly as $e^{-X/L}$, with $L \sim 10$–50 km and an altitude with scale heights of about 1 km or less; substantial fluxes of oceanic evaporates, however, have been shown to penetrate deeply inland during specific storm events. "Sodium storms" are observed at the South Pole in association with transport of warm, moist, marine-dominated air from the Weddell Sea embayment areas. The transport takes place primarily in the regimes of the troposphere, just above the temperature inversion that sets up in a layer several hundred meters deep above the ice-air boundary. Following the breakup of the circumpolar vortex in austral spring, remnants of storms spawned about the semipermanent cyclonic system storms in the South Atlantic and Weddell Sea infiltrate into the interior region of the continent, forming supercooled glaciated clouds as the moist air advances up the ice sheet. The cloudiness may aid the breakup of the near-surface radiative inversion (e.g., Hogan, et al. 1984) causing periods of a few days in length when midtropospheric marine sea salt-laden air teleconnects to the surface. In these episodes, Mie-scattering ($r > 0.1$ μ) particles and sodium (and other marine elements) increase dramatically for periods of a few days (Bodhaine et al. 1986).

One senses that the generation (source strength) and deep intrusion of sea salt into central ice sheets is very dependent on storminess. The concentration of sea salt-derived aerosol material at the South Pole (Na, Cl, Mg, etc.) is an order of magnitude stronger in winter than summer (Cunningham and Zoller 1981), apparently because of the greatly increased severity and frequency of storms. One might use sea salt in ice to derive "storm indices."

Extraterrestrial Sources

The estimated flux of cosmic dust on Earth as solid material reaching the surface has been estimated as 0.1–1 mt yr^{-1} (e.g., Langway 1967; Hodge et al. 1964; and many others), corresponding to a uniform deposition of 0.06–0.6 IFU (10^{-14}g cm^{-2} s^{-1}). If these are incorporated uniformly in snow with an annual water precipitation rate of 8 g cm^{-2} yr^{-1}, this would provide 2–20 ppb mass fraction loading in the snow, which in fact is the same order of magnitude as insoluble material actually observed in polar ice. It is now realized that much of what was once considered to be extraterrestrial debris, spheres for instance, are in fact terrestrial products of high temperature reactions. Estimates of cosmic influx to Earth of micrometeoric material, derived from measurements in space, though rather uncertain, suggest that the incoming flux of natural aerosol to the snow is negligible. Extraterrestrial sources of aerosol seem minor in comparison to all other sources of particles.

Sulfur Component

70–80% of the mass of Antarctic polar aerosol are nonsea salt sulfur compounds, frequently found as sulfates or bisulfates. There is a great deal of circumstantial evidence indicating that the sulfur in the aerosols is a product of gas-to-particle conversion process.

Probably the major SO_4^{2-} particulate, at least prior to the industrial revolution, are the products of oxidation of DMS from algae and phytoplankton in the oceans and bacteria in the marine boundary layer (Charlson et al. 1987). The relatively insoluble DMS enters the atmosphere and is oxidized to SO_2, methane sulfonic acid, and, finally, to the condensed phase of hydrated H_2SO_4. One important linkage, producing aqueous H_2SO_4 droplets a few hundredths of a micron diameter, is binary homogeneous nucleation of H_2SO_4 and H_2O molecules: a process which can operate at extremely low partial vapor pressures (i.e., $\simeq 10^{-9}$ Torr) of acid molecules. H_2SO_4 products may also be produced heterogeneously in clouds.

Regarding climatic fluctuations, of notable interest, is the sensitivity of the sulfur flux into the atmosphere to parameters such as oceanic circulation, salinity, sun intensity, etc. We are mostly ignorant of these dependencies. However, the SO_4^{2-} (and H^+) exhibit long period variations, albeit weak, throughout the period covered by the 2083 m Vostok ice core. Nonsea salt SO_4^{2-} was low (100 ± 20 ng g^{-1}) during the Holocene and Last Interglacial, about doubled to 200–300 ng g^{-1} during the Last Glacial Maximum, and was ~200–250 ng g^{-1} during the Penultimate Glacial Maximum (Legrand et al. 1988). The cause for this is uncertain, but it may reflect a higher biogenic activity of the ocean. One might gain insight by searching for methane sulfonic acid, as it would establish the importance of DMS as a source.

Volcanic sources

Major volcanic eruptions strongly disturb the ice sheet aerosol and soluble ionic species record. The volcanic layers in ice have to do with the debris location in the atmosphere, this being in the stable, cloud-free region 20–30 km altitude. Morton's plume rise theory, relating instantaneous energy release to the height of the convective plume produced, indicates that volcanoes which inject stratospheric debris have thermal energies of 10^{12}–10^{14} J, which is the order of magnitude deduced from seismometry. Ash, being a product of mechanical disruption, is in the relatively large size range and quickly falls out, although the minority remnant, around a micron and less in diameter and constituting probably only 10^{-3} to 10^{-5} mass fraction of the injected atmospheric ash, spreads out worldwide and ends up in the polar ice. The component so retained, being the very smallest fraction of the heterodispersion generated, may well have sufficiently suffered many chemical fractionation processes in production and in removal "winnowing" that the composition may have little to do with the parent magma.

Volcanoes produce gases, some of which react to produce particles, the most important of which are submicron droplets of sulfuric acid. The secondary aerosol creates acid bands in the ice, the interpretation and measurements of which have been thoroughly reviewed in a paper by Delmas et al. (1985). The bands from Krakatoa (1883) and Agung (1963), for example, are notable and sharp. Generally, the use of insoluble particles to detect volcanic horizons is less successful than the use of a technique sensing acid, such as electroconductivity, acid filtration, or ionic detection of SO_4^{2-}. The acidity of ice from a major volcanic eruption increases about an order of magnitude from the ice background acid concentration 2–4 μ eq l^{-1}. There is apparently little neutralization of the acid sulfate by NH_3.

The volcanic signals, being distinct, have great potential value to provide information about atmospheric transport and removal processes. So far not much attention has been directed toward this area of research. The residence time of submicron particles in the stratosphere is 1 to a few years, being sufficiently long to frequently homogenize the particles over a hemisphere and even over both hemispheres for large eruptions. Exactly how the particles leak out of the stratosphere is an open question; perhaps currents of ozone-rich and aerosol-laden stratospheric air communicate with the troposphere through stratospheric folds. Increased aridity is liable to occur in association with increased subsidence. The SO_4^{2-} tropospheric volcanic aerosol deriving from the stratospheric reservoir is then likely to build up and, when snowed out, may build enhanced peaks at the polar regions in comparison to the same stratospheric reservoir leaking into a stormy, high-precipitation, and cloudy, midlatitude troposphere. This may give the illusion of larger volcanoes during ice age periods.

Anthropogenic Sources

Anthropogenic material is frequently condensed from the gas phase in the diffusion region of combustion processes; since the concentrations in source regions is high, the aerosols coagulate and some enter into the Greenfield Gap removal-resistance size range (a few tenths of a micron). The result is that an anthropogenic component reaches remote regions since the atmospheric removal in the Greenfield Gap is very slow. Aerosols in this size range have "residence times" of tens of days.

Boutron and Patterson (1983), however, have recently shown that lead in recent Antarctic ice has not increased over that in prehistoric times more than 2- to 3-fold, confirming that remote polar areas of the Southern Hemisphere are still little affected by industrial lead pollution. The "pollution" components is <1 pg Pb|g of water in east Antarctica, and the corresponding fallout is 12 pg Pb cm^{-2} yr^{-1} (4×10^{-9} IFU). In Greenland ice the Pb deposition is 400 times larger, while the deposition of Pb in the North Pacific Westerlies and North Atlantic Westerlies are 4000 and 14,000 times larger. Lead is primarily injected into the Northern Hemisphere Westerlies, so eolian lead fluxes in the Southern Hemisphere are much smaller than in the north; this is congruent with the presence of the circumpolar atmospheric circulation barrier which hinders migration of anthropogenic lead poleward from lower latitudes. We begin to see interesting possibilities of reconstructing mass budgets for species on the basis of the concentrations observed in polar ice by calibrating the earth-atmosphere-ice system with tracers (such as industrial Pb), whose fluxes can be estimated quite accurately.

TRANSPORT CURRENTS

Nonmarine aerosols reach the polar ice from sources which are concentrated in the mid-latitudes. In the case of Greenland, backwash flow from Europe and the north westerly flow from North America frequently bring contaminants to the ice in a few days time. In the case of the Antarctic ice sheet, however, no such semidirect pathways from continental regions exist. Occasionally, direct teleconnections between the peninsular region and the Patagonia desert area of South America or between Australian deserts and western Antarctica may be set up, but such instances are difficult to verify and back trajectories over such vast distances (~6000–10,000 km), computed from a sparse meteorological network, have little meaning. One approach is to use the general circulation of the atmosphere in an attempt to understand the relation between sources and sinks.

Eddy Diffusion Models

Insight has been obtained into transport of aerosol to ice sheets with eddy diffusion models (e.g., Shaw 1979). One model approximated the Antarctic ice sheet as a cylinder of radius a and height H surrounded by an impressed source, and the aerosol mass loading, γ, is obtained through the solution to the diffusion equation,

$$D_H\left[\frac{1}{r'}\frac{d}{dr'}\left(r'\frac{d\gamma}{dr'}\right)\right] + D_V\frac{d^2\gamma}{dh^2} - S\cdot\gamma = 0, \qquad (1)$$

where D_H and D_V are horizontal (meridional) and vertical eddy diffusion coefficients and S is a sink term. The sink term, $S\gamma$, implicitly pertains to processes where the rate of loss is proportional to the mass concentration of aerosols.

In the case of an impressed source at the periphery of the continent, $\gamma(a)$, and suppressing the vertical diffusion, the solution of equation 1 is

$$\gamma(r') = \gamma(a)\, I_0(Kr')|I_0(Ka), \qquad (2)$$

where Io(x) is the modified Bessel function, with argument $Kr' = \sqrt{S/D_H^1} \cdot r' \equiv r'/\sqrt{D_H T_0}$, (the reciprocal of S represents a characteristic time T_0 for aerosol removal). This could be applied to the aerosol column, γH, where H is a characteristic scale height of species that vary with height, z, as $\exp(-z/H)$. In such a case, the term $S = T^{-1}$ can be expressed as v/H, where v is a "deposition velocity." For $D_H = 4.7 \times 10^{10}$ $cm^2\ s^{-1}$ (geometric mean of a large number of reported values) and for $a = 3300$ km (the radius of a cylinder with same area as the Antarctic ice sheet), the fraction and mass of aerosol at the ice sheet's center depends on the value of S and is listed in Table 1.

An example of this theory is the prediction of mean sea salt concentration at the altitude of South Pole (2700 m), in terms of sea salt with $r < 1\ \mu$ near the coasts (a typical concentration of which is 0.1 to 0.3×10^{-12} g cm^{-3}). Using a scale height of 5 km for vertical diffusion, the concentration is $\gamma_0 \exp(h|5)\, I_0(0)|I_0\,(a/\sqrt{D_H T_0}.) =$ 10 to 100 ng m^{-3} for T = 3 and 50 d,

Table 1 Ratio R of column particle load above 2.7 km altitudes over the seas of Antarctica to the column particle load at South Pole Station

S, d^{-1}	$T_0 = 1/S$, d	R
2.0	0.5	0.0008
1.0	1.0	0.004
0.5	2.0	0.053
0.2	5.0	0.22
0.1	10.0	0.44
0.05	20.0	0.62
0.025	40.0	0.79
0.125	80.0	0.88
0.0062	160.0	0.94
0.0031	320.00	0.97
0.00	∞	1.0

respectively, compared with mean observed sea salt mass concentration at South Pole of 10–30 ng m^{-3} in summer and 60–160 ng m^{-3} in winter (Tuncel et al., submitted). Another way of casting the problem is to deduce T from the observed ice sheet-coastal sea salt ratio: this provides T = 3 d for summer conditions, T = 100 d for winter. The long value of T for winter may not be too meaningful in the sense of an eddy diffusion model because the high values of Na arise from transient "sodium storms": (rapid transport) "events" carrying marine air rapidly into the interior of the continent.

The cylindrical geometry eddy diffusion model has been applied to the problem of estimating desert dust transport to Antarctica (Shaw 1979), where account was taken of the strong scavenging by the zone of precipitation (the "furious fifties") at about 55°S latitude surrounding the continent. The precipitation from these systems averages 114 g H_2O cm^{-2} yr^{-1}, whereas 1 mm of rain essentially cleanses the atmosphere. With representative rain-out efficiencies and a precipitation rate of 114 g cm^{-1} yr^{-1}, the southern precipitation would cleanse the atmosphere about a thousand times per year, implying a mean residence time of the order of hours for particulate material that enters the clouds. It can be concluded that the cloud and rain systems in the southern oceans must remove virtually all particles which attempt to pass through them. They would be nearly perfect barriers. Only a remnant above cloud top height (about a twentieth of the column mass load would be above 3 km for the case of an exponential height distribution of dust with a scale height of 1 km) would be available as a ring source feeding an interior cylinder provides the prediction that the mass concentration near the center of the ice sheet would be about a half percent of that at the zonally averaged source region latitude at $\simeq$ 32°S.

A year-round average of 5 ng m^{-3} dust loading over the ice sheets fed from the high latitude southern deserts is arrived at with this highly simplified eddy diffusion model approach; this would represent only a small fraction (2–3%) of the insoluble material found in interior ice sheet cores, from which it might be concluded that the arid regions of the Southern Hemisphere are only minor contributors. This surprising result may suggest (*a*) that much of the insoluble material is extraterrestrial (unlikely), and (*b*) that some essential aspect of the model is flawed. A probable reason is that the assumption of perfectly blocking ring-like storms is simply not valid. Indeed, mid- to low-level, lightly scavenged tropospheric meridional transport currents, capable of rapidly transporting crustal material from arid zones to the ice sheet, may occur in the Antarctic just as they do in the Arctic. Such transport is likely to be in association with large-scale anticyclonic systems; whether such injection pathways for desert dust to the ice sheet are commonly occurring features is not known. One might search for episodal dust events and associate them with meteorological patterns; presumably the transport would take place in anticyclonic air masses and go back to continents.

PARTICLE REMOVAL

Particles are removed from the atmosphere and deposited on the surface by two major mechanisms: (*a*) by processes involving precipitation or clouds and (*b*) by loss when turbulent eddies bring particles in the vicinity of the surface. In both instances, both diffusive losses and inertial impactive losses have to be considered. In the case of precipitation scavenging, nucleation is also a factor to be dealt with.

Inertial mechanisms involve the particles skidding across aerodynamic flow lines and impacting and sticking on objects; examples are simple sedimentation through the viscous atmosphere, impaction of particles on snow flakes, ice crystals or water droplets, or impaction on the surface from eddies. The Stokes number, S_t, (ratio of inertial to viscous constraining force) is the relevant parameter (e.g., Twomey 1977); when St $>$ 1 particles will be collected. These processes proceed at a rate proportional to r^2, where r is particle radius, so large particles are removed more rapidly than small ones.

Diffusive mechanisms of aerosol removal dominate for small particles. A diffusion coefficient, D, can be assigned to a particle with the Einstein relation, $D = kTB$, where k is Boltzmann's constant and B is the mechanical mobility for the particle moving in its viscous carrier gas. For the Stokes region, $B^{-1} = 6\pi r\eta$, where η is gas viscosity and v is velocity. We see that $D = kT/6\pi r\eta\rho$, for a spherical particle in a viscous gas, where we have included the Cunningham slip factor $\rho = 1 + A\ r\lambda$, where λ is the mean free collision pathlength of gas molecules and A is a constant of about unity.

Diffusion is a random walk process where the mean square distance traversed from the origin $\sqrt{X^2}$, by the particle executing Brownian motion, is X^2 = const N, where N is the number of collision steps taken by the particle and is proportional to t. The proportionality constant is the diffusion coefficient, $X^2 = Dt$. In this approach, the aerosol particle is treated as being a heavy molecule in thermal equilibrium with its carrier gas. Notice that a particle diffuses in unit time α distance a $1/r$ (for $r < \gamma$), hence small particles have greater probability of diffusing across streamlines and impacting objects than large. The rate of removal by diffusion is $\alpha\ r^{-2}$ for particles a few hundredths of a micron or smaller.

When the removal mechanisms are combined, one can write in a general way the rate of removal of particles from the atmosphere,

$$\frac{dn}{dt} = \left(ar^2 + \frac{b}{r^2}\right) n, \tag{3}$$

where the first term is due to inertial removal and the second is from diffusive removal of the smallest particles. Note that both very large and very small particles are removed rapidly and that the rate of removal undergoes a minimum at some radius r_m. Equation 3 can thus be written in a normalized form as

$$-\frac{dn}{dt} = \frac{R_m}{2}\left(\frac{r^2}{r_m^2} + \frac{r_m^2}{r^2}\right) n, \tag{4}$$

where R_m represents the minimum rate of removal occurring at $r = r_m$ (note that $dn/dt = R_m$ when $r = r_m$).

As aerosol laden air travels away from its source of aerosols the concentration of particles of radius r exponentially decreases as $\exp(-t/T)$, where the time constant T is

$$T = S^{-1} = \left\{\frac{R_m}{2}\left(\frac{r^2}{r_m^2} + \frac{r_m^2}{r^2}\right)\right\}^{-1}. \tag{5}$$

Values of r_m and R_m for particles in the free atmosphere or, alternately, the coefficients a and b depend very much on the turbulence spectrum of the air topography and surface characteristics over which the wind is blowing, the distribution of clouds in time and space, precipitation rate etc.

As a very general statement, the rates of both inertial and diffusive removal slow down in polar air masses due to the enhanced dynamical stability of the air, to the relatively small volume fraction occupied by cloudiness (in winter), and to the low rates of precipitation.

The atmospheric machine with its sundry complex mechanisms of removing particles builds up a mode of removal-resistant particles, according to equation 4, where aerosol-laden air masses travel for long enough times and distances away from sources and when photochemical gas-to-particle sources are not important.

The particle mode at $r = r_m$ is usually found to be a few tenths of a micron diameter and has variously been called the Greenfield-Gap mode, the accumulation mode, the lingering mode, etc. I prefer to call it the removal-resistant mode of particles. If it were not for the constant production of new particles, especially tiny particles a few hundredths or thousandths of a micron from gas-to-particle conversion, one would presumably find the aerosol over extremely remote places to be a strong, nearly monodisperse mode a few tenths of a micron in diameter.

At this time there is a great uncertainty in the size distribution of aerosols over the giant ice sheets of the planet; however there does seem to be a mode of particles centered roughly around a tenth of a micron and possessing from 1 to several micrograms m^{-3} of mass. In addition, one finds evidence of very tiny particles in polar air, at least in summer, and sometimes in quite high concentrations (e.g., 1000 cm^{-3}). These, presumably, are products of photolytically driven gas-to-particle conversion and seem to consist primarily of droplets of H_2SO_4. Such "Aitken" particles contribute no significant mass and would, at first thought, seem to be irrelevant vis-à-vis ice core analysis. This is not necessarily the case, however, as the Aitken particles coagulate with the larger particles both by direct collision and through cloud scavenging process. In the latter case, a cloud drop evaporates and leaves behind a consolidation of the original freezing or condensation nucleus and particles that diffused to the drop. Large particles, therefore, pick up a coating of smaller particle debris and an open question in atmospheric physics is what is the nature of the film coating large particles and does it act as a surfactant to retard droplet growth as some observations suggest (e.g., Bigg 1986)? This is not just an academic question because the particles brought to the surface by nucleation might be smaller than would otherwise be predicted.

I have attempted (Shaw 1979) to calculate the rates of removal for particles to the Antarctic polar plateau by different mechanisms using a two mode aerosol size distribution. Mass flux to the surface from the small Aitken mode was nil. For the large particle mode, mass loss by sedimentation and diffusion to the surface was negligible in comparison to the predominant "dry" process: impaction to the surface from turbulent eddies which provided a flux of 2×10^{-14} g cm^{-2} s^{-1}.

Wet removal was estimated to be dominated by impaction of particles on snowflakes, but later it was found that the aerosol size mode used for the calculation was overestimated by about a factor of 3. When the calculations were repeated for aerosols centered in a mode around 0.3 μm diameter, the dominant "wet" removal mechanism became nucleation of snowflakes and ice crystals (diamond dust). There is a great deal of additional work yet to be done on the question of the role of nucleation of aerosols in the atmosphere to form cloud drops, which then coagulate to form precipitation.

The calculated downward flux from the standard polar aerosol size distribution to the surface is a few $\times 10^{-14}$ g cm^{-2} s^{-1} for the polar plateau with an assumed snowfall of 7 g H_2O cm^{-2} yr^{-1} and an ice crystal fallout rate of 1 g H_2O cm^{-2} yr^{-1}. The ice crystals were parameterized with a mean radius of 30 μ and fell from a layer 500 m above the surface (T. Ohtake, personal communication).

The calculations of particle precipitation to ice sheets do provide the correct order of magnitude and suggest that removal by nucleation of ice crystals and snowflakes dominates. These calculations, however, should not be taken too seriously as there are a manifold of questions that come up that ought to be answered. In my mind, one of the main questions involves the physical chemistry of the cloud and ice crystal nucleation. Insoluble particles require larger supersaturations to nucleate a water droplet than soluble salts. For example, if an air mass is supersaturated with respect to water vapour by one half percent (a representative maximum value in natural cloud formation), all particles with diameters greater than 0.05 μm would be nucleated if the particles were soluble salts like $(NH_3)_2SO_4$, but if they are insoluble material, only the particles larger than 0.4 μm would nucleate. We see that all large mode soluble particles would nucleate and be incorporated in cloud, but a small fraction of only the soluble particles would be cloud-activated. The ratio of soluble to insoluble particles fluxes reaching the ice would differ considerably from the ratio of these species in air. Another fractionation process which may be important is that coming about by the variable efficiency of compounds to serve as ice nuclei, for example, montmorillonite clay is an active ice-nucleating agent and may preferentially be deposited in ice sheets by nucleating ice crystals. It is in fact found in snow crystals (Kumai 1976). We are at an early stage in being able to make calculations pertaining to particle removal from ice sheets; there is an enormous need for experimental work on these problems, especially in regard to the physical chemistry of the nucleation process, including an assessment of possible surfactant films on cloud- and ice-condensation nuclei.

Is There A Background Aerosol?

The question of a background aerosol has meaning only to the extent that one can define "background." Here I have in mind the term to mean a natural aerosol in the troposphere above the marine boundary layer, below the tropospause, and away from strong concentrated sources. This "system" is envisioned to possess a certain homogeneity and this requires that the particles' residence time be comparable or larger than large-scale mixing times in the atmosphere. This requirement may to some approximation be found to hold for particles around a few tenths of a micron in the region of r_m; one would expect much more variability for particles both much larger or smaller than r_m. Indeed, Gayley and Ram (1985) and Ram et al. (submitted) report evidence for a rather constant size distribution for particles in ice around the vicinity of r_m and independent of depth (time). This could be construed as evidence for the idea of a "background" aerosol whose size distributions are dictated by the removal machinery of the atmosphere and are rather independent of the size distribution of the original source.

REFERENCES

Bigg, E. 1986. Discrepancy between observations and prediction of concentrations of cloud condensation nuclei. *Atmos. Res.* **20**: 82–86.

Bodhaine, B.; DeLuisi, J.; Harris, J.; Houmere, P.; and Bauman, S. 1986. Aerosol measurements at the South Pole. *Tellus* **38B**: 223–235.

Boutron, C., and Patterson, C. 1983. The occurrence of lead in antarctic recent snow, firn deposited over the last two centuries and prehistories ice. *Geochim. Cosmo. Acta* **47(8)** 1355–1368.

Charlson, R.; Lovelock, J.; Andreae, M.; and Warren, S. 1987. Oceanic phytoplankton, atmospheric sulfur, cloud albedo and climate. *Nature* **326**: 655–661.

Cunningham, W., and Zoller, W. 1981. The chemical composition of remote area aerosols. *J. Aerosol Sci.* **12**: 367–384.

De Angelis, M.; Barkov, N.; and Petrov, V. 1987. Aerosol concentrations over the last climatic cycle (160 K yr) from an Antarctic ice core. *Nature* **325**: 318–321.

De Angelis, M.; Legrand, M.; Petit, J.; Barkov, N.; and Korotkevitch, Ye. 1984. Soluble and insoluble impurities along the 950 m deep Vostok ice core, Antarctica climatic implications. *J. Atmos. Chem.* **1**: 215–239.

Delmas, R.; Legrand, M.; Aristarain, A.; and Zanolini, F. 1985. Volcanic deposits in antarctic snow and ice. *J. Geophys. Res.* **90**: 12,901–12,920.

Gayley, R., and Ram, M. 1985. Atmospheric dust in polar ice and the background aerosol. *J. Geophys. Res.* **90**: 12,921–12,925.

Hodge, P.; Wright, F.; and Langway, C. Jr. 1964. Studies of particles for extra-terrestrial origin. 3. Analysis of dust particles from polar ice deposits. *J. Geophys. Res.* **69**: 2919–2931.

Hogan, A.; Samson, J.; Kebscull, K.; Townsend, R.; Barnard, S.; and Murphy, B. 1984. On the interaction of aerosol with meteorology. *J. Rech. Atmos.* **18**: 41.

Kondratyev, K.; Ivanov, U.; Pozdnyakev, D.; and Prokofyev, M. 1985. Natural and anthropogenic aerosols: a comparative analysis. *Pont. Acad. Sci. Scr. Var.* **56**: 281–303.

Kumai, M. 1976. Identification of nuclei and concentrations of chemical species in snow crystal samples at the South Pole. *J. Atmos. Sci.* **33**: 833–841.

Langway, C., Jr. 1967. Stratigraphic analysis of a deep ice core from Greenland. Report RR77. Hanover, NH: Cold Reg. Res. and Eng. Lab.

Legrand, M.; Lorius, C.; Barkov, N.; and Petrov, V. 1988. Atmospheric chemistry changes over the last climatic cycle (160,000 yr) from Antarctic ice. *Atmos. Envir.*, in press.

Mörikofer, W. 1941. Über die Trübung der Atmosphäre durch Wüstenstaub und Schneetreiben. *Helv. Phys. Acta* **16**: 537–548.

Prospero, L., and Carlson, T. 1972. Vertical and areal distribution of Saharan dust over western equatorial North Atlantic Ocean. *J. Geophys. Res.* **77**: 5255–5265.

Shaw, G. 1979. Considerations on the origin and properties of the Antarctic aerosol. *Rev. Geophys. Space Phys.* **17**: 1983–1998.

Tuncel, G.; Aras, N.; and Zoller, W. 1987. Temporal variations and sources of elements at the South Pole atmosphere: non-enriched and moderately enriched elements. *J. Geophys. Res.*, in press.

Twomey, S. 1977. Atmospheric Aerosols. New York: Elsevier.

The Environmental Record in Glaciers and Ice Sheets
eds. H. Oeschger and C.C. Langway, Jr., pp. 29–51
John Wiley & Sons Limited

Mechanisms of Wet and Dry Deposition of Atmospheric Contaminants to Snow Surfaces

C. I. Davidson

Departments of Civil Engineering and Engineering & Public Policy
Carnegie Mellon University
Pittsburgh, PA 15213, U.S.A.

Abstract. Proper interpretation of glacial record data requires an understanding of contaminant transport mechanisms from the atmosphere to the surfaces of glaciers. Such transport may occur by wet or dry deposition. Wet deposition generally dominates; the process of nucleation and the role of riming are especially important. Because the complexities of deposition are only poorly understood, ice core data can merely provide rough estimates of airborne contaminant levels in previous times. Improving these estimates requires better deposition model input data on the characteristics of atmospheric contaminants and on characteristics of clouds in the polar regions.

INTRODUCTION

Trace amounts of contaminants in glaciers have been valuable for studying changes in atmospheric contaminants over time. For example, gases trapped in ice cores have provided information on historical variations in amounts of these gases in the Earth's atmosphere. Microparticles in the ice have been linked to large-scale dust storms, volcanic eruptions, biological activity, and other particle-producing phenomena. High concentrations of sulfate have provided additional evidence of volcanic activity, while nitrate, sulfate, and certain trace metals have indicated the extent of anthropogenic pollution.

It is often assumed that concentrations of contaminants in an ice core are proportional to concentrations in the atmosphere at the time of deposition. Such an assumption may not be justified, however. For example, the diversity of particle sizes and composition in the atmosphere may not be accurately reflected by characteristics of microparticles in glaciers. Chemical species in either gas or particulate form may be altered during transport from air to snow, and rates of transport often vary greatly with time. Because of these problems, interpretation of ice core data is not always straightforward.

Making the best use of the glacial record requires that mechanisms of transport from the atmosphere to the snow surface be accounted for. The purpose of this paper is to discuss these transport mechanisms to provide a link between atmospheric contaminants and ice core data. The paper is divided into three sections. First, our current understanding of the process of dry deposition from the atmosphere onto natural snow surfaces is discussed. Wet deposition is discussed next, with particular attention to scavenging by snow. Finally, methods of accounting for both dry and wet deposition processes in interpreting glacial record data are presented. Processes occurring after deposition, e.g., ion migration in the snowpack or effects of snow densification, may also influence contaminants in ice core but are beyond the scope of this paper.

DRY DEPOSITION ONTO SNOW

The literature contains a considerable amount of information on dry deposition of atmospheric gases and particles. However, much of this information is for vegetated surfaces; only a limited number of studies have considered deposition to snow. This section summarizes current knowledge of dry deposition onto snow surfaces. First, the physical mechanisms of deposition are described. Mathematical models developed for snow are then presented. Finally, experimental data are summarized.

Mechanisms of Dry Deposition

Dry deposition may be defined as the transport of particulate and gaseous contaminants from the atmosphere onto surfaces in the absence of precipitation. It is worthwhile to consider dry deposition as part of an overall atmosphere-surface exchange: gases are sometimes reversibly adsorbed onto surfaces only to be reemitted, while particles may be deposited and subsequently resuspended. We often refer to dry deposition as the net result of a balance between downward and upward fluxes.

The process of dry deposition is typically divided into three steps. First, contaminants are carried through the lowest layers of the atmosphere and into the quasi-laminar sublayer just above the surface. This sublayer is a region of relatively stagnant air and is a consequence of the no-slip condition: the air an infinitesimal distance above a stationary surface will also be stationary. During the second step, the contaminants are transported across the sublayer. Finally, the depositing particles or gas molecules interact with the surface. These three steps are known as aerodynamic transport, boundary layer transport, and surface interaction, respectively.

Aerodynamic transport. Transport from the atmosphere into the quasi-laminar sublayer can occur by eddy diffusion and sedimentation. The former mechanism refers to contaminant movement by turbulent wind eddies from regions of high concentration to regions of lower concentration. This movement reflects the stochastic nature of the wind: turbulent motion is random, and hence there is a tendency for contaminants transported by the eddies to move away from regions of high concentration to form a more uniform system without strong concentration gradients. When the surface is a net sink for the contaminant, the concentration will always be low near the ground, and eddy diffusion will cause a continual flux of contaminant toward the surface. Both particles and gases experience eddy diffusion.

Sedimentation is significant only for particles with diameters greater than a few tenths of a μm. A particle accelerating downward under the influence of gravity will experience an aerodynamic drag force which increases as the velocity increases. The drag force opposes gravity and retards the acceleration. Eventually, the magnitude of the drag force reaches that of the gravitational force and the acceleration decreases to zero. The particle is then falling at the sedimentation velocity, which is a function of the size, shape, and density of the particle.

Boundary layer transport. Transport across the quasi-laminar sublayer can occur by diffusion, interception, inertial motion, and sedimentation. These mechanisms are illustrated in Fig. 1.

Diffusion of contaminants results from the motion of air eddies as well as from Brownian motion. Air eddy motion in the sublayer is much weaker than in the free atmosphere, although turbulent eddies may assist transport to within a few μm of a surface. Brownian diffusion transports particles and gases along a concentration gradient in a manner similar to that of eddy diffusion. However, the driving force is the random thermal energy of the molecules of air and molecules of contaminant species rather than turbulent energy. Brownian diffusion is important for gases and submicron particles, and becomes significant only very close to the surface where turbulent eddies are virtually nonexistent. Transport by Brownian and eddy diffusion is a function of atmospheric conditions, characteristics of the depositing contaminants, and the magnitude of contaminant concentration gradients.

Interception applies only to particles with diameters greater than about 0.1 μm. The mechanism occurs when particles moving with the mean air motion pass sufficiently close to an obstacle to collide with it. Thus interception occurs only at places where the sublayer is smaller than the size of the particle, e.g., where there are rapid changes in the direction of

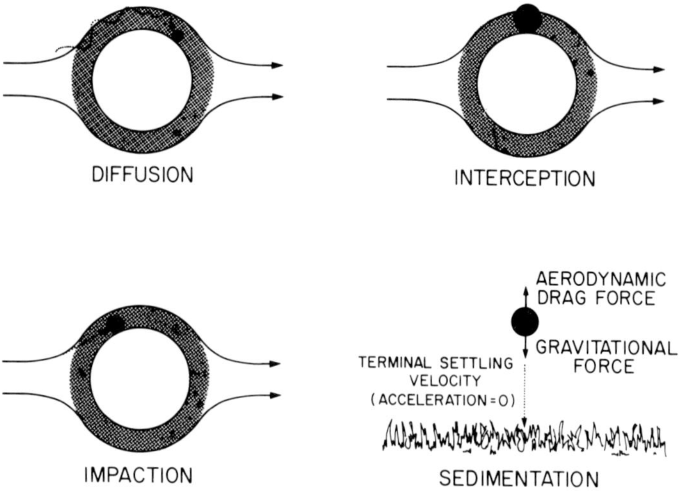

Fig. 1—Mechanisms of transport across the quasi-laminar sublayer. For illustrative purposes, the surface shown is a top view of a circular cylinder, although roughness elements on a snow surface can be highly irregular (reproduced from Davidson and Wu 1988).

air flow and where the sublayer may be narrow. Irregularly shaped ice crystals on a snow surface may collect particles by this mechanism, although it is not likely to be of major importance except where ice needles protrude into the airstream.

Inertial deposition similarly applies to particles with diameters greater than about 0.1 μm. This category includes two mechanisms: inertial impaction and turbulent inertial deposition. Impaction occurs when particles cannot follow rapid changes in the direction of the mean air flow, and their inertia carries them across the sublayer and onto the surface. Unlike interception, a particle subject to impaction leaves the air streamline and crosses the sublayer with inertial energy imparted from the mean air flow. Surface snow characteristics that promote interception, e.g., ice and snow crystals that protrude into the airstream, also promote impaction.

Turbulent inertial deposition applies to particles carried close to the surface by air eddies. The particles are carried across the relatively quiescent air of the sublayer by inertial energy imparted from the eddies. This mechanism differs from impaction in that the transport energy is derived from the component of the turbulent air flow that is normal to the direction of mean flow, rather than from the mean flow. Turbulent inertial deposition is likely to influence deposition of particles with diameters greater than 0.1 μm onto snow, even when the surface is relatively smooth.

Sedimentation through the quasi-laminar sublayer is significant for surfaces with components oriented horizontally. Although especially important for supermicron particles, the mechanism can assist deposition of smaller particles once they are within a few μm of the surface. Sedimentation is often a dominant mechanism of transport through the sublayer for particles with diameters greater than a few tenths of a μm.

Electrostatic forces, thermophoresis, and diffusiophoresis may also affect deposition to varying degrees. These are usually considered to be of lesser importance than the mechanisms described above, although all may be potentially significant for a snow surface. For discussion of these mechanisms, the reader is referred to Davidson and Wu (1988).

Surface interactions. The final step in the deposition process occurs as particles or gases reach the surface. Particles may simply adhere to the surface or react chemically, producing irreversible changes in the deposited material. Gases may adsorb reversibly onto a surface or may undergo chemical reaction. Interaction with the surface may also involve resuspension of incoming particles or reemission of adsorbing vapors.

In the case of a snow surface, resuspension of particles may be significant. This is particularly likely in the case of windblown snow, where the integrity of the surface is poorly defined. Reemission of vapors may also occur.

Mathematical Models

Most modeling efforts begin by defining the dry deposition velocity ν_d,

$$\nu_d(z) = -\frac{F_d}{C_a(z)}, \tag{1}$$

where F_d is the flux or deposition rate per unit area from dry deposition (g/cm^2 s) and C_a is the airborne concentration of contaminant (g/cm^3). The minus sign is needed since downward flux has a negative value but deposition velocity is defined as a positive quantity. Because C_a is a function of height z above the surface, ν_d is also a function of height. F_d is assumed to be constant over an appropriate range of heights.

The deposition velocity ν_d is usually defined in terms of resistances to transport,

$$\nu_d(z) = \frac{1}{r_{total}} = \frac{1}{r_a(z)+r_b+r_c} + \frac{1}{r_g}, \tag{2}$$

where r_a, r_b, r_c, and r_g represent resistances to aerodynamic transport, boundary layer transport, surface transport, and gravitational transport, respectively. The gravitational resistance is a separate term since r_g is

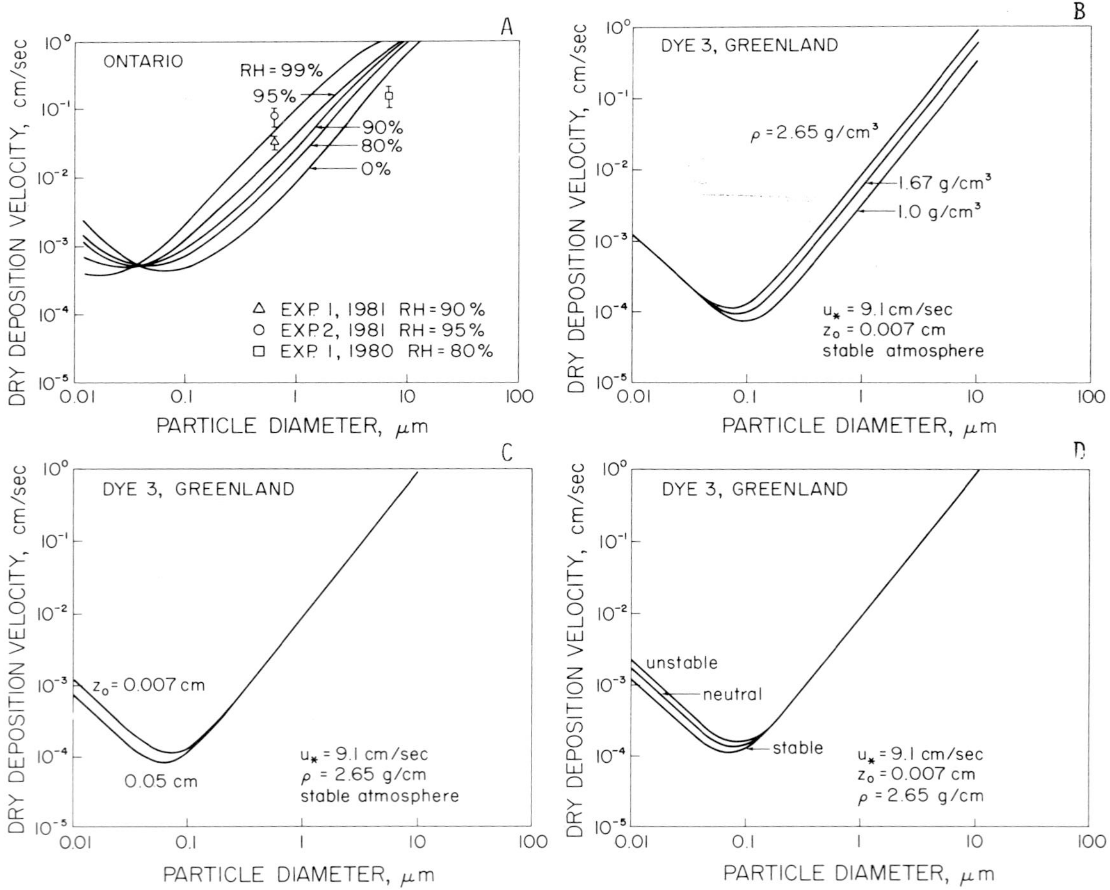

Fig. 2—Deposition velocity versus particle diameter over snow for several conditions, based on a 10 cm reference height. Graph A is redrawn from Ibrahim et al. (1983) and includes experimental data obtained with monodisperse ammonium sulfate aerosol.

considered to be in parallel with the sum of the series resistances ($r_a + r_b + r_c$). r_a is derived by analogy with heat transport in a turbulent atmosphere and uses an eddy diffusivity proportional to height. Corrections for buoyant forces in a nonadiabatic atmosphere are often necessary. r_b is described by eddy and Brownian diffusion, interception, inertial motion, and sedimentation. r_c is usually assumed to be zero for particles, implying perfect retention by the surface, although this resistance may be significant in many circumstances. For gases, r_c is often large enough to be rate-limiting.

Most dry deposition methods have been developed for vegetated surfaces; very little information is available in the literature on modeling dry deposition to snow. However, Ibrahim et al. (1983) have proposed a framework which accounts for the individual mechanisms of deposition onto a snow surface. Separate expressions have been developed for aerodynamic and boundary layer transport, with perfect retention of particles assumed at the surface. The snow is assumed to be composed of ice needles spaced 10 per cm^2, each 0.01 cm in diameter and 0.2 cm in length.

Besides these mechanisms, hygroscopic growth in the region just above the surface is included. Ibrahim et al. estimate that the relative humidity just above the snow surface exceeds 90% when surface temperatures exceed −10°C. They have applied their model to ammonium sulfate aerosol, which deliquesces (dissolves in adsorbed water) at relative humidities in the range 80–90%, depending on temperature. Thus ammonium sulfate particles approaching the snow surface will grow as they encounter the region of high humidity, which will result in enhanced deposition.

For the purpose of this paper, the model has been applied to conditions measured at Dye 3, Greenland. Wind data obtained during summer 1980 at Dye 3 showed that the roughness height z_0 of the snow surface was 0.007 cm, although this value may vary considerably. Note that z_0 is not a measure of the size of snow crystals, but rather is a mathematical term representing the y-intercept on a plot of windspeed (x-axis) versus the logarithm of height. The average friction velocity u_*, which is related to the wind shear stress at the surface, was 9.1 cm/s during these measurements.

Results of the modeling effort are shown in Fig. 2. Graph A gives families of curves taken from Ibrahim et al. (1983), showing the influence of relative humidity and the resulting hygroscopic growth on ν_d. The remaining graphs apply to conditions at Dye 3 at 0% relative humidity. Graph B shows the influence of particle density on ν_d; the value of 1.67 g/cm^3 corresponds to ammonium sulfate, while 2.65 g/cm^3 is typical of soil dust. The effect of roughness height z_0 is shown in graph C. Finally, graph D shows the influence of atmospheric stability, based on a Monin-Obukhov length L equal to −100 m, ∞, and +100 m for an unstable, neutral, and stable atmosphere, respectively.

All of the curves show minima at particle diameters in the range 0.05–0.1 μm. Because Brownian diffusivity increases as particle size decreases, particles smaller than 0.05 μm have large diffusivities and hence rapid transport across the quasi-laminar sublayer. For particles larger than 0.1 μm, interception and impaction become increasingly important, so ν_d increases. Particles larger than a few tenths of a μm in diameter are deposited almost entirely by sedimentation. In general, high deposition velocities are promoted by unstable atmospheric conditions, high windspeeds, and rough surfaces.

Experimental Data

Several studies have reported deposition data for snow. Some of these are field experiments investigating the transport of ambient aerosols and gases, while others have examined deposition of artificially generated aerosol in the field. Chamber studies involving deposition to snow under controlled conditions have also been conducted. Many of these studies are summarized in Table 1.

Overall, the deposition velocities in Table 1 show wide ranges of values. The values depend on characteristics of the depositing species, the type of experiment (field or chamber), and, in the case of the field experiments, the atmospheric conditions. Despite the variations, some interesting conclusions can be reached from the data.

The overall average ν_d for SO_2, for the nine sets of field experiments including this species (excluding the very high value of 1.6 cm/s for an unstable atmosphere), is 0.17 ± 0.15 cm/s; in contrast, values of ν_d for SO_4^{2-} reported in Table 1 are somewhat smaller. This suggests that more sulfur may reach the surface of glaciers by dry deposition when the dominant form of sulfur in the atmosphere is SO_2 than when the dominant form is SO_4^{2-}. Such may be the case during the polar winter when oxidation to SO_4^{2-} is slow. A particularly large deposition velocity is seen for ambient HNO_3; other investigators have shown that this species reacts efficiently with most types of surfaces including snow, so that overall dry deposition is essentially determined by the rate at which eddy diffusion transports the gas to the surface (aerodynamic transport). More nitrogen will reach glaciers by dry deposition when atmospheric HNO_3 is abundant than when other nitrogen-containing species are dominant.

Of particular interest are the values of ν_d for trace metals. The small deposition velocities for Cu and Pb reflect the predominantly submicron particle sizes associated with these species. Emissions of both metals on a global scale are dominated by man-made combustion processes. Values for the other metals, many of which are derived from erosion of the Earth's

Table 1 Examples of dry deposition velocities for snow reported in the literature

Species	Dry Deposition Velocity, cm/s	Details of Experiments	Reference
SO_2	0.25 ± 0.20	stable atmosphere	Barrie & Walmsley (1978)
SO_2	0.057 ± 0.025	cold	Cadle et al. (1985)
	0.15 ± 0.13	warm	
SO_2	0.082 ± 0.062	surface snow	Dasch & Cadle (1986)
	0.12 ± 0.11	bucket containing snow/water	
SO_2	0.13	ν_d assumes no SO_4^{2-} deposition	Dovland & Eliassen (1976)
SO_2	0.1	chamber with snow	Granat & Johansson (1983)
SO_2	1.6	unstable atmosphere	Whelpdale & Shaw (1974)
	0.52	neutral atmosphere	
	0.05	stable atmosphere	
SO_4^{2-}	0.03	surface snow in Greenland	Davidson et al. (1985)
SO_4^{2-}	0.039 (Exp. 1, 1981)	stable atmosphere, 0.7 μm	Ibrahim et al. (1983)
	0.096 (Exp. 2, 1981)	unstable atmosphere, 0.7 μm	
	0.16 (Exp. 1, 1980)	unstable atmosphere, 7 μm	
NH_4^+	0.10 ± 0.11	surface snow	Cadle et al. (1985)
NH_4^+	0.083 ± 0.083	surface snow	Dasch & Cadle (1986)
	0.13 ± 0.17	bucket containing snow/water	
HNO_3	1.4 ± 1.0	surface snow	Cadle et al. (1985)
HNO_3	0–0.67	chamber with snow	Johansson & Granat (1986)
Cl^-	4.3 ± 6.1	surface snow	Dasch & Cadle (1986)
	5.1 ± 4.0	bucket containing snow/water	
Ca	2.1 ± 1.8	surface snow	Cadle et al. (1985)
Mg	1.5 ± 1.3		
Na	0.44 ± 0.48		
K	0.51 ± 0.60		
Ca	2.0 ± 1.8	surface snow	Dasch & Cadle (1986)
	2.7 ± 2.6	bucket containing snow/water	
Cu	0.08 ± 0.04	surface snow in Greenland	Davidson et al. (1985)
Al	0.2 ± 0.06		
Fe	0.6 ± 0.09		
K	0.05 ± 0.02		
Mg	0.2 ± 0.06		
Mn	0.3 ± 0.1		
Na	0.2 ± 0.02		
Pb	0.16	surface snow	Dovland & Eliassen (1976)

crust, are generally greater due to larger particle sizes. The importance of size distributions is also seen when comparing v_d for eastern Michigan (Cadle et al. 1985; Dasch and Cadle 1986) and for Greenland in Table 1: the difference reflects decreasing particle sizes during long-range transport.

WET DEPOSITION

This mechanism refers to the removal of gaseous and particulate contaminants from the atmosphere by precipitation. Most wet deposition studies concern scavenging by rain; in this paper, we are primarily interested in removal by snow. This section begins by describing the physical mechanisms of contaminant scavenging by snow. Then mathematical modeling of the wet deposition process is considered. Finally, relevant experimental data are summarized.

Mechanisms of Wet Deposition

Junge (1977) describes three categories of precipitation scavenging in the Arctic. These are nucleation, in-cloud scavenging by existing cloud droplets and ice crystals, and below-cloud scavenging. The various steps in the scavenging process are illustrated in Figure 3.

Nucleation scavenging. This process refers to contaminant particles serving as condensation nuclei or ice nuclei. In the first category, hygroscopic particles may become centers of growth for water droplets when the supersaturation with respect to liquid water becomes sufficiently high (step 1 in Fig. 3). Clouds can contain liquid droplets at temperatures down to −30°C, known as supercooled water. Atmospheric sulfate particles, abundant in remote as well as urban areas, generally make good condensation nuclei. Hence condensation is likely to be important even in the polar regions.

If temperatures become sufficiently cold during transport of the cloud, ice nuclei may become activated (step 2). Unlike condensation nuclei, these consist of nonhygroscopic particles with crystal lattice structures and shapes conducive to the formation of ice. Aerosols derived from crustal erosion generally make good ice nuclei (Kumai 1967; Mason 1975; Pruppacher and Klett 1980). Upon initiation of the ice phase in clouds, the water droplets formed by condensation begin to evaporate and the resulting water vapor diffuses to the growing ice crystals. This occurs because the supersaturation with respect to liquid water is always less than that with respect to ice. Note that many of the condensation nuclei may be returned to the atmosphere during this evaporation. In cases where clouds form in very cold temeratures, ice nucleation may be initiated without the condensation step.

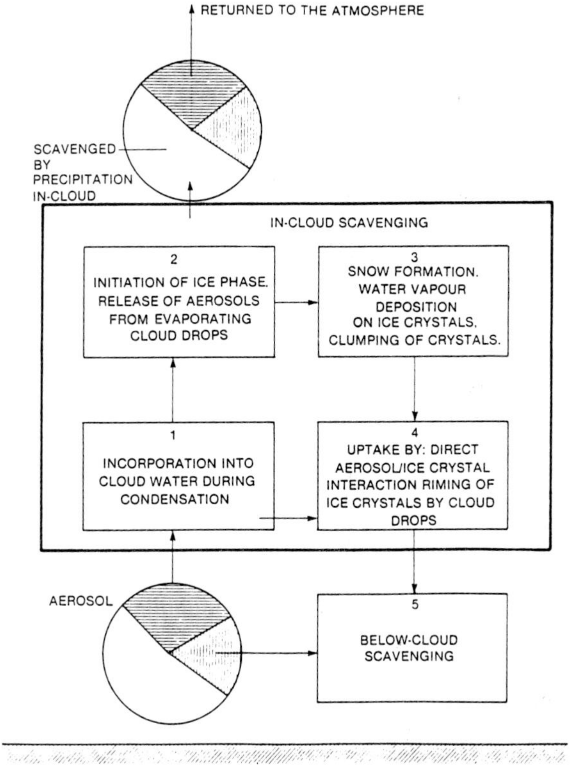

Fig. 3—Steps in the precipitation scavenging process (reproduced from Barrie 1985a).

Once the contaminant particles become active as nuclei, the resulting ice crystals and water droplets can coalesce to form snowflakes and raindrops (steps 3 and 4). The crystals and droplets are collectively referred to as cloud water and are typically associated with diameters less than 50 μm. Snowflakes and raindrops are collectively known as hydrometeors and have diameters of 100–2000 μm. Slinn (1984) notes that a single hydrometeor may consist of roughly 10^6 original ice crystals or water droplets, each with a contaminant particle as a nucleus, illustrating the effectiveness of precipitation scavenging as a removal mechanism. In some cases, this coalescence can involve growth of ice crystals by accretion of supercooled liquid water droplets as opposed to ice crystal growth only by diffusion of water vapor, a process known as riming (step 4). Scott (1981), Borys et al. (1983), and Barrie (1985b) have demonstrated the importance of riming in scavenging contaminants.

Collection by existing cloud droplets and ice crystals. In-cloud scavenging occurs as contaminant gases and particles collide with cloud water or hydrometeors (step 4). Each ice crystal, water droplet, or hydrometeor is surrounded by a quasi-laminar sublayer which the contaminants must traverse if collision is to occur. The mechanisms of transport are similar to those in the dry deposition process, namely diffusion, interception, impaction, and sedimentation. In addition, phoretic forces caused by evaporation and electrical effects may be significant.

It is important to note that the in-cloud scavenging rate for particles depends primarily on the rate of transport across the sublayer: it is usually assumed that all particles reaching the surface will be collected (see Slinn, 1984, for additional discussion). The scavenging rate for gases, on the other hand, depends on retention by the cloud water or hydrometeor, which is a function of the solubility or adsorptivity of the gas. Transport across the quasi-laminar sublayer is not likely to be rate-limiting due to the high diffusivities of most gases.

Below-cloud scavenging. Gases and particles can also be scavenged by falling precipitation (step 5). Each hydrometeor falling toward the Earth's surface is surrounded by a quasi-laminar sublayer across which contaminants may be transported. Thus the mechanisms of below-cloud scavenging are similar to those within the cloud.

Because most of the growth of ice crystals, water droplets, and hydrometeors occurs within clouds, below-cloud scavenging is not likely to be as important as the previous two mechanisms. Junge (1977) and Shaw (1980) conclude that nucleation scavenging is probably the dominant mode of wet deposition of particles in the polar regions. It is likely that in-cloud scavenging dominates for wet deposition of gases.

Mathematical Models

It is possible to define a wet deposition velocity as the ratio of the wet flux F_w to the airborne concentration of a contaminant, analogous to the dry deposition velocity. However, it is usually more instructive in the case of wet deposition to express the loss of airborne contaminant as a first order process defined by the scavenging rate $\psi(s^{-1})$:

$$\frac{dC_a}{dt} = -\psi C_a \tag{3}$$

$$F_w = -\int_0^\infty \psi C_a \, dz = \rho_w C_s P_0. \tag{4}$$

The parameter ρ_w in equation 4 represents the density of snow meltwater (g/cm^3), C_s is the contaminant concentration in the falling snow (g/g snow), and P_0 is the precipitation rate at the ground in cm of snow meltwater/s.

To evaluate ψ, Slinn (1984) proposes a simple expression for the scavenging rate which applies both within and below clouds:

$$\psi = \int_0^\infty \epsilon\,(a,\lambda) A \nu_s(\ell)\, N(\ell)\, d\ell \tag{5}$$

where $\epsilon(a,\lambda)$ is efficiency of collision between a contaminant particle of radius a and a hydrometeor of characteristic capture length λ. A, ν_s, and N represent the effective cross-sectional area, sedimentation velocity, and number concentration of hydrometeors of size ℓ. Note that ℓ and λ may be identical for a spherical droplet, but are likely to be different for irregularly shaped snowflakes or ice crystals. The collision efficiency ϵ is computed by accounting for the various transport mechanisms as in the case of dry deposition. Slinn (1984) suggests a semiempirical expression for ϵ based on values of λ which are adjusted to provide a reasonable fit to available laboratory and field data. Graphs of ϵ versus contaminant particle radius are shown in Fig. 4. The shapes of the curves are similar to those in Fig. 2, illustrating similarities in the mechanisms of wet and dry deposition.

One can further simplify equation 5 by assuming a characteristic hydrometeor size ℓ_c (based on volume/area) and incorporating a total precipitation rate P:

$$\psi = \gamma\epsilon(a,\lambda)P/\ell_c \tag{6}$$

where γ is a dimensionless constant of order unity. Equation 6 is graphed in Fig. 5 for the case of $\epsilon = 1$ and $\gamma = \pi/4$ for several crystal types, and for other values of the product $\gamma\epsilon$ chosen to fit the available experimental data.

Experimental Data

Only a limited number of field studies have considered precipitation scavenging of airborne gases and particles by snow. Results have typically been expressed in terms of a scavenging ratio W, equal to $\rho_a C_s/C_a$, where ρ_a is the density of air. Note that W represents the combined effects of all of the precipitation scavenging mechanisms discussed above, and is related to ψ through equation 4. Examples of typical values of W for snow are given in Table 2.

The studies summarized in Table 2 can be used to explore precipitation scavenging under field conditions. The possible role of crustal aerosol in ice nucleation is suggested by the large values of W for Al, Fe, K, Mg, Mn, and Na. Much smaller values are reported for Ag, As, Pb, and SO_4^{2-}, associated mainly with combustion-derived submicron particles (Milford and Davidson 1985). The large values for Cd and Cu are surprising considering the small particle sizes for these metals; possible reasons are discussed in

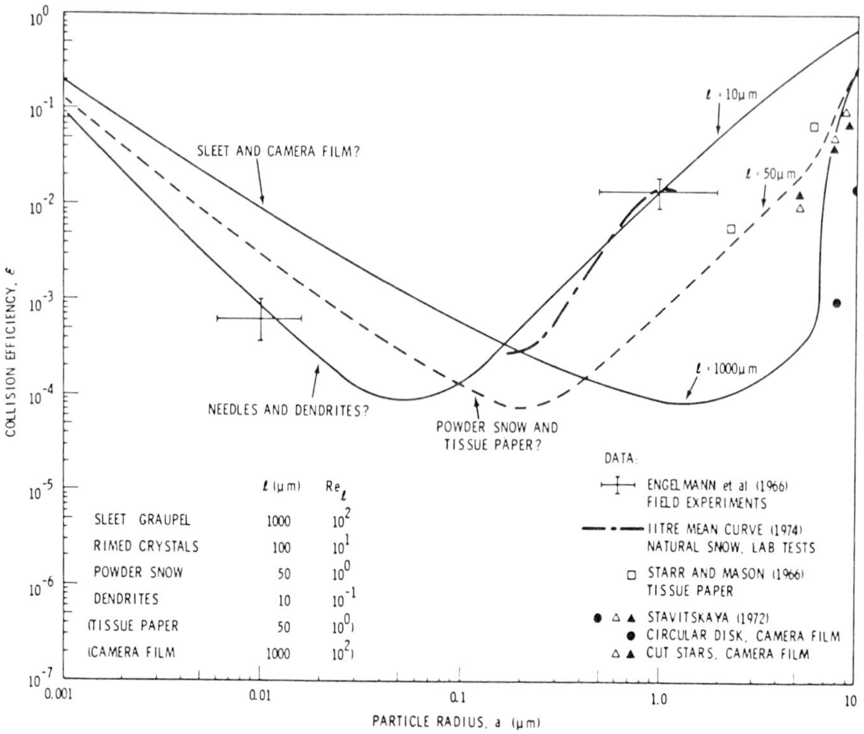

Fig. 4—Collision efficiency ε as a function of particle radius a for various types of collectors: hydrometeors, tissue paper fibers, and cut pieces of camera film (reproduced from Slinn 1984). The Reynolds number Re_ℓ is defined as $\nu_s \ell/\nu$, where ν_s is the sedimentation velocity of the collector and ν is the kinematic viscosity of air.

the original reference. Note that the Dye 3 scavenging ratios are based on sampling when precipitating clouds were at or near the surface of the ice sheet. The inverse dependence of W on precipitation amount is reflected in the data of Engelmann (1970) and of Barrie (1985b). Figure 6, taken from the latter reference, illustrates an empirical relation derived from scavenging ratios for snow at six eastern Canadian sites at times when most of the airborne sulfur was in the form of particulate SO_4^{2-}. Finally, the importance of riming is shown in figure 7, taken from Scott (1981). Barrie (1985a) notes that the high scavenging ratios for ^{210}Pb and ^{90}Sr, reported by Lambert et al. (1983) in Table 2, suggest the importance of riming in Antarctica. The low value of W for SO_4^{2-} in Greenland in the table is consistent with the lack of riming during these winter/spring storms. It must be cautioned that the scavenging ratio data in Table 2 represent a wide range of temperatures, which complicates attempts to compare values among the studies.

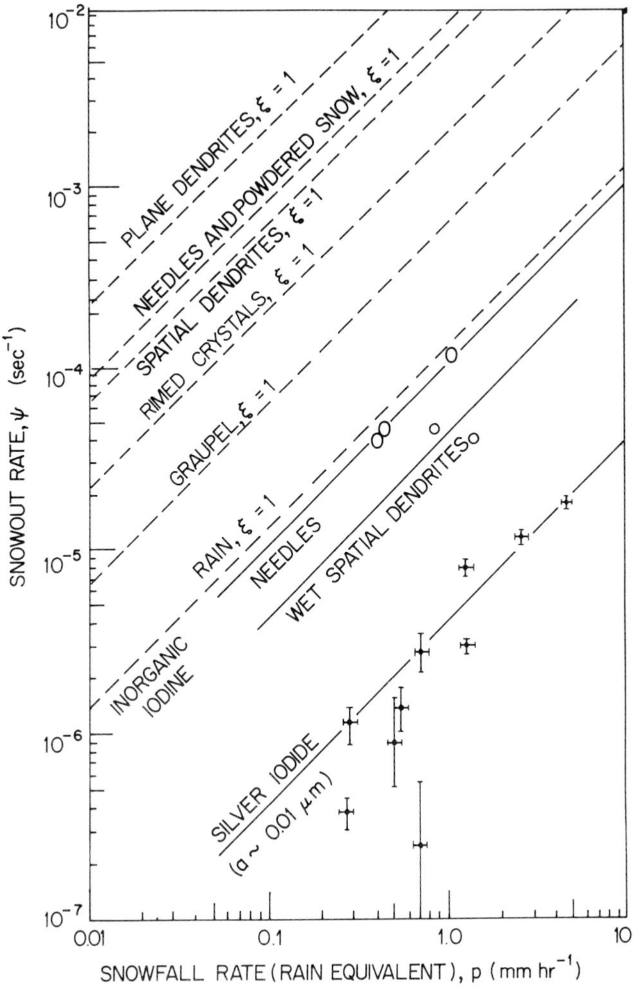

Fig. 5—Scavenging rate ψ, also termed the snowout rate, as a function of precipitation rate P (reproduced from Slinn 1984).

ACCOUNTING FOR WET AND DRY DEPOSITION WHEN INTERPRETING THE GLACIAL RECORD

In the preceding sections, many of the processes by which atmospheric contaminants are incorporated into snow were described. We now explore the extent to which our current knowledge of deposition can apply to polar glaciers. First I consider application of the modeling results and field data presented above to conditions in the polar regions. Then I explore the use

Table 2 Examples of scavenging ratios for snow reported in the literature

Species	Scavenging Ratio (g contaminant/g snow) / (g contaminant/g air)	Details of Experiment	Reference
Gross Beta Activity (Bomb Debris)	1100 620 290	P = 0.15 mm/day P = 1 mm/day P = 10 mm/day snow in Germany	Engelmann (1970)
^{131}I	100–2700		Engelmann (1970)
^{210}Pb	535–724	Antarctica	Lambert et al. (1983)
^{90}Sr	877–1451		
Al	1300 ± 130	Dye 3, Greenland	Davidson et al. (1985)
Fe	1700 ± 400		
K	1200 ± 620		
Mg	1800 ± 1100		
Mn	1600 ± 860		
Na	2000 ± 1200		
Ag	< 250		
As	< 490		
Cd	1100 ± 520		
Cu	1800 ± 860		
Pb	160 ± 70		
NO_3^-	980 ± 780		
SO_4^{2-}	180 ± 120		
SO_4^{2-}	150–3000 20–400	rimed snow unrimed snow in Michigan	Scott (1981)
SO_4^{2-}	150–7000	rimed and unrimed snow in Eastern Canada	Barrie (1985b)

of recent shallow ice core and snowpit data to help understand contaminant transport to polar glaciers.

Application of Modeling Results and Field Deposition Data

The dry and wet deposition models discussed above are of only limited use in helping interpret ice core data. This is because of weaknesses in the models as well as uncertainties in the input data for polar conditions.

For dry deposition, uncertainties exist in defining the three major resistances to transport. For example, micrometeorological data for estimating r_a in the polar regions are scarce, while details of the surface snow characteristics influencing r_b and r_c are also limited. Particle size distributions for many chemical species of interest are not available. Furthermore, rates

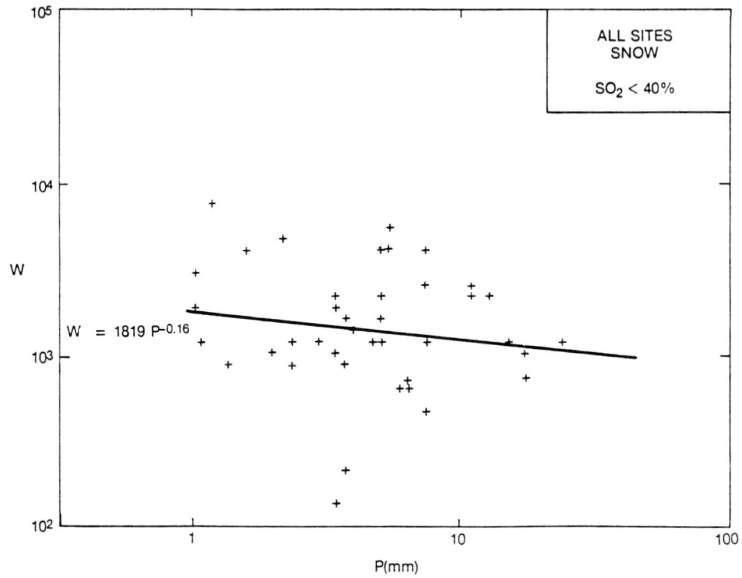

Fig. 6—Scavenging rate W versus total precipitation on an event basis for snow in eastern Canada (reproduced from Barrie 1985b).

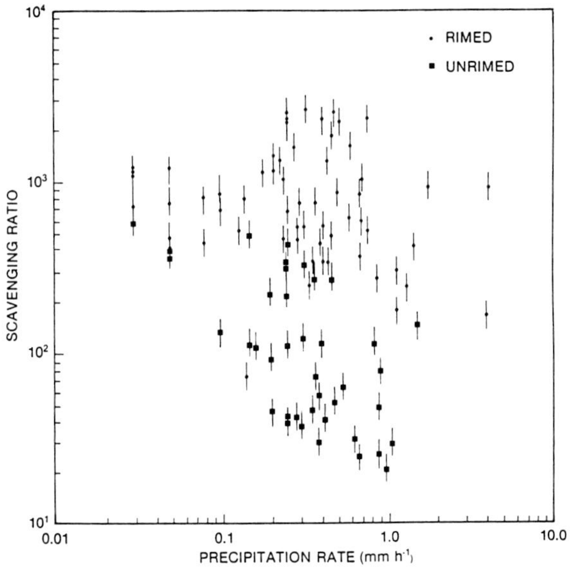

Fig. 7—Scavenging rate W versus precipitation rate for rimed snow and for unrimed snow (reproduced from Barrie 1985a, based on the data of Scott 1981).

of hygroscopic particle growth in the humid layer just above the surface are not known for key species at temperatures encountered in the polar regions. Dry deposition of gases may depend on the presence of liquid water at the surface, a variable which is not well understood for glaciers. Such deposition is likely to vary greatly from one gaseous species to another, depending on gas solubility and other parameters.

There is also considerable uncertainty in using wet deposition models, since little information exists on characteristics of clouds. This is especially true of cold clouds resulting from ice nucleation. Proper use of the models requires input on airborne concentrations of gaseous and particulate species of interest in and below clouds, nucleation properties of the contaminant particles, size distributions of ice crystals, water droplets, and hydrometeors in clouds, the extent of riming, and other factors. Such information for the mid-latitudes is scarce, and for the polar regions is virtually absent. Slinn (1984) has discussed possible errors in using ground-level airborne concentrations as inputs to scavenging models where concentration data at cloud-level are needed. However, this problem may not be severe in areas far from sources where concentrations may be more uniform with height.

It is clear that the existing models cannot be used to predict accurately rates of dry and wet deposition to glaciers. However, in combination with field deposition data, the models can provide valuable information on the relative importance of wet and dry deposition, and can identify areas where research is needed. For example, using available size distribution data for Pb in remote areas (Milford and Davidson 1985) with the dry deposition models of Fig. 2 yields dry deposition velocities of 0.02–0.05 cm/s. This may be compared with an annual average wet deposition velocity of 0.2 cm/s, calculated from the scavenging ratio for Pb in Table 2 with P_0 = 50 cm/yr at Dye 3. A similar dry/wet deposition ratio also applies to the crustal elements at Dye 3, since both dry and wet deposition rates are increased in rough proportion for these larger particles. However, dry deposition will be more important at locations where P_0 is smaller. For example, Davidson et al. (1985) have estimated that dry deposition accounts for at least 35% of the total deposition of trace elements at Thule, Greenland. Furthermore, Legrand (1987) estimates that dry deposition accounts for 60–70% of the total deposition of particulate impurities in central Antarctica, due to very small precipitation rates there. For events of short duration, such as emissions from the Chernobyl nuclear reactor accident, deposition may depend on the amount of precipitation at the receptor site during the brief periods of high airborne concentration. Radioactive cesium from Chernobyl reached the southern Greenland ice sheet mainly by dry deposition; concentrations in the snow would have been much greater if atmospheric transport of the radioactivity had coincided with precipitation over Greenland (Davidson et al. 1987a). It is thus likely that the relative importance of dry and wet deposition varies greatly from one location to another and is dependent upon the contaminant species of interest.

Application of Shallow Ice Core and Snowpit Data from the Arctic

The utility of shallow ice cores and snowpits as a means of studying deposition has only recently been recognized. Two studies have attempted to compare airborne concentrations from long-term monitoring programs with snow chemistry records.

The study of Barrie et al. (1985) has compared acidity levels in Agassiz ice cap with acidity in the atmosphere at Alert, 240 km away. Both sites are on Ellesmere Island in the Canadian Arctic. Data over a three-year period show much stronger seasonal variations in the air compared with the snow, attributed to seasonal variations in deposition. An overall average scavenging ratio of 240 has been calculated from the data.

Davidson et al. (1987b; 1988) have examined seasonal variations in airborne SO_4^{2-}, NO_3^-, and Cl^- concentrations in the Arctic by statistically analyzing airborne data from sites in Norway, Greenland, and Canada. The results have been compared with Dye 3 snowpit data for the interval 1982–1987. An example of the findings for SO_4^{2-} is shown in Fig. 8, using snowpit data obtained by Mayewski et al. (1987) and Davidson et al. (1988) for the five-year period. Concentrations in air show a strong peak in late February, while those in the snow show a weaker peak in early March. These differences in seasonal patterns are attributed in part to the effect of riming, which results in more efficient scavenging during warmer weather. Note that increasing scavenging efficiencies coincide with decreasing airborne concentrations in the spring, yielding smaller variations in concentration in the snow than in the air. It is also significant that the scavenging ratio peaks in the summer: values of W in June, July and August average 470, while values during the rest of the year average 210. Such variations in deposition are partially responsible for observed seasonal variations in concentrations of SO_4^{2-} and other contaminants in snowpits and ice cores, such as those reported by Herron (1982), Neftel et al. (1985), Finkel et al. (1986), and Mayewski et al. (1986).

Despite these complexities, glacial record data show consistencies indicating that past atmospheric concentrations are, to some extent, represented in ice cores. In order to quantify the relations between concentrations in air and those in glaciers, however, atmospheric monitoring programs need to be established at ice coring sites. Such programs should include simultaneous sampling of contaminants in air and snow during the variety of meteorological conditions likely to be encountered throughout a full year. Air sampling is needed both at the surface and at cloud level. Snowpit studies examining changes in contaminant levels in the snow after deposition are also needed

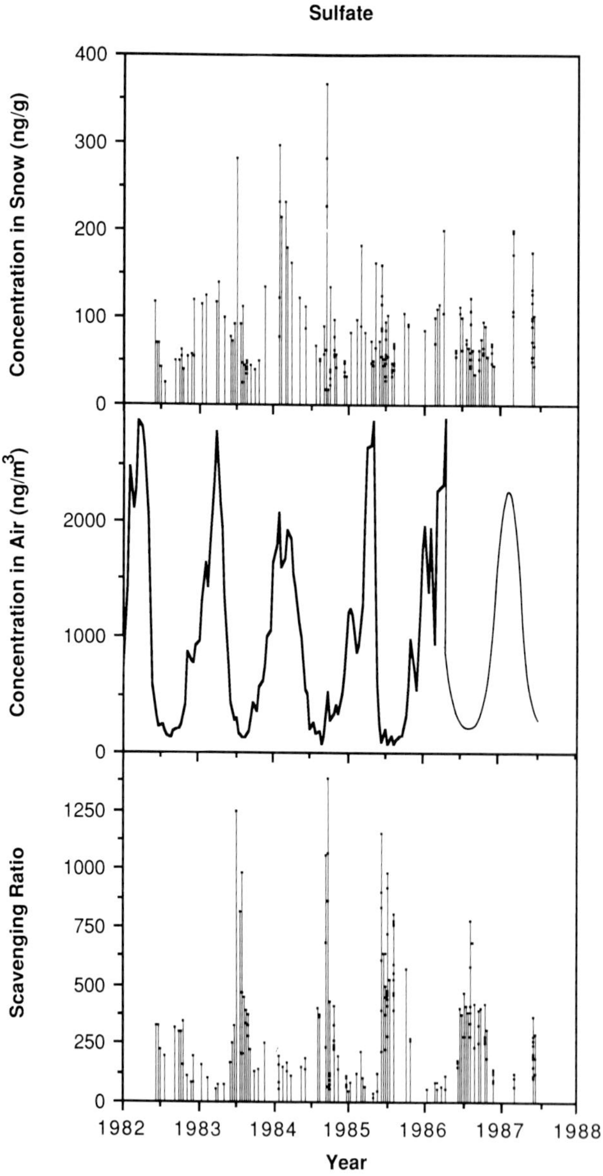

Fig. 8—Sulfate concentrations in snow and air, and scavenging ratio W, based on Dye 3 snowpit data (modified from Davidson et al. 1988).

to assess the integrity of the glacial record. Laboratory studies of fundamental mechanisms of wet and dry deposition are especially needed to improve our understanding of air-to-snow transport processes influencing contaminants in ice cores.

SUMMARY

Transport of atmospheric gases and particles to glaciers occurs by dry and wet deposition. Dry deposition can be considered a three-step process. First, gases and particles are brought from the free atmosphere down to the quasi-laminar sublayer just above the snow surface. The contaminants are then transported across the sublayer. Finally, the gases or particles interact with the surface and may become incorporated into the snow. Wet deposition also involves a number of steps. Water droplets and ice crystals in clouds are formed when contaminant particles are activated as condensation nuclei and ice nuclei. The droplets and crystals grow by diffusion of water vapor and by collisions with other droplets and crystals, resulting in raindrops and snowflakes which eventually fall to the Earth's surface. Contaminant particles and gases can be scavenged during any of these steps.

The relative importance of dry and wet deposition can vary greatly among contaminant species and at different locations. Wet deposition is likely to dominate at sites where annual precipitation rates are appreciable. At sites with particularly low annual precipitation, and where the snowfall is mostly unrimed, dry deposition may dominate.

Although wet and dry deposition data for snow surfaces exist for some species, data applicable to the polar regions are extremely limited, and for many species are completely lacking. As a result, we cannot yet calculate with reasonable accuracy atmospheric concentrations based on ice core data. At this point, we are only beginning to understand the complexities of air-to-snow transport which are vital to proper interpretation of the glacial record.

Acknowledgments. This work was supported by National Science Foundation Grant DPP-8618223. Calculations leading to the dry deposition models for Dye 3 were performed by Richard Honrath. Valuable comments on the manuscript were received from several participants of this Dahlem Workshop.

REFERENCES

Barrie, L.A. 1985a. Atmospheric particles: their physical/chemical characteristics and deposition processes relevant to the chemical composition of glaciers. *Ann. Glaciol.* **7**: 100–108.

Barrie, L.A. 1985b. Scavenging ratios, wet deposition, and in-cloud oxidation: an application to the oxides of sulphur and nitrogen. *J. Geophys. Res.* **90**: 5789–5799.

Barrie, L.A.; Fisher, D.; and Koerner, R.M. 1985. Twentieth century trends in Arctic air pollution revealed by conductivity and acidity observations in snow and ice in the Canadian High Arctic. *Atmos. Envir.* **19**: 2055–2063.

Barrie, L.A., and Walmsley, J.L. 1978. A study of sulfur dioxide deposition velocities to snow in Northern Canada. *Atmos. Envir.* **12**: 2321–2332.

Borys, R.D.; Demott, P.J.; Hindman, E.D.; and Feng, D. 1983. The significance of snow crystal and mountain-surface riming to the removal of atmospheric trace constituents from cold clouds. Proc. Fourth International Conference on Precipitation Scavenging, Dry Deposition, and Resuspension, Santa Monica, CA, 1982, pp. 181–189. New York: Elsevier.

Cadle, S.H.; Dasch, J.M.; and Mulawa, P.A. 1985. Atmospheric concentrations and the deposition velocity to snow of nitric acid, sulfur dioxide and various particulate species. *Atmos. Envir.* **19**: 1819–1827.

Dasch, J.M., and Cadle, S.H. 1986. Dry deposition to snow in an urban area. *Water Air Soil Poll.* **29**: 297–308.

Davidson, C.I.; Harrington, J.R.; Stephenson, M.J.; Monaghan, M.C.; Pudykiewicz, J.; and Schell, W.R. 1987a. Radioactive cesium from the Chernobyl accident in the Greenland ice sheet. *Science* **237**: 633–634.

Davidson, C.I.; Honrath, R.E.; Kadane, J.B.; Tsay, R.S.; Mayewski, P.A.; Lyons, W.B.; and Heidam, N.Z. 1987b. The scavenging of atmospheric sulfate by Arctic snow. *Atmos. Envir.* **21**: 871–882.

Davidson, C.I.; Santhanam, S.; Fortmann, R.C.; and Olson, M.P. 1985. Atmospheric transport and deposition of trace elements onto the Greenland ice sheet. *Atmos. Envir.* **19**: 2065–2081.

Davidson, C.I.; Small, M.J.; Harrington, J.R.; Stephenson, M.J.; Boscoe, F.P.; and Gandley, R.E. 1988. The transport of atmospheric sulfate, nitrate, and chloride to the Greenland ice sheet. *Atmos. Envir.*, in press.

Davidson, C.I., and Wu, Y.L. 1988. Dry deposition of particles and vapors. In: Acid Precipitation, ed. D.C. Adriano, vol. 2. New York: Springer, in press.

Dovland, H., and Eliassen, A. 1976. Dry deposition on a snow surface. *Atmos. Envir.* **10**: 783–785.

Engelmann, R.J.; Perkins, R.W.; Hagen, D.I.; and Haller, W.A. 1966. Washout Coefficients for Selected Gases and Particulates. Report BNWL-SA-657. Richland, WA: Battelle Pacific Northwest Laboratories.

Engelmann, R.J. 1970. Scavenging predictions using ratios of concentrations in air and precipitation. In: Precipitation Scavenging (1970), eds. R.J. Engelmann and W.G.N. Slinn, pp. 475–486. CONF-700601. Springfield, VA: NTIS.

Finkel, R.C.; Langway, C.C., Jr.; and Clausen, H.B. 1986. Changes in precipitation chemistry at Dye 3, Greenland. *J. Geophys. Res.* **91**: 9849–9855.

Granat, L., and Johansson, C. 1983. Dry deposition of SO_2 and NO_x in winter. *Atmos. Envir.* **17**: 191–192.

Herron, M.M. 1982. Impurity sources of F^-, Cl^-, NO_3^-, and SO_4^{2-} in Greenland and Antarctic precipitation. *J. Geophys. Res.* **87**: 3052–3060.

Ibrahim, M.; Barrie, L.A.; and Fanaki, F. 1983. An experimental and theoretical investigation of the dry deposition of particles to snow, pine trees, and artificial collectors. *Atmos. Envir.* **17**: 781–788.

IITRE 1974. See Knutson, E.O., and Stockham, J.D. 1977. Aerosol scavenging by snow: comparisons of single-flake and entire-snowfall results. In: Precipitation Scavenging (1974), eds. R.C. Semonin and R.W. Beadle, pp. 195–207. CONF-741003. Springfield, VA: NTIS.

Johansson, C., and Granat, L. 1986. An experimental study of the dry deposition of gaseous nitric acid to snow. *Atmos. Envir.* **20**: 1165–1170.

Junge, C.E. 1977. Processes responsible for the trace content in precipitations. In: Isotopes and Impurities in Snow and Ice. Proc. IUGG Symp., Grenoble, 1975. *IAHS-AISH Publ.* **118**: 63–77.

Kumai, M. 1967. Fog modification on the Greenland Ice Cap. Proc. First National Conference on Weather Modification, Albany, NY, pp. 414–422.

Lambert, G.; Ardouin, B.; and Mesbah-Bendezu, A. 1983. Atmosphere to snow transfers in Antarctica. Proc. Fourth International Conference on Precipitation Scavenging, Dry Deposition, and Resuspension, Santa Monica, CA. 1982, pp. 1353–1360. New York: Elsevier.

Legrand, M. 1987. Chemistry of Antarctic snow and ice. *J. de Phys.* **48**: C1-77–C1-86.

Mason, B.J. 1975. Clouds, Rain and Rainmaking, pp. 66–67. Cambridge: Cambridge Univ. Press.

Mayewski, P.A.; Lyons, W.B.; Spencer, M.J.; Twickler, M.; Dansgaard, W.; Koci, B.; Davidson, C.I.; and Honrath, R.E. 1986. Sulfate and nitrate concentrations from a south Greenland ice core. *Science* **232**: 975–977.

Mayewski, P.A.; Spencer, M.J.; Lyons, W.B.; and Twickler, M.S. 1987. Seasonal and spatial trends in south Greenland snow chemistry. *Atmos. Envir.* **21**: 863–869.

Milford, J.B., and Davidson, C.I. 1985. The sizes of particulate trace elements in the atmosphere: a review. *J. Air Poll. Control Assn.* **35**: 1249–1260.

Neftel, A.; Beer, J.; Oeschger, H.; Zurcher, F.; and Finkel, R.C. 1985. Sulfate and nitrate concentrations in snow from south Greenland. *Nature* **314**: 611–613.

Pruppacher, H.R., and Klett, J.D. 1980. Microphysics of Clouds and Precipitation, pp. 248–268. Dordrecht: Reidel.

Scott, B.C. 1981. Sulfate washout ratios in winter storms. *J. App. Met.* **20**: 619–625.

Shaw, G.E. 1980. Optical, chemical and physical properties of aerosols over the Antarctic ice sheet. *Atmos. Envir.* **14**: 911–922.

Slinn, W.G.N. 1984. Precipitation scavenging. In: Atmospheric Sciences and Power Production, ed. D. Randerson. Technical Information Center, Oak Ridge, TN: U.S. Department of Energy.

Starr, J.R., and Mason, B.J. 1966. The capture of airborne particles by water drops and simulated snow crystals. *Quart. J. Roy. Meteor. Soc.* **92**: 490–499.

Stavitskaya, A.V. 1972. Capture of water-aerosol drops by flat obstacles in the form of star-shaped crystals. *Izv. Akad. Nauk. Atmos. Ocean. Phys.* **8**: 768–774.

Whelpdale, D.M., and Shaw, R.W. 1974. Sulphur dioxide removal by turbulent transfer over grass, snow, and water surfaces. *Tellus* **26**: 196–204.

The Environmental Record in Glaciers and Ice Sheets
eds. H. Oeschger and C.C. Langway, Jr., pp. 53–67
John Wiley & Sons Limited

The Transformation of Snow to Ice and the Occlusion of Gases

Jakob Schwander

Physics Institute
University of Bern
3012 Bern, Switzerland

Abstract. The gases enclosed in the bubbles of glacier ice represent samples of the atmosphere approximately from the time of bubble formation. The age of this air is different from the surrounding ice. Also, the age cannot be given by a simple value but is described by an age distribution. This age distribution is determined by the mixing in the permeable firn layer and the air trapping rate at the firn-ice transition. At present it is possible to estimate the age distribution, using model calculations, in special cases. More data, especially on the spatial variability of the air permeability and diffusivity in the firn, are needed to permit modeling of more general cases. Tracer experiments would be a further possibility to assess the age of the air in the bubbles of ice. Due to physical and chemical processes, the concentrations of the gases in the bubbles may differ from the atmospheric concentrations. Apparently the gas compositions in bubbles of very cold glaciers (no surface melting during summer) do not differ from samples of the atmosphere.

INTRODUCTION

Glacier ice contains a certain amount of air entrapped as small bubbles during the transformation of snow to ice. The main potential of the analysis of the air enclosed in ice is the possibility to reconstruct the composition of the ancient atmosphere. Polar ice covers a time period of more than 100,000 years but not all ice is equally well suited to investigate the earlier atmospheric gas composition. A number of processes may cause the air extracted from ice cores to have a different composition than the atmosphere at the time the bubbles were formed. Figure 1 shows schematically the main processes that determine the amount and composition of gases in the glacier ice. Due to the varying solubility of the different gaseous species, the presence of liquid water has an important influence on the composition of

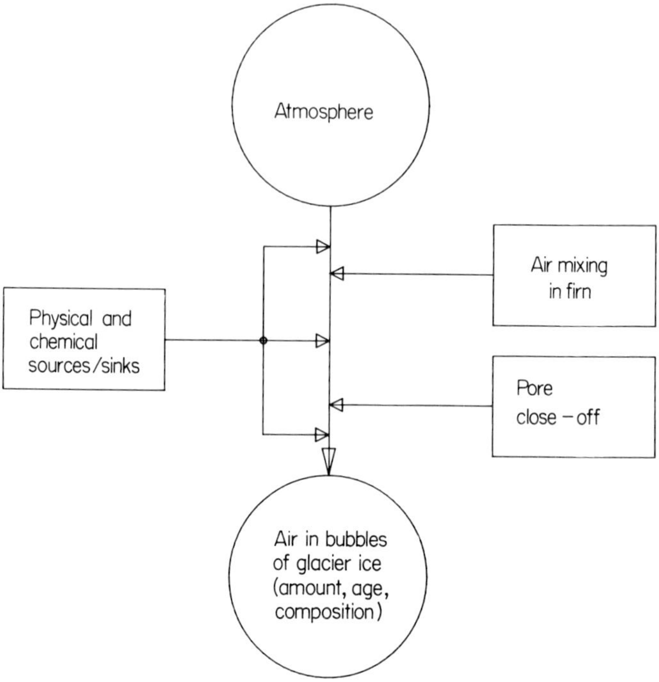

Fig. 1—Schematic representation of the main factors influencing the composition of the gases entrapped in the bubbles of glacier ice.

the air in the bubbles. We therefore distinguish three different types of glacier zones (a particular glacier may be composed of combinations of the following zones):

1. "cold" glacier zone without surface melting (dry snow zone),[1]
2. "cold" glacier zone with surface melting (percolation zone),[1]
3. "temperate" glacier zone.

The temperature of a "cold" glacier zone is below the melting point. Melting, if any, is limited to a layer near the surface. The meltwater percolates through the uppermost snow layers and spreads out laterally when it encounters a more dense layer, where it refreezes to form ice layers or ice lenses. Generally, surface melting occurs regularly in areas with mean annual air temperatures above −25°C. However, even in Central Greenland, where the mean annual temperature is around −30°C, surface melting occurs occasionally. Real dry snow zones are encountered only in Central Antarctica

[1] After C. Benson (1962)

and near the summits of high mountains at high latitudes. However, the firn and ice between melt layers of a cold glacier zone with relatively little ice layers behaves much like ice of a dry snow glacier.

A "temperate" glacier zone is at the pressure melting point throughout, except for a surface layer in which the temperature is below 0°C for part of the year. A network of water carrying capillary veins exists between the ice grains. The gas content and the gas composition of temperate ice are not constant since the gases are washed out by water flowing through its veins. In addition, chemical reactions in the aqueous environment may act as sinks or sources for some gases. Although the analysis of air from a temperate glacier or from a glacier with numerous ice layers may provide some glaciological or climatological information, it is not suited to assess the composition of the ancient atmosphere.

In this chapter I wish to outline processes and their respective relevance that influence the amount, the composition, and the age of the air entrapped in glacier ice. Due to the minor significance of temperate glaciers or ice layer-rich polar ice for gas analysis, the discussion will be restricted to the cold glacier ice.

THE TRANSFORMATION OF SNOW TO ICE

The snow deposited in the accumulation area of a glacier or an ice sheet is compressed and sintered as a result of water vapor diffusion and plastic deformation of the ice grains under the weight of subsequently fallen snow layers. The resulting porous ice structure is called firn. During the first stage of the firn metamorphosis the most important process for the density increase is settlement, that is, rearrangement of the firn grains to reach a closer packing. Above a density of about 550 kg/m^3 a further rearrangement of the ice grains no longer leads to a significant increase of the density. Sintering and plastic deformation become then the dominant processes. Around a density of 800 kg/m^3 the pores are gradually pinched off to form individual bubbles. This is the transition to glacier ice which is, by definition, impermeable to air in a macroscopic sense. Glacier ice is thus a polycrystalline ice aggregate that normally contains air bubbles, but also a variety of chemical inclusions.

The firn metamorphosis is also characterized by a growth of the average grain size. Near the surface the metamorphosis of snow is initially very fast, as a result of large temperature gradients. Below 10 m depth, where seasonal temperature variations are small, the crystal growth is approximately constant and follows the equation (Gow 1975)

$$D^2 - D_0^2 = k \times t, \tag{1}$$

where D^2 and D_0^2 are the mean cross-sectional areas of one firn grain at time t and 0, respectively. The growth factor k has a temperature dependence according to the Arrhenius equation with an activation energy of ca. 42 kJ/mol. At −25°C the growth rate is on the order of 10^{-2} mm^2/yr. The mean cross-sectional area at 10 m depth ranges from 0.25 to 1 mm^2 and is on the order of 1–4 mm^2 at the firn-ice transition.

In the context of air occlusion, the change of the firn structure during the transition from snow to glacier ice can be characterized by several macroscopic parameters: (*a*) density, (*b*) porosity (open and closed),[2] (*c*) air permeability, and (*d*) diffusivity of gases. Parameters (*c*) and (*d*) are closely related to parameter (*b*).

Herron and Langway (1980) developed an empirical formula to describe the densification of firn:

$$\rho(z) = \frac{\rho_i \times X}{1 + X} \tag{2}$$

where

$$X = \begin{cases} \exp[\rho_i \times k_0 \times z + \ln\{\rho_0/(\rho_i - \rho_0)\}] & (\text{if } \rho(z) < 550\ \text{kg/m}^3) \\ \exp[\rho_i \times k_1 \times (z - z_{550})/A^{0.5} + \ln\{550/(\rho_i - 550)\}] & (\text{if } \rho(z) > 550\ \text{kg/m}^3) \end{cases}$$

$z_{550} = 1/(\rho_i \times k_0)[\ln\{550/(\rho_i - 550)\} - \ln\{\rho_0/(\rho_i - \rho_0)\}]$
(= depth where density is 550 kg/m^3)
$k_0 = 0.011 \times \exp[-10160/(RT)]$
$k_1 = 0.575 \times \exp[-21400/(RT)]$
$R = 8.134$ J/(K mol)
z: depth below surface (meter)
T: mean annual temperature (Kelvin)
$\rho(z)$: density at depth z (kg/m^3)
ρ_i: density of pure ice (= 918 kg/m^3)
ρ_0: surface density
A: accumulation rate (meter water per year).

Although ρ_0 varies somewhat from site to site, Eq. 2 yields good results if one takes an average value of 350 kg/m^3. Equation 2 is based on data from the accumulation areas in Greenland and Antarctica (mean annual temperatures between −15 and −57°C) and is thus only valid for relatively cold and flat accumulation areas. The densification rate of the firn is mainly controlled by the mean temperature and the accumulation rate. For sites in

[2] The porosity of a sample is defined as the volume fraction that is not occupied by ice and is the sum of open and closed porosity. Open porosity is the volume fraction of the open space that is connected with the surface of the sample. From this definition it is clear that in general the ratio of the closed to the open porosity of a sample depends on the size of that sample.

Greenland and Antarctica the depth where the firn-ice transition density is attained ranges from about 50–120 m. The corresponding ages of the ice range from about 50–3000 years.

The density dependence of parameters (*b*) to (*d*) for firn of a cold flat area proves to be similar for sites with mean annual temperatures between −20°C and −30°C (Schwander et al., in preparation). The following empirical equations are found:

$$\text{closed porosity } s_{cl} \simeq \begin{cases} s \times \exp[75(\rho/830-1)] & \text{if } 0<\rho<830 \text{ kg/m}^3 \\ s & \text{if } \rho>830 \text{ kg/m}^3 \end{cases} \quad (3)$$

$$\text{permeability} \quad k \simeq 7\times10^{-4}\times s_{op}^2 \{1-\exp[4.455(\rho/810-1)]\} \quad (4)$$

$$\text{diffusivity} \quad D \simeq D_0(1.7s_{op} - 0.2) \quad (5)$$

where s is the total porosity (= $1-\rho/\rho_i$), s_{op} is the open porosity, and D_0 is the diffusion coefficient of the considered gas in air.

In addition, the following equation holds by definition:

$$\text{open porosity } s_{op} = s - s_{cl}. \quad (6)$$

This means that, with Eqs. 2 to 6, we can estimate the depth profiles of parameters (*a*) to (*d*) for cold accumulation areas with annual temperatures between −20°C and −30°C, and as a first approximation also for colder sites, as I do not expect that the Eqs. 3 to 5 change much at temperatures below −30°C.

Snow properties show seasonal variations. For example, winter snow is usually of higher density than summer snow. Since these seasonal properties are at least to some degree conserved throughout the whole firn layer, this causes seasonal variations of the parameters superposed on the general trend with depth. An example showing the behavior for the density, closed porosity, and air permeability is shown in Fig. 2. In addition to the general trend with depth and the more or less regular seasonal pattern, inhomogeneities are observed where ice layers or wind crusts[3] have been formed. The ice layers do not cover a large region continuously but are more or less frequently interrupted by holes or dislocations. According to my knowledge, there exist no extensive investigations on the geometry of

[3] Wind crusts are tiny layers (ca. 1 mm thick) that appear more transparent when looking perpendicular through a firn or ice core. The higher transparency is due to less pores or bubbles in that layer. Wind crusts are formed by wind compaction at the snow surface (Lliboutry 1964), so that they follow the irregular topography. As a consequence the wind crusts are oriented at different inclinations in an ice core and are often interrupted.

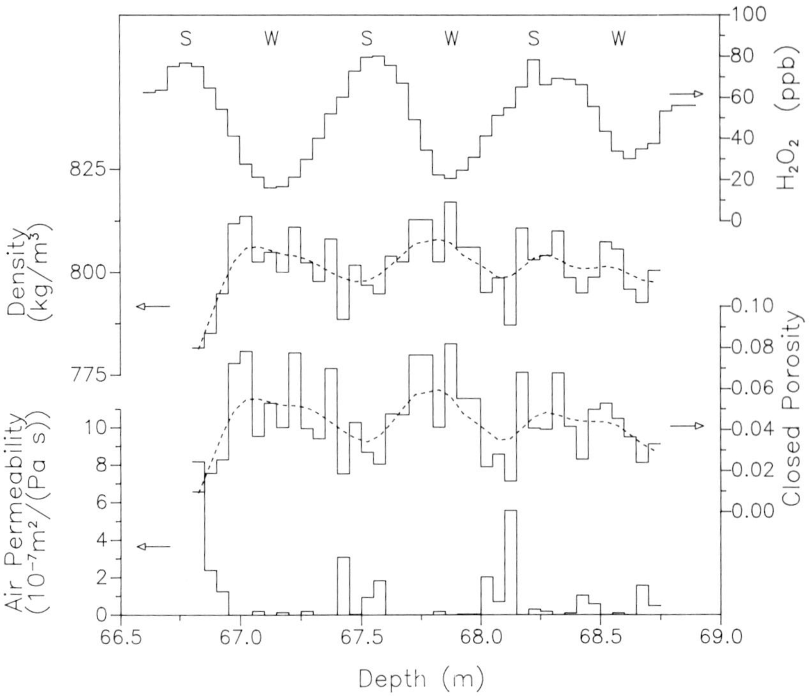

Fig. 2—Seasonal variability of H_2O_2, density, closed porosity and air permeability at Siple Station. Summer (S) and winter layers (W) are marked according to the H_2O_2 record (the lowest values in winter).

ice layers. Ice layers influence the air mixing in the firn. Wind crusts, however, do not significantly influence the air exchange, as air permeability and diffusivity measurements showed (Schwander et al., in preparation). The parameters described so far (density, porosity, permeability, ice layers) depend on climatic parameters such as temperature, snow accumulation rate, wind, solar irradiation, etc. This infers that profiles of the firn parameters are not necessarily stationary for a given site but are subject to secular trends due to changes of the climate.

THE OCCLUSION OF AIR

The first bubbles are already formed in the snowflakes. Especially rimed snowflakes contain numerous micro-bubbles. Most of them are lost during recrystallization. The amount of air in glacier ice originating from such micro-bubbles is not known at present, but there are indications that it is less than one percent (Schwander 1980). A second way in which bubbles are formed is by the refreezing of percolating meltwater. The gases that are dissolved in the meltwater can only partly escape during refreezing. The rest is trapped in the ice layer or ice lens.

The major part of the bubbles is formed when the firn pores are pinched off, that is, during the firn-ice transition. The occlusion of air at the firn-ice transition has been investigated in experiments (Schwander and Stauffer 1984; Schwander et al., in preparation) and in theory (Stauffer, Schwander et al. 1985; Enting 1985; Enting and Mansbridge 1985; Enting 1987). While giving a qualitative insight in the close-off processes, the theory cannot numerically determine the amount of enclosed air in function of depth.

The closed pore volume was measured on samples from Siple Station. The major increase of the total bubble volume takes place between 65 and 80 m depth corresponding to the density interval 795–830 kg/m^3. We must bear in mind that the profile of the closed porosity has been measured on relatively small samples and depends on the size and shape of those samples. For example, if a sample at 80 m depth still shows an appreciable open porosity it does not necessarily mean that its open pore system was connected through the whole firn layer to the atmosphere. There might exist layers at shallower depths that are already completely impermeable to air. In particular, it has been observed that winter snow layers reach pore close-off more quickly than summer layers and isolate the latter from the atmosphere before their permeability decreases to zero. The total amount of air that is finally trapped in cold glacier ice is approximately 100 cm^3 (STP)/kg. Raynaud and Lebel (1979) give an empirical equation relating the total gas content with the mean temperature T_c(K) and the mean pressure p_c(hPa) at pore close-off:

$$V(\text{cm}^3(\text{STP}) \text{ per kg ice}) = p_c(0.20–15.4/T_c). \qquad (7)$$

AIR MIXING AND AGE OF THE AIR

The gases enclosed in the bubbles of glacier ice are of a different age than the surrounding ice. The knowledge of this age relationship is especially necessary for the accurate chronological interpretation of fluctuations and trends of trace-gas concentrations as obtained from ice core studies: for example, the increase in trace gases due to human activity (Neftel et al. 1985; Stauffer, Fischer et al. 1985; Khalil and Rasmussen 1987), or the changes in the CO_2 concentration during and at the end of the last glaciation in relation to other climatic parameters recorded in the ice (Staffelbach et al. 1988).

Besides the enclosure process at the firn-ice transition, the gas exchange in the firn is an important factor that determines the age relation between ice and occluded air. There are two methods for investigating the air mixing in the firn. One method is to carry out tracer experiments. Either the spread of a tracer deposited on the snow surface could be measured or the air in the firn pores or in bubbles of shallow ice cores could be analyzed for atmospheric trace substances (CFCs, CO_2, ^{85}Kr, ^{14}C) whose concentration courses in the atmosphere had been measured for the last decades. At present no data from tracer methods are yet available. The other method is to measure the diffusion coefficients on firn samples in the laboratory and to calculate the air mixing in the firn layer (Schwander et al. 1988). In the following I expand the discussions of this paper.

I define by age of the ice, τ_i, the time elapsed since the snowfall that was the origin of that ice and by age of an air molecule, τ_a, the time elapsed since it crossed for the last time the atmosphere/snow boundary. The age of the air at a certain depth cannot be expressed by a simple value but must be described by an age distribution. We have seen above that, in the case of a cold glacier, the major part of the air is sealed off in a relatively small depth range corresponding to the density range 795–830 kg/m^3. We now define two normalized age distributions:

$P_p(z,\tau)$: age distribution of the air in the open pore space at depth z,
$P_b(z,\tau)$: age distribution of the air in the bubbles at depth z,

where $\tau = \tau_i - \tau_a$. τ is thus the age difference between the ice and the air. If V_z is the closed-off air volume per kg ice at depth z, then the amount of air per kg ice, $dV_{z'}(\tau)$, with an age difference between τ and $\tau+d\tau$, trapped between z' and $z' + dz$ is

$$dV_{z'}(\tau) = P_p(z',\tau)\, d\tau \frac{dV_{z'}}{dz'}\, dz'. \tag{8}$$

The total amount of air per kg ice, $V_z(\tau)$, with an age difference between τ and $\tau+d\tau$, trapped at depth z is therefore

$$V_z(\tau) = \int_0^{V_z} dV_{z'}(\tau) = \int_0^z Pp(z',\tau)\, d\tau\, r(z')\, dz'. \tag{9}$$

where $r(z') = dV_{z'}/dz'$, i.e., the amount of air per kg ice trapped per depth unit.

Normalization of Eq. 9 yields the age distribution in the bubbles:

$$P_b(z,\tau) = 1/V_z \int_0^z r(z') \times Pp(z',\tau)\, dz'. \tag{10}$$

Under stationary conditions, P_b becomes independent of the depth z for the ice below the firn-ice transition:

$$P_b(z,\tau) = P_b(\tau) \quad (z > z_b), \tag{11}$$

where z_b is the lower bound of close-off interval.

For the hypothetical case of instantaneous mixing of the air in the firn down to the transition zone, the air would be of zero age throughout the open pore system and therefore P_p would be a δ-function and the age distribution of the air in the ice would be solely controlled by r(z). The mean age difference between air and ice would be approximately the age of the ice where 50% of the air is closed-off and the width of the age distribution would correspond to the time that an ice particle needs to cross the close-off interval. For lack of knowledge on the ventilation of the firn, this case has usually been assumed in previous works, if any corrections for the age difference had been made at all. In reality, the mixing is not instantaneous but is controlled by diffusion. We must distinguish between molecular diffusion and macroscopic mixing or eddy diffusion. The latter occurs when air is forced to flow through the firn. Total diffusion constants, in function of the air flow velocity, have been measured on firn samples from Siple Station (Schwander et al. 1988). The results are shown in Fig. 3. To estimate the contribution by eddy diffusion we must have an idea of the speed of air in a firn layer. Air flow in the firn is caused by several reasons (Schwander et al., in preparation): by the wind, by barometric pressure variations, and by the densification of the firn.

The situation of large ice sheets with no ice layers is first discussed. Effects by laminar air flow can be estimated with Bernouilli's formula. The wind causes the static pressure at hilltops or saddels to be lowered compared with still air conditions. The surface of the polar ice sheets is, however, rather flat. We computed the dynamic pressure effect for a surface undulation on the order of 50 m per kilometer. At a wind speed of 10 m/s the maximum pressure deviation from still air conditions is about 10 Pa. The resulting air flow through the firn is very small and leads only to a minor contribution to the air mixing.

Barometric pressure variations cause the air in the firn to be moved up and down. Based on measurements of the porosity, the air permeability and the power spectrum of the atmospheric pressure variations (Gossard 1960), it is possible to compute the vertical air velocity at different depths. The mean resulting air speed for a firn with a firn-ice transition depth at 65 m is about 10^{-5} m/s near the surface and 10^{-7} m/s at 60 m depth.

Owing to the densification of the firn, air is expelled from its open pore space. The resulting air flow is on the order of 10^{-8} m/s. Unlike the air movement from the atmospheric pressure variations, this flow leads to a net transport of the air upward relative to the firn. The mass transport due to this small speed, however, is completely negligible compared to the one by molecular diffusion.

In the assumed case of ice layer-free cold firn, the total mean air flow is estimated to be less than 10^{-5} m/s at any depth in the firn. From Fig. 3 we see that this low air flow does not increase the diffusivity significantly compared to zero air flow velocity. The mixing of the air in the firn is thus

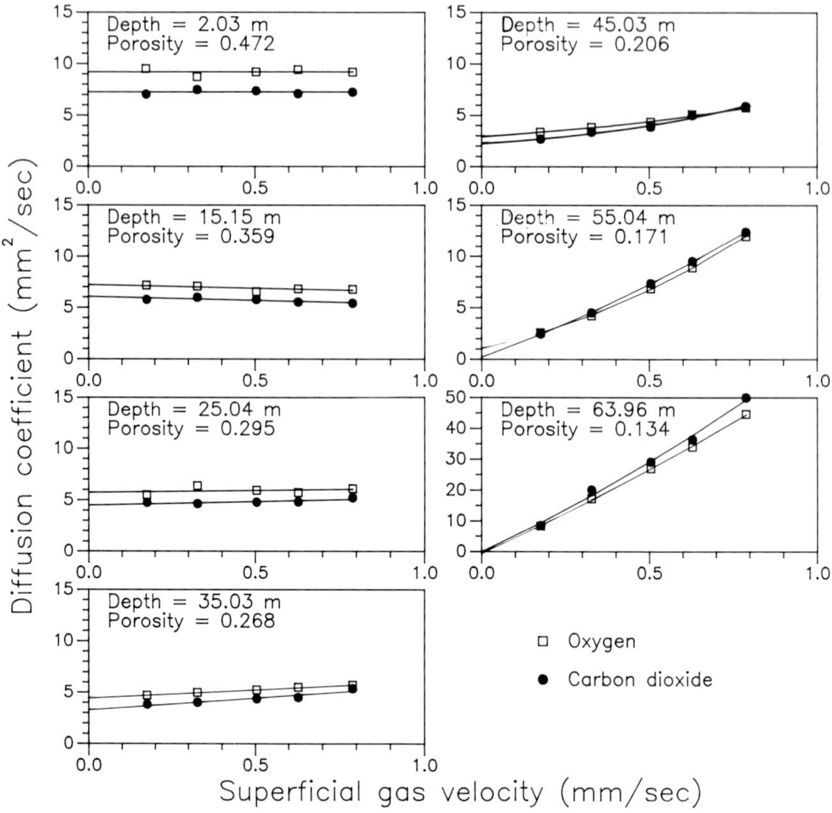

Fig. 3—Diffusion coefficients of O_2 and CO_2 (in N_2) measured on firn samples from Siple Station versus the superficial gas velocity (from Schwander et al. 1988).

essentially controlled by molecular diffusion. It is therefore possible to compute P_p based on the measured porosity and gas diffusivity in the firn. Within the firn-ice transition zone the determination of P_p is, however, made difficult by the fact that there the porosity and diffusivity show rather large variations. We must distinguish between low and high accumulation areas. In high accumulation areas (accumulation $\geq$ 200 kg/(m^2 year)) the sequence of summer and winter layers is usually well conserved. Relatively dense winter layers with air permeabilities and diffusivities near zero are already encountered near the top of the firn-ice transition, namely at mean densities of about 795 kg/m^3. Such layers impede further downward mixing significantly. This means that the effective sealing-off of the air from the atmosphere takes place in a much narrower depth range and earlier than it is suggested from measurements on small samples (Schwander and Stauffer 1984). I estimate that the density range for the effective sealing-off is about

795–800 kg/m^3. As the time a certain firn layer needs to cross this density range is relatively short compared to the mixing time by diffusion of the whole firn layer, the age distribution in the bubbles is about equal to the age distribution in the pores at the top of the close-off interval:

$$P_b(\tau) \simeq P_p(z_t,\tau), \tag{12}$$

where z_t is the upper bound of close-off interval. Fig. 4 shows the calculated age distributions for CO_2 at the top of the firn-ice transition versus the dimensionless age $\tau_a^* = \tau_a \times D/(z_t)^2$ at three different polar sites. D is the diffusion constant of CO_2 in air, at the temperature and pressure of the site. Although the climate is rather different at the three sites[4] these age distributions are similar and hence probably of universal character. They can be used to estimate the age distribution at the depth of pore close-off of most cold, ice layer-free accumulation areas. The mean age difference between ice and air, $<\tau>$, is then given by

$$<\tau> = \tau_{i,t} - <\tau_{a,t}> \tag{13}$$

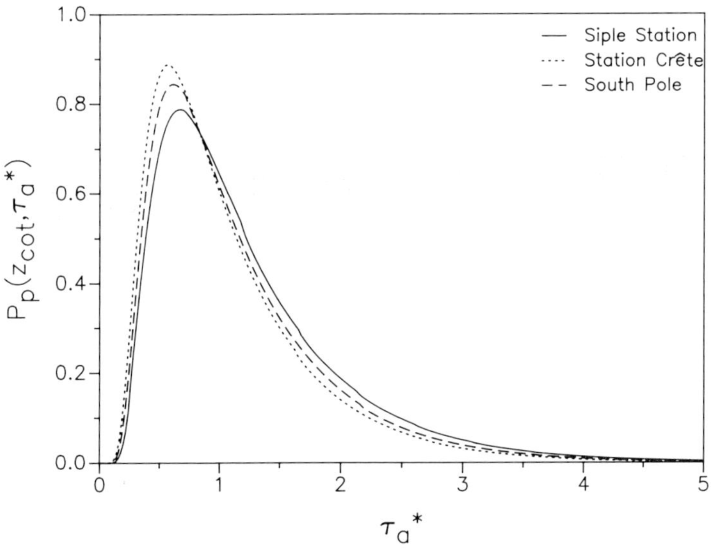

Fig. 4—Calculated age distributions of CO_2 at the top of the firn-ice transition versus a dimensionless age (see text) for Siple Station, Station Crête, and South Pole.

[4]Siple Station, West Antarctica: mean temperature = −24°C, accumulation = 500 kg/(m^2 year); Station Crête, Greenland: −30°C, 250 kg/(m^2 year); South Pole = −51°C, 80 kg/(m^2 year).

where $\tau_{i,t}$ is the age of the ice and $<\tau_{a,t}>$ is the mean age of the air at the top of the firn-ice transition.

If the accumulation is smaller, the seasonal snow layers are interrupted due to surface erosion and therefore horizontally continuous air tight layers do not exist within the transition zone. The pore close-off extends over a larger depth interval and according to Eq. 10 the age distribution becomes broader. However, the variability of the diffusivity in the transition zone makes a quantitative description difficult.

If ice layers are present in the firn, the following effects have to be considered: First, diffusion must be treated three-dimensionally. The diffusion is hampered since the path between any two points separated by an ice layer is increased. Second, squeezing out of air by densification becomes more important. The upward speed of the air through the remaining open pores may be significant. Nearly uninterrupted ice layers with only a small percentage of open pores may seal off lower layers at irregular intervals and lead to a stepwise trapping of the air.

CHANGES OF THE AIR COMPOSITION IN THE FIRN AND IN THE ICE

The most important goal of the analysis of gases entrapped in ice cores is to get information on the concentration of the investigated gases in the past atmosphere. It is therefore desirable that the concentration of the gases whose concentration is to be measured in the air bubbles is as close as possible to the atmospheric concentration at the time of bubble formation. At present, there exists only one way to check experimentally whether the concentration in the bubbles reflects exactly the atmospheric one. It consists of analyzing relatively young ice containing air of a time when the concentration of the investigated gas species had already been measured directly on atmospheric air samples. It seems that the gases in the bubbles of ice from areas with a mean annual temperature below −25°C are almost unchanged samples of the atmosphere (Neftel et al. 1985; Stauffer, Fischer et al. 1985). This direct comparison is, however, limited by the uncertainty of the age of the air in the bubbles.

There are different causes leading to a deviation from the atmospheric composition in the firn pores and the bubbles in the ice: physisorption and chemisorption, chemical reactions, and separation by gravity. As these mechanisms are specific for each gas species, an extensive discussion of them would exceed the scope of this paper. The most important effects are:

1. Physi- and chemisorption. Most probably adsorption on the surface of ice crystals plays a minor role for most gases. If liquid water is present, the solution of gases therein is important, especially for the very soluble gases such as NH_3, CO_2, and N_2O. If melting at the surface occurs only

during warm summer days the site may still be suitable for the investigation of less soluble gases, such as methane for example. The earlier mentioned micro-bubbles may also lead to a deviation of the mean gas concentration in the bubbles from the atmospheric concentration. Micro-bubbles are formed when supercooled water droplets freeze rapidly onto the snow crystals. The more soluble gases are thus probably enriched in them. The contribution of micro-bubbles to the total air content is, however, estimated to be less than 1% and the effect is accordingly small.

2. Chemical reactions. Solid phase reactions could play a role only over very long time spans. For example, some indications point to a slow decomposition of H_2O_2 in ice (Neftel et al. 1986). Reactions changing the gas composition in the bubbles cannot be completely ruled out.
3. Separation by gravity. According to the barometric formula

$$p_{iz} = p_{i0} \times \exp(M_i gz/RT), \tag{14}$$

where z: depth below surface
p_{iz}: partial pressure of gas i at depth z
p_{i0}: partial pressure of gas i at the surface
M_i: molecular weight of gas i
g: Earth's acceleration of gravity
R: gas constant (= 8.134 J/(K mol))
T: temperature (Kelvin).

The heavier air components are enriched at depth due to sedimentation. If c(0) is the concentration at the snow surface and c(100) is the concentration at 100 m depth, we obtain for CO_2, for example, $c(100)/c(0) \simeq 1.006$, and for Kr, $c(100)/c(0) \simeq 1.026$. Since the time to reach equilibrium concentrations in firn is of the order of the mixing time by diffusion, which is generally short compared to the age of the ice at pore close-off, we expect that the air in the firn is close to gravity equilibrium. Thus, for very precise concentration measurements this effect has to be taken into account.

An effect that may redistribute the gaseous components in the ice at greater depths is the formation of air hydrates (clathrates) (Shoji and Langway 1982). Under high hydrostatic pressure conditions, as they exist within the large polar ice sheets, all bubbles disappear and are at least partly transformed into clathrates. It is still uncertain whether all of the gas is transformed into clathrates or whether part of the gas is simply dissolved in the ice. After core recovery, air bubbles slowly reemerge but part of the clathrates are stable for 20 years or longer (Shoji and Langway 1987). It is imaginable that during clathrate formation or bubble retransformation the different gas species do not behave alike. It has been shown, however, that the CO_2 concentration in air extracted with a dry extraction technique from relatively fresh ice cores is not influenced by the degree of bubble reappearance (Neftel et al. 1983).

CONCLUSIONS

To derive the composition of the ancient atmosphere from the analysis of ice cores we should know the age of the extracted air and we should know how well the gas composition of the bubbles reflects the atmospheric composition at the time of bubble formation.

The age difference between ice and air and the age distribution of the air is at present still quantitatively not well known. This is due to a lack of knowledge of the three-dimensional variability of the porosity and diffusivity at the firn-ice transition and, in the case of ice layers, due to the insufficient assessment of their superficial extent. Presently we are able to estimate, based on a one-dimensional diffusion model, the age distribution of the air in the pores of firn without ice layers. Further investigations of the transition zone and of the geometry of the ice layers should help us to understand better the sealing-off of the air. The measurement of tracers in the firn pores and in the bubbles of shallow ice cores would be a possibility to determine the age distribution experimentally.

Ice formed at very low temperatures contains the best preserved gas records. On the other hand, sites with low temperatures usually have a low annual snow accumulation, which leads to a broader age distribution of the air in the bubbles and consequently time resolution is poor. In warmer depositional areas, meltwater becomes involved in snow/ice formation processes and often changes the composition of the gases, especially the highly soluble components.

REFERENCES

Benson, C. 1962. Stratigraphic studies in the snow and firn of the Greenland ice sheet. U.S. Army SIPRE Res. Report, No. 70.

Enting, I.G. 1985. A lattice statistics model for the age distribution of air bubbles in polar ice. *Nature* **315(6021)**: 654–655.

Enting, I.G. 1987. On the application of lattice statistics to bubble trapping in ice. *Tellus* **39B(1–2)**: 100–113.

Enting, I.G., and Mansbridge, J.V. 1985. The effective age of bubbles of polar ice. *Pure Appl. Geophys.* **123**: 777–790.

Gossard, E.E. 1960. Spectra of atmospheric scalars. *J. Geophys. Res.* **65(10)**: 3339–3351.

Gow, A.J. 1975. Time-temperature dependence of sintering in perennial isothermal snowpacks. In: Proc. of Snow Mechanics Symposium, Grindelwald, 1974, ed. E.R. LaChapelle. *IAHS-AISH Publ.* **114**: 25–41.

Herron, M.M., and Langway, C.C., Jr. 1980. Firn densification: an empirical model. *J. Glaciol.* **25(93)**: 373–385.

Khalil, M.A.K., and Rasmussen, R.A. 1987. Atmospheric methane: trends over the last 10,000 years. *Atmos. Envir.* **21(11)**: 2445–2452.

Lliboutry, L. 1964. Traité de Glaciologie, vol. 1. Paris: Masson et Cie.

Neftel, A.; Jacob, P.; and Klockow, D. 1986. Long-term record of H_2O_2 in polar ice cores. *Tellus* **38B**: 262–270.

Neftel, A.; Moor, E.; Oeschger, H.; and Stauffer, B. 1985. Evidence from polar ice cores for the increase in atmospheric CO_2 in the past two centuries. *Nature* **315(6014)**: 45–47.

Neftel, A.; Oeschger, H.; Schwander, J.; and Stauffer, B. 1983. CO_2 concentration in bubbles of natural ice. *J. Physical Chem.* **87**: 4116–4120.

Raynaud, D., and Lebel, B. 1979. Total gas content and surface elevation of polar ice sheets. *Nature* **281(5729)**: 289–291.

Schwander, J. 1980. CO_2-Gehalt von Schnee-, Firn- und Eisproben. Lizentiat (Diploma), Univ. of Bern.

Schwander, J., and Stauffer, B. 1984. Age difference between polar ice and the air trapped in its bubbles. *Nature* **311(5981)**: 45–47.

Schwander, J.; Stauffer, B.; and Sigg, A. 1988. Air mixing in firn and the age of the air at pore close-off. *Ann. Glaciol.* **10**: 141–145.

Shoji, H., and Langway, C.C., Jr. 1982. Air hydrate inclusions in fresh ice core. *Nature* **298(5874)**: 548–550.

Shoji, H., and Langway, C.C., Jr. 1987. Microscopic observations of the air hydrate-bubble. Transformation process in glacier ice. *J. Physique* **48(3-Cl)**: 551–556.

Staffelbach, T.; Stauffer, B.; and Oeschger, H. 1988. A detailed analysis of the rapid changes in ice core parameters during the last ice age. *Ann. Glaciol.* **10**: 167–170.

Stauffer, B.; Fischer, G.; Neftel, A.; and Oeschger, H. 1985. Increase of atmospheric methane recorded in Antarctic ice core. *Science* **229**: 1386–1388.

Stauffer, B.; Schwander, J.; and Oeschger, H. 1985. Enclosure of air during metamorphosis of dry firn to ice. *Ann. Glaciol.* **6**: 108–112.

The Environmental Record in Glaciers and Ice Sheets
eds. H. Oeschger and C.C. Langway, Jr., pp. 69–83
John Wiley & Sons Limited

Environmental Records in Alpine Glaciers

D. Wagenbach

Institut für Umweltphysik, Universität Heidelberg
6900 Heidelberg, F.R. Germany

Abstract. The potential of alpine glaciers to provide environmental records which may be compared with the polar records is considered. The principal differences of glaciochemical and climatic records from alpine glaciers and those from polar ice sheets are outlined and their representativeness is discussed. A review of longer records available from alpine glaciers in nonpolar regions shows that the large glacier sites in the tropics and in northwest Canada provide primarily useful paleoclimatic records while a clear record of atmospheric pollution is available only from the small glacier sites in the European Alps. The specific problems in interpreting glaciochemical and isotopic records from these small glacier sites are therefore outlined in somewhat more detail.

INTRODUCTION

The rather long tradition of glaciological research in the Alps is illustrated by the fact that nearly a hundred years ago more than 200 m deep boreholes were drilled to measure glacier thickness. Nowadays, however, ice core drilling has been emphasized on the two vast ice sheets of Greenland and Antarctica, since here the glaciometeorological conditions permit reconstruction of long-term isotopic and glaciochemical records providing paleoclimatic information of global relevance (Langway et al. 1985). Ice core studies of alpine glaciers have been limited. Consequently, the chemical composition of the atmospheric aerosol fifty years ago is more precisely known for the geographic South Pole and for Southern Greenland than it is for the continental source areas.

The records of alpine glaciers, although covering rather short time scales (at best several thousands of years) and being representative more or less for a regional scale, can, nevertheless, provide important information different from that drawn from the polar ice cores. This is particularly true for filling the gap in the geographical distribution of glaciochemical and climatic records and also for a retrospective look at man's impact on his local atmospheric environment.

The lack of environmental records from midlatitude glaciers is mainly due to the fact that the glaciometeorological and topographical conditions of most glaciers are rarely good enough to provide well preserved and datable long-term records. One of the most important and also most difficult problems is to minimize the influence of meltwater percolation on the chemical and isotopic stratigraphy. This is only guaranteed in the dry snow or in the infiltration-recrystallization zone of cold glaciers (temperature well below the pressure melting point).

This paper discusses the potential of alpine glaciers to provide glaciochemical and paleoclimatic records from a nonpolar region. Emphasis is on the cold high-altitude glaciers, which may be found for example in the Alps (Haeberli 1976; Haeberli and Alean 1985), since this type most likely permits reconstruction of the anthropogenic influence on the atmospheric chemistry on a regional or medium (continental) scale.

DIFFERENCES IN THE RECORDS OF ALPINE AND POLAR GLACIERS AND HOW WE CAN BENEFIT FROM THESE DIFFERENCES

The notion "alpine glacier" is used here to describe the high-altitude accumulation zone of mountain glaciers from nonpolar regions (latitude 60°N, 60°S). In contrast, the polar ice sheets (Greenland, Antarctica) and the larger ice caps or ice domes located at their margin (i.e., Devon Ice Cap, Law Dome) are here referred to as polar glaciers (referring to their geographical situation rather than to their temperature regime). A retrospective look at past atmospheric trace gas concentrations such as CO_2, CH_4, and N_2O has so far only been possible from polar ice cores (see Schwander and Rasmussen, both this volume). Due to the long atmospheric lifetime of these gases (compared to intercontinental atmospheric mixing), these records in midlatitude glaciers (if any reliable existed) would not provide much supplementary information. Presently, no reliable records of trace gases with much shorter atmospheric lifetimes (like H_2O_2 or formaldehyde) have been obtained from alpine glaciers. Therefore, only those glaciochemical records related to aerosol particles or their atmospheric precursors will be discussed here.

The features relevant to ice core studies of alpine glaciers are significantly different from those of the polar glaciers. Some of the most significant features are:

1. The nontemperate part of an alpine glacier accumulation zone usually has a relatively small volume, compared to polar glaciers. Thus under ideal conditions, records of only several thousand years can be analyzed.

According to the usual small-scale glacier geometry, the flow pattern is very complex, and permeable firn accounts for a considerable fraction of the total glacier thickness.
2. Most sites are influenced by formation and percolation of meltwater resulting in a stratigraphy of various melt features such as ice layers, ice lenses, and vertical ice channels. The chemical and isotopic stratigraphy may be significantly altered by these phenomena.
3. The net snow accumulation rate will generally be higher than on polar ice sheets and may reach up to 2 m of water equivalent per year. A specific effect of the small-scale alpine glacier is the overall net loss of surface snow by wind-induced erosion which results in a significantly lower net mean snow accumulation compared to the local precipitation.

Initially, most of these peculiarities would tend to disqualify an alpine glacier from being a reliable archive.

Generally speaking, the existence of polar glaciers is connected with their polar locations, whereas alpine glaciers only survive above a certain altitude which depends on latitude and the regional precipitation rate. Consequently, different information is to be expected from the records of polar and alpine glaciers. Since there is no significant production of trace substances on the polar ice sheets, the clean air status at ground level is controlled by long-range air mass transport from the surrounding oceans and continents, as well as by the subsequent downward mixing of atmospheric constituents through the local boundary layer. On the other hand, the alpine glaciers are always located within the source area of various atmospheric trace substances. On the average, the clean air status of alpine glaciers is controlled by the local vertical mixing pattern. It is also important to note that the distance to the nearest ground level source of trace gases and particles may be only ~10 km away.

The impurity record related to atmospheric aerosol particles of polar glaciers will be representative primarily for the polar background aerosol on a hemispheric scale. Apart from episodic haze events, which are now recognized as typical for the high Arctic (Barrie 1986), this polar background aerosol is expected to consist of a mixture of aged continental aerosol and of primary and secondary marine aerosol consisting of sea-salt and organo-sulfur species.

The impurity content of high-altitude alpine glaciers is determined by the varying influence of several aerosol bodies, depending on the position of the snowfield relative to the mixing height and on local wind conditions. Important aerosol bodies contributing to the impurity content of alpine glaciers are:

1. the continental background aerosol representative for the aged aerosol of the free continental troposphere (which are sometimes disturbed by

high altitude injections of large amounts of desert soil dust),
2. the local or regional aerosol originating from the underlying mixing layer (probably heavily polluted nowadays),
3. the aerosol derived from the immediate vicinity of the glacier (most likely large-sized mineral dust particles from exposed proglacier soil areas).

The representativeness of the glaciochemical records from alpine glaciers ranges from a large-scale composite with true background conditions to simply the regional scale, resulting from prevailing intense vertical mixing, or even to the very local scale at episodic events of specific mountain winds.

It is the regional influence which causes serious problems in evaluating to what degree the glaciochemical record in an alpine glacier is representative on a larger scale. On the other hand, the regional record or pattern is of great importance and is very sensitive to all kinds of particulate matter or reactive trace gases originating from near continental sources. A change in the midlatitude source strength of these species is directly recorded in alpine glaciers. This is of primary importance for the reconstruction of the anthropogenic impact on the atmospheric cycle of C, S, N, and halide-containing constituents as well as on the cycles of a series of minor trace elements (such as heavy metals) in populated regions. On a regional scale, the anthropogenic impact on the midlatitude glaciers may cover a much longer period of time than the anthropogenic influence does on polar regions, like the high Arctic. Note that one of the longest-lasting relevant anthropogenic emissions may be due to the permanent increase in agricultural activities.

In mid-latitude glaciers a three orders of magnitude higher level of anthropogenically produced heavy metals (e.g., Pb, Zn, V, Cu) is expected in comparison to polar glaciers. In view of the serious contamination problems arising during sampling and analysis of these species, it would be much easier to gain an unequivocal time trend from alpine glaciers (Batifol and Boutron 1984). Due to this contamination problem, most of the deposition records of heavy metals and related trace elements extracted so far from polar glaciers (with the exception of Pb) may be questioned (see Peel, this volume).

Apart from the obvious question about the history of atmospheric pollution in man's environment, there is also a considerable lack of knowledge concerning the natural cycle of the species derived from land biota and crustal weathering such as organic acid, particulate organic carbon compounds, and mineral dust. The organic acid and particulate organic C compounds may give better insight into past air chemistry, while the source strength of mineral dust may be related to aridity, vegetation cover, and surface wind speed, thus offering some paleoclimatic or paleometeorological information on the climatic zone of the alpine glacier investigated.

It may sound rather trivial, but nevertheless, the fact that polar and alpine glaciers belong to quite different climatic zones and circulation regimes is their most important distinguishing feature. A good deal of important information, in addition to the already established glaciochemical and paleoclimatic records in polar glaciers, may perhaps only be gained by extending the apparatus of polar ice core studies to nonpolar latitudes. The following two important aspects of the potential paleoclimatic records from alpine glaciers can be discerned:

1. The paleoclimatic records derived from glacio-proxy data (such as $\delta^{18}O$, δD, accumulation rate (if possible), frequency of melt features, and deposition rate of mineral dust and cosmic ray produced radioisotopes (^{10}Be)) may be examined for conspicuous coincidences with (climatic) records reported from appropriate polar glaciers to verify their hemispheric or even global relevance.
2. The working hypothesis translating the proxy data mentioned into climatic records may be tested for reliability by comparing the corresponding historical records most likely available from nonpolar sites. The cold periods, known as The Little Ice age, which are reported from extended regions of Central and Northern Europe after the end of the Middle Ages, are considered to be one of the prominent examples in this context.

CURRENT INFORMATION DRAWN FROM ALPINE GLACIER RECORDS

To date only a few environmental records from alpine glaciers have been established which deserve mentioning, due to their covering a substantial period of time. There are only a few sites where ice cores with potential records of more than 100 years have been retrieved, offering as well the necessary information on glacier geometry, surface flow pattern, and local glaciometeorological conditions. These drill sites are found at Mt. Logan (Canada), Quelccaya Ice Cap (Peru), and, as lately discovered, the more recent sites at Dunde Ice Cap (China), Colle Gnifetti (Alps), and, to a lesser extent, Mt. Blanc (Alps) (see figure 1).

Mt. Logan

The Mt. Logan ice field plateau (60°N, 140°W) in the Canadian St. Elias Mountains, where extensive research has been carried out (Holdsworth and Peak 1985), shows unique favorable conditions for gaining well preserved records. The site is characterized as follows: altitude in the range of 5300 m,

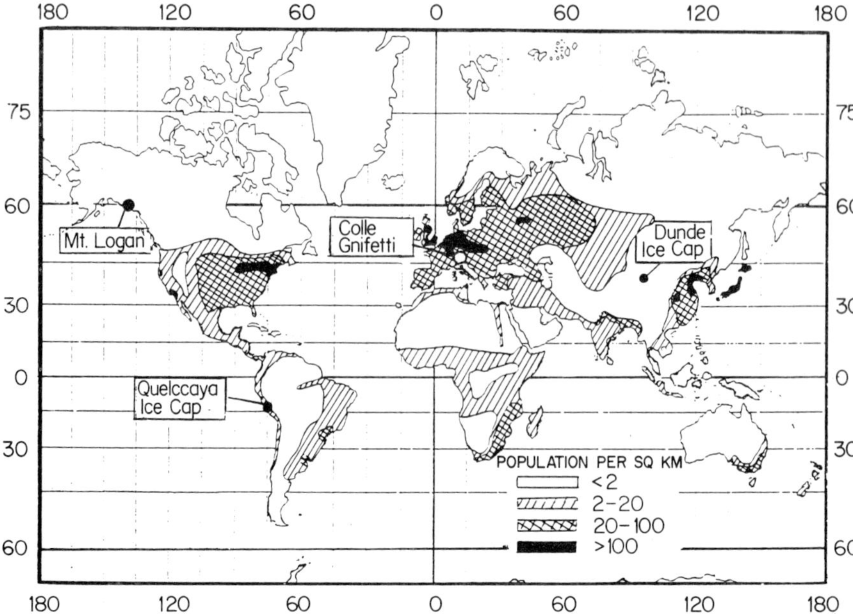

Fig. 1.—Location of alpine drill sites discussed in the text and their relation to the global distribution of population density which is assumed to be equivalent to fossil fuel consumption (the smaller energy consumption in the low latitudes has been taken into account by reducing the population index into the next lower category).

total area approximately 20 km^2, glacier thickness up to about 350 m, net accumulation rate about 0.35 m water per year, firn transition 65 m below surface, and firn temperature close to −29°C. A real advantage here is that the annual variation of the $\delta^{18}O$ record can still be identified in most cases despite the wind exposed situation typical for high-altitude alpine sites. As demonstrated from pH, liquid conductivity, and sulfate and nitrate analysis (Holdsworth and Peak 1985) on parts of a 103 m ice core, covering ~300 years, the Mt. Logan site seems to be consistently unaffected by man-made increase in mineral acid, which is not true for polar glaciers in the Northern Hemisphere. This site may, therefore, be representative primarily for the aerosol body related to the free upper troposphere of this region and for air masses originating from the northern Pacific. Sodium concentration in Mt. Logan snow is reported to be in the range of 1.4 ppb (Delmas et al. 1985), one of the lowest values ever measured in precipitation samples and thus a minor influence of primary marine or continental aerosol is expected here. The Mt. Logan ice core data show the excellent potential of this site to provide, probably on a 1000 year time scale, well preserved records including isotopic thermometry and aerosol species related to the volcanic or stratospheric source as well as to the upper tropospheric background aerosol. The possibility of extracting the history of man's impact on the precipitation chemistry seems to be limited, however.

Quelccaya Ice Cap

Numerous research programs have been carried out on the tropical Quelccaya Ice Cap in the Peruvian Andes (14°S, 71°W) (Thompson et al. 1985). This relatively large glacier plateau is characterized as follows: summit elevation of 5650 m, total area approximately 55 km^2, glacier thickness about 160 m, flat bedrock topography and net snow accumulation in the range of 1 m water per year. Since this glacier is temperate, records of soluble impurity are expected to be difficult to interpret. It is, however, a specific feature of these tropical accumulation zones that the stratigraphy of the relatively thick annual layers will be well preserved, owing to the sharp contrast between the precipitation rate in the dry winter and in the wet summer half-year. Annual layers which reflect about 80% precipitation during the wet season are separated by intense dust layers providing a visual stratigraphy.

In combination with annual variations of $\delta^{18}O$ (being not linked to air temperature), microparticle, and liquid conductivity, a very precise time scale has been established by annual layer counting for two ice cores drilled to the bedrock, covering up to 1500 years. These data have been extensively used for climatological interpretation in terms of past precipitation rate. It was found that the recent occurrence of El Niño events is likely to be connected with a decrease in Quelccaya accumulation rate by almost 30% (Thompson et al. 1984). Due to the relatively simple glacier geometry, a reasonable correction of annual layer thinning has been applied, providing a 1500 years trend of the glacier mass balance. The period from A.D. 1500–1720 was thereby identified as the wettest during the last 1500 years (Thompson et al. 1985). Furthermore, the long-term $\delta^{18}O$ record in both ice cores was found to be apparently parallel to the corresponding air temperature trend in the Northern Hemisphere during the last 400 years. From this a global occurrence of The Little Ice Age was deduced (Thompson et al. 1986). Note that the stable isotope pattern of tropical glaciers is more complex and less well understood than that of the polar or nontemperated midlatitude glaciers. Part of the isotopic variation may arise from postdepositional changes and also from substantial differences in water vapor source and advection during the wet and dry periods. Therefore, a straightforward deduction of mean air temperature changes from the glacial isotope variation is not possible here.

Dunde Ice Cap

Recently, the Dunde Ice Cap (38°N, 96°E), located within the Tibetan Plateau, has been selected for ice core studies primarily related to the paleoclimatology of this subtropical region (Thompson et al. 1988). Size, altitude, and stratigraphical features (predominance of accumulation from the wet season) are found to be quite similar to those at the Quelccaya Ice Cap. During the summer of 1987, three ice cores to the bedrock were successfully recovered from this nontemperated ice cap, and it is believed that these ice cores may contain 10–13 m of glacial stage ice (L.G. Thompson, personal communication). At present it would be difficult to assess the potential of this site for detailed glaciochemical studies. However, due to the rather high firn temperature (associated with the formation of melt layers), the relatively small accumulation rate, and the extremely large mineral dust concentration, no straightforward interpretation of the records of major ions or noncrustal trace elements is supposed to be possible.

The record of mineral dust deposition seems to be more promising here in view of the Asian source areas of mineral dust (such as the Gobi desert), which are known to influence strongly the atmospheric dust burden of the North Pacific and the elioan dust fraction of deep-sea sediment cores, accordingly. The glaciochemical records of the Dunde Ice Cap associated with the mineral dust components are therefore likely to play an indispensable role in decoding the history of the mineral dust source strength of this region. Note that the dust level in the upper troposphere, as recorded in the Dunde Ice Cap, is of high relevance to the long-range dust transport.

Colle Gnifetti

Being different from the relatively large glacier plateaus described above, the truly midlatitudinal Colle Gnifetti site is represented by a small col (4450 m a.s.l.) at the summit of the second highest peak of the Alps (Monte Rosa). Ice core studies and related work were started here in 1976 (Haeberli et al. 1983; Oeschger et al. 1977; Schotterer et al; 1985), showing a maximum glacier thickness of about 130 m, a net snow accumulation rate of 0.2–0.4 m water per year, and a firn temperature close to −15°C. This site is located at the boundary between the dry zone and the infiltration–recrystallization zone (Alean et al. 1984). Due to the rather small col area of not more than 0.3 km^2, the annual snow accumulation is controlled by wind erosion, which prevents any regular seasonal snow deposit. Consequently, dating by annual layers cannot be applied.

Records for the last several hundred years may be extracted from the two Colle Gnifetti cores drilled to bedrock (Haeberli et al. 1988). The following features of this midlatitude site have been approximately deduced from quasi-continuous records from the last 100 years:

1. Short-term variation and glacier inventory of mineral dust are dominated by the deposition of Saharan dust, providing some evidence that on an episodic base this site is substantially influenced by subtropical air masses (Wagenbach et al. 1983).
2. There is a significant decrease in the background level of liquid conductivity, as well as in the main acidity, sulfate, and nitrate level with depths in the cores (Wagenback et al. 1988), which possibly reflects the change caused by man-made emission of the mineral acid precursors SO_2 and NO_x (see Table 1). Space-charge measurements confirm that the relative maximum concentrations of the acid species are found in the snow deposited around the mid-seventies (Schotterer et al. 1985).
3. Regional climatic conditions are deduced from a good correlation of the smoothed $\delta^{18}O$ record with the corresponding records of summer air temperature reported from a Swiss mountain station (U. Schotterer, personal communication), and also from a more complex influence of alpine summer air temperature on the stratigraphy of melt layers (Schotterer el al. 1978).

This midlatitude glacier will be particularly useful to recover well-preserved glaciochemical records of anthropogenic impurities, dating back to the preindustrial period. The same is to be expected from comparable sites in the adjacent Mt. Blanc region. Since 1950 an increase of up to a factor of two in the trace metals Pb, V, and Cd has been reported from this region (Briat 1978). Also, some detailed analyses of major ions have been carried out in recent snow samples (Roueaux and Delmas 1986), which will be continued on a 70 m ice core (R.J. Delmas, personal communication). Considerable shortcomings of the Colle Gnifetti site, however, arise from its small-scale nature introducing chaotic irregularities of snow deposition

Table 1 Twentieth century change of mean sulfate (total) and nitrate level in ice cores from central Europe (Colle Gnifetti, Wagenbach et al. 1988) and South Greenland (Dye 3, Neftel et al. 1985)

	Sulfate [ppb]		Nitrate [ppb]	
Period	Colle Gnifetti	Dye 3	Colle Gnifetti	Dye 3
Present day	480	90	280	86
Turn of the century	110	36	60	44
Present day / Turn of the century	4.4	2.5	4.7	2
Preindustrial*	–	24	–	52

* see Finkel et al. 1986

in space and time (due to snow erosion) and a rapid change in glaciometeorological conditions upstream to the saddle drill site. These features, probably typical for most high-altitude glacier sites of the midlatitudes, will cause a series of inevitable ambiguities in the ice core data interpretation. Therefore, a short outline of this problem, as experienced in some detail at the Colle Gniffetti site is given below.

SMALL-SCALE EFFECTS ON HIGH-ALTITUDE ALPINE DRILL SITES

Snow Deposition

At wind exposed sites significantly smaller than the characteristic transport length of drifting snow particles, the annual net snow accumulation will be controlled by surface wind conditions, surface slope, and snow consolidation effects (such as the formation of wind crusts and melt layers) rather than by the regional precipitation rate. When this is the case, the dry winter snow will preferentially be lost and the following consequences are to be considered:

1. Representative deposition fluxes, as frequently calculated from mean snow impurity concentration and mean snow acumulation rate, will always be underestimated and can therefore not be compared to those at other sites.
2. Dry deposition of aerosol particles and reactive trace gas species becomes more important to total deposition because newly fallen snow is most likely removed totally from the sampling site.
3. The air concentration at high mountain sites, as already mentioned, is controlled by the vertical mixing intensity, which causes a pronounced annual cycle of the land-derived species with a maximum level during the summer half-year (see figure 2). The annual layer average of snow impurities and the stable isotopes $\delta^{18}O$, δD will therefore depend mainly on the relative amount of winter (or summer) snow contributing to this annual layer. Hence, there is not only the already mentioned uncertainty of the spatial representativeness of alpine ice core data but also the uncertainty concerning the season primarily represented by the snow deposited. In this context, pollen analysis would be a powerful tool to overcome these problems. At the Colle Gnifetti it is assumed that the net snow accumulation accounts for approximately only one third of the local precipitation rate, showing a pronounced predominance of summer snow (see figure 3). This should be considered in view of any interpretation of the chemical and climatic proxy data of these cores.

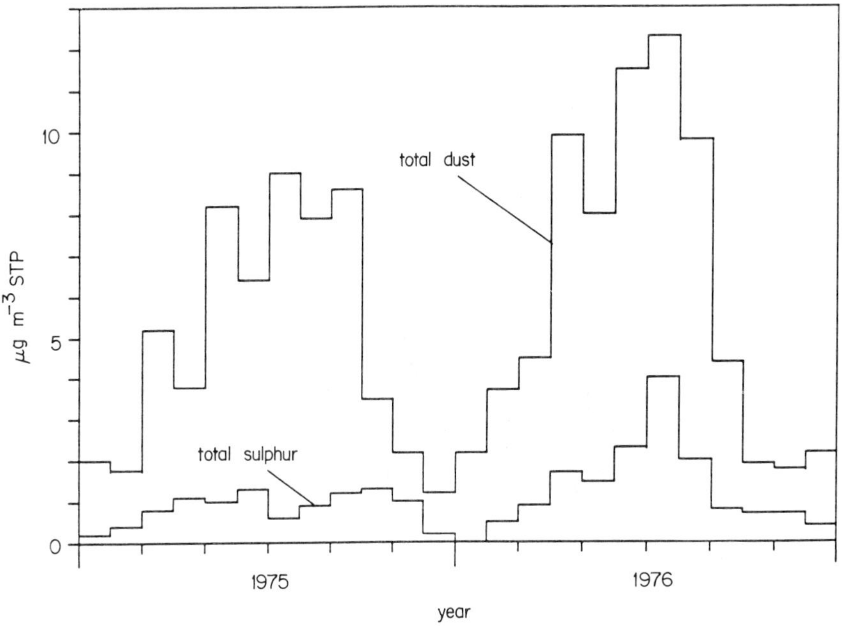

Fig. 2.—Seasonal cycle of monthly mean total dust and total sulfur in aerosol samples collected at the Swiss Jungfraujoch high-altitude research station, 3580 m a.s.l. (adopted from Bundesamt für Umweltschutz 1980). The lower levels observed during winter are mainly attributed to the low vertical mixing intensity during this season. Note that due to the annual cycle of fossil fuel burning, a winter maximum of total sulfur (and related species) is found at ground level.

Ice Flow Regime

A rapid systematic variation in surface slope, accumulation rate, incoming solar radiation, snow density, and local wind condition along the surface flow line upstream to the saddle point is to be expected at drill sites forming a small col (see for example figure 4). Among other things, the deposition processes of atmospheric impurities and, above all, the formation of meltwater are affected by these parameters (Alean et al. 1985; Beck 1985), leading to corresponding systematic variations in the properties along an ice core drilled there. Thus, to establish long-term records from high alpine ice cores, the lateral effects must be known and must be removed from the records analyzed, by means of an appropriate flow model. (Due to the small-scale nature of these glaciers a steady state assumption is essential for these flow models).

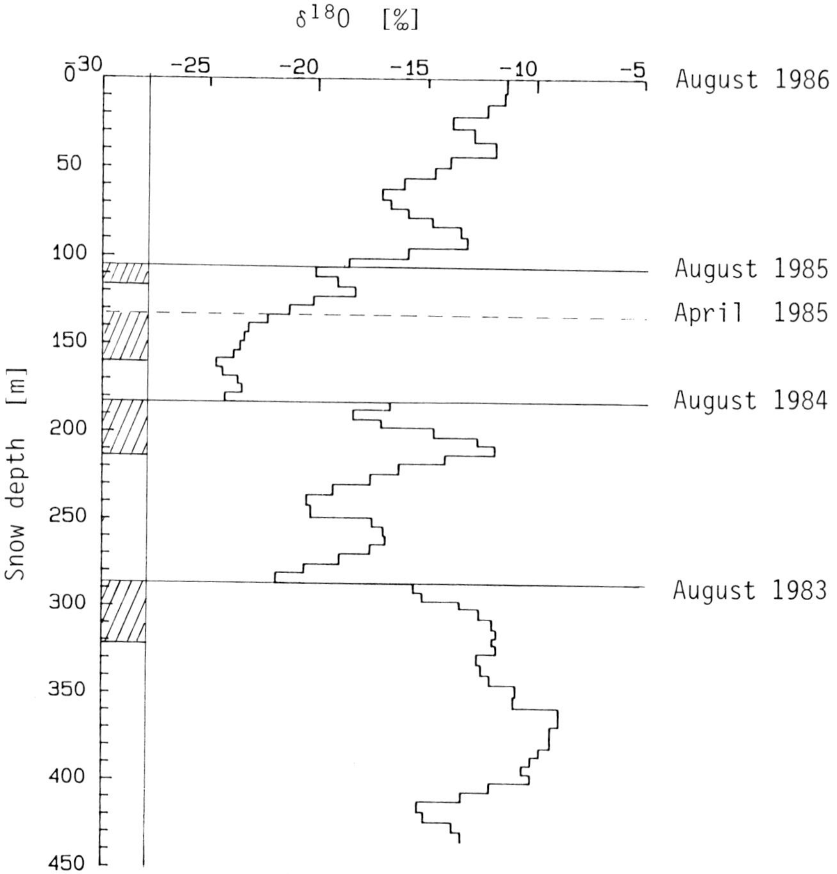

Fig. 3—$\delta^{18}O$ variation in surface snow on a small wind exposed drill site (Colle Gnifetti). The depth profile refers to a sequence of five snow pits, each dug at the date indicated; overlapping sections are marked by hatched depth intervals. As confirmed by parallel snow gauge readings, the obvious lack of an annual $\delta^{18}O$ variation is primarily due to the absence of any net accumulation between autumn and early spring.

FUTURE WORK

Until now, results gained from ice core analyses made from alpine glaciers are less continuous than those from polar glaciers; this is particularly true for small high-altitude sites. To reduce some of the problems encountered in ice core studies from alpine glaciers, special emphasis should be given to:

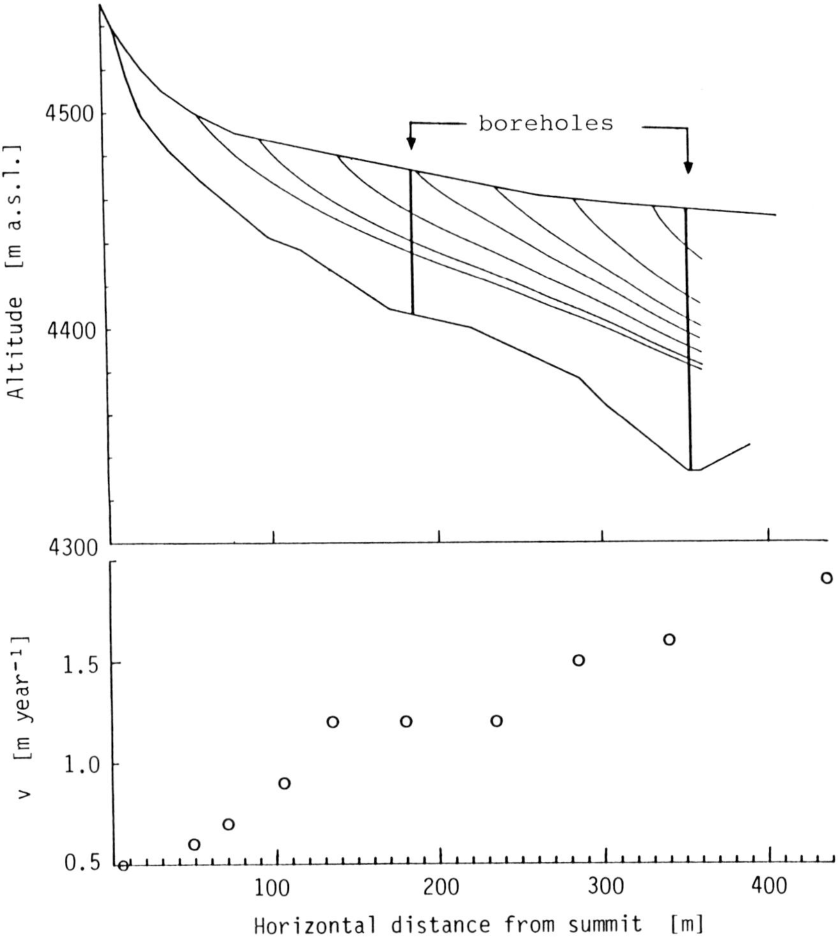

Fig. 4—Calculated back trajectories of particle path and vertical component of surface velocity measured along the surface flow line through two boreholes at a col forming drill site (Colle Gnifetti) (adapted from Haeberli et al. 1988). As indicated, material in the lower half of the ice cores has been deposited within the slope region which is characterized by a significant lower rate of net snow accumulation and of incoming solar radiation.

1. the investigation of the stable isotope pattern of mountain precipitation and surface snow (particularly important for low-latitude glaciers),
2. the influence of a high mineral dust concentration on the ion chemistry of ice core samples,

3. supplementary long-term field studies at the drill sites investigating glacier flow, the seasonal pattern of snow deposition and atmospheric impurity concentration, and the local glaciometeorological conditions, and
4. the exploration of drill sites on additional cold glaciers in the midlatitudes to allow a more comprehensive reconstruction of the anthropogenic impact on the continental aerosol chemistry.

Acknowledgments. Part of this work has been supported by the German Federal Ministry for Research and Technology (grant no. FZ 01QF07291).

REFERENCES

Alean, J.; Haeberli, W.; and Schaedler, B. 1984. Snow accumulation, firn temperature and solar radiation in the area of the Colle Gnifetti core drilling site (Monte Rosa, Swiss Alps): distribution patterns and interrelationships. *Z. Gletscher. Glazial* **19, 2**: 131–147.

Barrie, L.A. 1986. Arctic air pollution: an overview of current knowledge. *Atmos. Envir.* **20, No. 4**: 643–663.

Batifol, F.M.,, and Boutron, C.F. 1984. Atmospheric heavy metals in high altitude surface snows from Mont Blanc, French Alps. *Atmos. Envir.* **18, No. 11**; 2507–2515.

Beck, N. 1985. Laboruntersuchungen zur Bildung von oberflächigen Schmelzschichten in Trockenschnee. Master Thesis, Inst. für Umweltphysik, Univ. of Heidelberg.

Briat, M. 1978. Evaluation of levels of Pb, V, Cd, Zn and Cu in the snow of Mont Blanc during the last 25 years. *Atmos. Poll.* **1**: 225–227.

Bundesamt für Umweltschutz. 1980. Bern, Switzerland. EMPA Report Nr. 42380.

Delmas, R.J.; Legrand, M.; and Holdsworth, G. 1985. Snow chemistry on Mount Logan, Yukon Territory, Canada. *Ann. Glaciol.* **7**: 213.

Finkel, R.C.; Langway, C.C., Jr.; and Clausen, H.B. 1986. Changes in precipitation chemistry at Dye 3, Greenland. *J. Geophys. Res.* **91, D9**: 9849–9855.

Haeberli, W. 1976. Eistemperaturen in den Alpen. *Z. Gletscher. Glazial.* **11, 2**: 203–220.

Haeberli, W., and Alean, J. 1985. Temperature and accumulation of high altitude firn in the Alps. *Ann. Glaciol.* **6**: 161–163.

Haeberli, W.; Schmid, W.; and Wagenbach, D. 1988. On the geometry, flow and age of firn and ice at the Colle Gnifetti core drilling site (Monte Rosa, Swiss Alps). *Z. Gletscher. Glazial.*, in press.

Haeberli, W.; Schotterer, U.; Wagenbach, D.; Haeberli-Schwitter, H.; and Bortenschlager, S. 1983. Accumulation characteristics on a cold, high-Alpine firn saddle from a snow-pit study on Colle Gnifetti, Monte Rosa, Swiss Alps. *J. Glaciol.* **29, 102**: 260–271.

Holdsworth, G., and Peak, E. 1985. Acid content of snow from a mid-troposphere sampling site on Mount Logan, Yukon Territory, Canada. *Ann. Glaciol.* **7**: 153–160.

Langway, C.C., Jr; Oeschger, H.; and Dansgaard, W., eds. 1985. Greenland Ice Core: Geophysics, Geochemistry and the Environment. *Geophys. Monog.* **33**. Washington, D.C.: Amer. Geophys. Union.

Neftel, A; Beer, J.; Oeschger, H.; Zürcher, F.; and Finkel, R.C. 1985. Sulphate and nitrate concentrations in snow from South Greenland 1895–1978. *Nature* **314**: 611–613.

Oeschger, H.; Schotterer, U.; Haeberli, W.; and Röthlisberger, H. 1977. First results from alpine core drilling projects. *Z. Gletscher. Glazial.* **13, H1/2**: 193–208.

Rouseaux, F., and Delmas, R.J. 1986. Chemical composition of bulk atmospheric deposition to snow at Col de la Brenva (Mont Blanc area). In: NATO Advanced Research Workshop, Acid Deposition Processes at High Elevation Sites, Edinburgh, ed. M.H. Unsworth. Dordrecht: Reidel.

Schotterer, U.; Haeberli, W.; Good, W.; Oeschger, H.; and Röthlisberger, H. 1978. Datierung von kaltem Firn und Eis in einem Bohrkern vom Colle Gnifetti, Monte Rosa. *Jahrb. Schweiz. Naturforsch. Ges.*

Schotterer, U.; Oeschger, H.; Wagenbach, D.; and Münnich, K.O. 1985. Information on paleo-precipitation on a high altitude glacier Monte Rosa, Switzerland. *Gletscher. Glazial.* **21**: 379–388.

Thompson, L.G.; Mosley-Thompson, E.; and Arnao, B.M. 1984. El Niño-Southern oscillation events recorded in the stratigraphy of the tropical Quelccaya Fee Cap, Peru. *Science* **226**: 50–53.

Thompson, L.G.; Mosley-Thompson, E.; Bolzon, J.F.; and Koci, B.R. 1985. A 1500-year record of tropical precipitation in ice cores from the Quelccaya Ice Cap, Peru. *Science* **229**: 971–973.

Thompson, L.G.; Mosley-Thompson, E.; Dansgaard, W.; and Grootes, P.M. 1986. The little Ice Age as reported in the stratigraphy of the tropical Quelccaya Ice Cap. *Science* **234**: 361–364.

Thompson, L.G.; Xiaoling, W.; Mosley-Thompson, E.; and Zichu, X. 1988. Climatic records from the Dunde Ice Cap, China. *Ann. Glaciol.* **10**: 80–84.

Wagenbach, D.; Görlach, U.; Haffa, K.; Junghans, H.G.; Münnich, K.O.; and Schotterer, U. 1983. A long-term aerosol deposition record in a high altitude alpine glacier. WMO Technical Conference on Observation and Measurements of Atmospheric Contaminants (TECOMAC) Vienna. *WMO Report No.* **647**: 623–631.

Wagenbach, D.; Münnich, K.O.; Schotterer, U.; and Oeschger, H. 1988. The anthropogenic impact on snow chemistry at Colle Gnifetti, Swiss Alps. *Ann. Glaciol.* **10**: 183–187.

Standing, left to right:
Gode Gravenhorst, Stuart Penkett, Ulrich Schotterer, Jakob Schwander, Glenn Shaw, Dietmar Wagenbach, Robert Delmas

Seated, left to right:
Christoph Brühl, Jim White, Karl Otto Münnich, Peter Brimblecombe, Cliff Davidson

The Environmental Record in Glaciers and Ice Sheets
eds. H. Oeschger and C.C. Langway, Jr., pp. 85–98
John Wiley & Sons Limited

Group Report
How Do Glaciers Record Environmental Processes and Preserve Information?

J.W.C. White, Rapporteur
P. Brimblecombe
C. Brühl
C.I. Davidson
R.J. Delmas
G. Gravenhorst
K.O. Münnich
S.A. Penkett
U. Schotterer
J. Schwander
G.E. Shaw
D. Wagenbach

INTRODUCTION

Measuring a parameter in glacial ice asks the question: What does this information reveal about the past history of the Earth's environment? When asking that question, however, we must also consider how well the recorder is functioning and how sure we are of its reliability.

If someone were to design an ideal recorder of the Earth's past environmental and atmospheric conditions, the following criteria would probably be required:

1. it should sample the particulates and gases in the atmosphere,
2. it should also record the changing climatic conditions,
3. it should collect this information in datable, consecutive layers without mixing to blur the information,
4. it should operate on a variety of time scales so that one recorder in one location offers a detailed picture in time, while another records over long time scales, and
5. it should be devoid of postdepositional chemical reactions, either organic or inorganic, so that the information is maintained without change.

While we cannot expect to find an ideal recorder in nature, glacial ice is proving to be a remarkably close approximation. As with any real recorder, the perception of perfection depends on how closely one looks at the recorder. When information is viewed on long time scales, ca. 100 000 years, and with a time resolution of decades to centuries, glacial ice comes very close to acting as an ideal recorder. As one looks closer, decreasing the time scale, increasing the time resolution, and placing more specific demands upon the recorder, the imperfections begin to become apparent and more important to the interpretation of the record.

This report summarizes the discussions of the working group on how glaciers record environmental processes and preserve the information. The topic is broad, and the discussions ranged over all of the parameters currently being measured in glacial ice, as well as what species should be added to our shopping list of parameters to measure in the future. In such discussions there is a danger of focusing on the imperfections of the recorder. After all, scientists are concerned with extracting more precise and detailed information from the ice and hence spend much of their time focusing on the problems which occasionally block this quest.

We must not lose sight, however, of the phenomenal successes achieved in extracting environmental records from glacial ice. These successes are not recounted here, but can be found throughout this volume.

ORGANIZATION OF THE REPORT

Tackling such a large subject requires organization. Consequently, we began our discussions with how well records are preserved in ice then moved from the records successively back to the atmosphere through a series of important transfers, namely *(a)* the firn to ice transition, *(b)* the surface snow to firn transition, *(c)* deposition from the atmosphere to the snow surface, and, finally, *(d)* long range transport from sources to the air above the glacier. For each step we attempted to address the major species of interest: gases (inert and reactive), particulates (soluble and insoluble), stable isotopes, and physical changes in the ice. We also attempted to consider glacier locations, specifically Arctic, Antarctic and low and midlatitude glaciers. Each area has its own strengths and weaknesses and so must be considered separately. In addition, we were interested in the possibility of using records in low and midlatitude glaciers as a bridge between the large polar ice sheets. The success of such an approach largely depends on how well the glacial record from these different regions complement one another.

HOW IS THE RECORD PRESERVED IN THE ICE?

The answer to the above question is, briefly, "very well." Nevertheless, a number of current problems in interpreting the record were identified.

For isotopes, the $\delta^{18}O$ and δD seasonal cycles are smoothed by vapor diffusion (Johnsen 1977), although this can be modeled. This process is also a function of accumulation rate so that if seasonal cycles are desired, one must go to an area of higher snow accumulation. In this case, seasonal cycles in stable isotopes can be preserved as far back as ca. 8000 yrs and perhaps longer.

There was much discussion on the possible effects of soluble particles accumulating in liquid brines at crystal triple junctions, such as that recently reported for H_2SO_4 (Mulvaney et al. 1988). There is no evidence, however, that other species are collecting at these points and reacting chemically to change their concentrations in the ice. There is also no indication of these brines from space charge (conductivity) measurements made deep on the Byrd core, where crystal sizes as large as 5 cm should have been large enough to permit observation of such brines.

There was also much discussion of the possibility that H_2O_2 may react chemically in the ice (Neftel et al. 1968). For the other gases, the main problem is not how to interpret the record but how to physically extract the core, i.e., technical problems in drilling (cracking the core must be avoided) and in collecting the gas from the ice. The numerous successes in reconstructing CO_2 concentrations indicates that many of the problems thought to affect this gas, i.e., melt layers, *in situ* production of CO_2 from carbonates, etc., can either be overcome, avoided, or do not affect the record.

The fact that the $\delta^{18}O$ of CO_2 reflects equilibrium with the ice indicates that atomic exchange does occur (Siegenthaler et al. 1988). This, however, is a concern for isotopes in gases—and then only if the atom is exchangeable at such low temperatures—and not for bulk concentrations.

In general, the group felt that chemical reactions of particulates and gases in ice at low temperatures are largely unknown and deserve future study. This is particularly important as temperatures in glacial ice span a large range, from −50°C in central Antarctica to close to 0°C in many nonpolar glaciers. While we have no evidence that large-scale chemical reactions occur which change concentrations of gases or particulates in ice cores, the group recommended that low temperature ice chemistry be investigated to improve our overall basic understanding.

HOW IS THE RECORD PRESERVED DURING THE FIRN TO ICE TRANSITION?

The group agreed that the firn to ice transition is important only for the atmospheric gases which are trapped in the ice. The only potential problem which could be foreseen for particulates was the possible reaction of particulate components for which gas phase species also exist: for example, HNO_3 affecting NO_3^- measurements. Whether such reactions occur is a question that could be addressed by laboratory studies if and when strong enough suspicion arises that such reactions are important.

For gases, the firn to ice transition presents complications which must be considered in assigning dates to ice segments, and particularly when comparing gas concentrations to particulate and stable isotope concentrations. These complications include molecular diffusion of gas in the firn (the gas exchange time in polar firn is on the order of 5 to 30 years) and variability in the closure rate of bubbles as firn goes to ice. These processes, however, are theoretically well understood (Enting and Mansbridge 1985; Schwander et al. 1988).

The main concern of the group was the effect of melt layers in the ice. These melt layers can affect the concentrations of gases observed in the ice in two basic ways. First, some gases, notably CO_2, will preferentially dissolve in the liquid, resulting in melt layers with high gas concentrations. Thus, melt layers can alter the gas record in the ice. This phenomenon can be put to a positive use where partial surface melting in the summer is common. In this case, gas concentrations show seasonal cycles which add to our pool of dating tools as well as, in conjunction with other seasonal markers, provide information about whether melting has occurred.

The second problem with melt layers involves their effect on modeling diffusion of gases in the firn. Extensive melt layers could form blocks of gas movement in the firn. The importance of this effect is unknown and depends on how extensive and continuous the melt layers are. It is primarily a concern for studies of anthropogenic gas concentrations where detailed time information is required, such as CO_2 and CH_4.

The group felt that this concern could be readily addressed in field experiments. For example, the ^3He produced by tritium deposited during the extensive nuclear testing in the 1960s could be used to study gas diffusion in areas with extensive melt layers present. In this case, the tritium concentrations measured in the snow provide the input function and its distribution with depth.

In addition, the group recommended that purposeful tracer experiments be done both as a check on the diffusion models and as a method for determining the importance of ice layers on gas diffusion in the firn. Gases with extremely low detection limits, such as SF_6, could be injected at specified levels in the firn to study their horizontal and lateral diffusion.

Finally, the group concluded that more attention should be focused on ice physics. The growth of ice crystals depends on the temperature at the firn to ice transition zone, shown recently by Petit et al. (1987) on the Vostok core, and paleotemperatures reconstructed from the size of ice crystals are in excellent agreement with paleotemperatures reconstructed from δD values. Furthermore, the bubble close-off density also depends on the temperature at the firn–ice transition. This affects, in addition to the altitude effect (barometric pressure), the total gas content of the ice.

HOW IS THE RECORD PRESERVED DURING THE SURFACE SNOW TO FIRN TRANSITION?

The group identified several mechanisms which could potentially cause problems in interpreting the environmental record in glaciers. These processes include redistribution of deposited snow by wind, sublimation of ice, direct condensation of water vapor on the snow surface, chemical transformation at the snow/air interface, and surface melting.

It was generally agreed that these processes would become important as more detailed (in time) and more diverse information is demanded from glacial ice. With the possible exception of surface melting, these factors do not appear to significantly affect deep ice cores where the time resolution is on the order of decades to centuries. Also, most of these processes are closely linked to the glaciometeorological parameters of the drill site, primarily the net snow accumulation rate, the mean firn temperature, and the surface wind conditions. Their relevance on the interpretation of the environmental record in ice cores is thus site dependent.

Most of the group agreed that redistribution of surface snow by winds was a potentially important but puzzling phenomenon—puzzling because its potential seems unrealized. Despite the consensus that snow could be moved large but mostly unknown distances, the group could identify few specific examples where this effect has seriously distorted an ice core record. For example, despite the relatively low accumulation rate at Vostok, the climate record is clearly preserved. The group suggested that sharp stratigraphic markers, such as fallout from the recent accident at the Chernobyl power plant, could be examined in surface snow pits as a way of examining redistribution. Davidson pointed out that the Chernobyl spike at Dye 3 is still very sharp, despite the obvious presence of blowing snow (Davidson et al. 1987).

Sublimation of snow, primarily during the summer, was considered to be a potential problem for reconstructing seasonal cycles. The loss of snow results in an apparent increase and/or decrease in concentration of some species. This was thought to be potentially important for all particulates. It may also affect the comparison of seasonal cycles between species. There is some indication that the stable isotope content may be affected by sublimation, particularly in alpine glaciers. However, it was noted that despite sublimation and vapor deposition, the mean $\delta^{18}O$ values of snow studied in coastal Antarctica by Peel and co-workers (1988) were not affected.

The condensation of vapor directly on the snow surface is not considered to be a potential problem except for those areas where the accumulation rate of snow is very small or approaches zero. Its effect on particulate concentrations, for example, may be significant. At this time, there are no examples of this effect, so it must remain a possibility to be considered if and when problems do arise.

The chemical transformation (primary photochemical reactions) of species at the snow surface is another potential, but as yet unproven, problem. It was suggested that this could affect H_2O_2 concentrations in surface snow, as well as Cl concentrations through the remobilization of chlorine. The latter possibility deserves serious consideration in light of the recent results concerning the large variations in the $^{10}Be/^{36}Cl$ ratio in the ice cores (Suter et al. 1988). The group suggested that for all important volatile species laboratory studies should be done to check for this effect.

The impact of melt layers on the environmental record in glacial ice was discussed in the previous section with respect to gases. In the present context, it was suggested that surface melting might also have an impact on the concentration of reactive trace gases. It was also suggested that stable isotopes, particularly deuterium excess, be investigated as a possible tool for identifying melt layers after the bubbles have disappeared and are no longer identifiable in the ice. This technique relies upon some degree of evaporation from the liquid melt which could be reflected in shifts in the deuterium excess of the ice. It was pointed out, however, that there was apparently no shift in the $\delta^{18}O$ of ice in melt layers observed in Greenland, an observation which suggests that melting and subsequent refreezing as the liquid sinks in the snow occurs without significant evaporation.

HOW IS THE RECORD PRESERVED DURING THE TRANSFORMATION FROM THE ATMOSPHERE TO SURFACE SNOW?

This part of the atmosphere to ice pathway is particularly important if we are to accurately reconstruct past concentrations of aerosols and trace gases in the atmosphere from their concentrations in glacial ice. Two mechanisms are important: wet deposition (including precipitation scavenging) and dry deposition. The overall goal is to formulate a transfer function linking the concentration of a species in the air, C_{air}, to its concentration first in surface snow, C_{snow}, and ultimately in glacial ice, C_{ice}. This last step clearly includes some of the processes discussed in the previous sections of this report. The group agreed that a two-pronged approach was best suited to attack this problem: *(a)* improvement in our knowledge of the key physical and chemical mechanisms involved in wet and dry deposition, and *(b)* the development of empirical and semi-empirical mathematical models linking C_{air} to C_{snow} and C_{ice}. When reading the following discussion it is important to keep in mind that concentration changes of particulates and trace gases in ice observed in ice cores (i.e., the large changes in many species between the Holocene and Last Glacial) have already yielded a wealth of information on past changes in environmental conditions. It is clear, however, that this information will be much more valuable when reliable links between C_{air} and C_{ice} are formulated. This is particularly true for the intercomparison of concentrations of particulates and gases in the same ice core and between Arctic, Antarctic, and low and midlatitude glacial ice.

Wet Deposition and Precipitation Scavenging

The key mechanisms involved in wet deposition are nucleation scavenging of particulates, the in-cloud transport of particles and gases to existing droplets and ice crystals, and possible below-cloud scavenging of particulates and gases by falling precipitation. The importance of below-cloud scavenging is uncertain, but there was general consensus that this is not an important process in the polar regions.

For particles, nucleation scavenging appears to dominate. For gases, we need to determine for the important species which of the steps involved in transporting the gases to cloud droplets and ice crystals (diffusion, adsorption, and absorption) are the most important. For particulates, we need to determine for the important species which of the nucleation mechanisms are important (condensation nuclei, ice nucleation, and riming). In this regard, particular attention should be focused on ice nucleation in glacial clouds (clouds in which no liquid phase is present), as such clouds may dominate in the interiors of ice sheets, particularly in Antarctica. We are currently ignorant of the important processes occurring in such clouds.

The group proposed a series of experiments which can help answer these questions:

1. laboratory experiments to examine adsorption/absorption of gases onto cloud droplets and ice crystals while measuring all relevant parameters (e.g., temperature, cloud droplet and ice crystal size distributions, air velocities);

2. laboratory experiments to examine nucleation efficiency of particles while measuring all relevant parameters (all above parameters plus size distributions and crystal lattice structure of particles—both lab-generated and field aerosol);
3. field experiments to examine adsorption/absorption of gases under field conditions: measure concentration in air (preferably at or even above cloud level), concentration in precipitation, and relevant influential parameters;
4. field experiments to examine nucleation efficiency under field condition.

Dry Deposition

The key mechanisms involved in dry deposition are: *(a)* transport from the turbulent atmosphere to the viscous surface boundary layer, *(b)* diffusion across the viscous boundary layer, and *(c)* interaction with the surface. For gases, the snow and firn act somewhat like a grab sampler, with the exception of highly reactive gases such as HNO_3, in which case transport from the free atmosphere may be rate limiting. For particles, transport across the viscous surface boundary layer and/or interaction with the surface is probably the rate-limiting step on which research efforts should be concentrated.

For gases with low reactivities, we need to determine the mechanisms and rates of interaction with the surface. For gases with high reactivities, we need to identify the mechanisms and rates of transport from the turbulent atmosphere to the viscous surface boundary layer and then across the boundary layer. For the important particulate species, we need to determine the mechanisms and rates of transport across the surface boundary layer and the nature of the surface interactions. Also for particulates, we need to address the problem of how we can separate the effects of sublimation and dry deposition. Both of these processes concentrate the particulates in the surface snow, but the relative importance of these processes may vary for the important species as a function of climatic conditions.

The group again proposed a series of experiments which can help answer these questions:

1. wind tunnel experiments to examine dry deposition of gases and particles while measuring all relevant parameters (e.g., wind speed, turbulence, intensity, stability, characteristics of surface, characteristics of depositing species such as solubilities of gases, size distribution and hygroscopicities of particles);
2. field experiments to examine dry deposition under field conditions while measuring all relevant parameters.

Field Studies of the Combined Effects of Wet and Dry Deposition

The overall goal in linking C_{air} to C_{ice} is to derive the transfer function

$$C_{ice} = f(C_{air}).$$

While the function f is likely to be nonlinear, it would be helpful to begin with the "zeroth order" approximation and write

$$C_{ice} = W \times C_{air},$$

where W, sometimes called the scavenging ratio, is a function with a wide variety of parameters. To keep the problem solvable, the parameters should be derivable from analyses done on the ice core. The most important parameters are snow accumulation rate, temperature, concentration of species which may interact with one another to influence C_{air}, and (for the insoluble particles) the size distribution.

As a first step in exploring how each of these parameters influences the value of W, we can perform experiments comparing measurements of the concentrations of species in the air, in falling snow, and in surface snow. The above equation can be used with C_{snow} in place of C_{ice}, using data from these comparisons to evaluate W. Other parameters should also be measured, such as air temperature, cloud characteristics, precipitation rates, and characteristics of the surface snow. The group felt that it is important to measure C_{air} as a function of elevation to provide information at the cloud level.

If these measurements could be conducted over a complete year they would provide a valuable reference year for the snow, which could then be followed in successive years to study possible changes occurring during burial and firnification. Key species suggested by the group include:

–excess sulfate and methyl sulfonic acid,
–nitrate aerosol and HNO_3,
–chloride: NaCl, HCl,
–H_2O_2,
–^{210}Pb, $\delta^{18}O$, deuterium excess, ^3H, ^{137}Cs (Chernobyl),
–trace metals: Pb, Al,
–pollen, diatoms,
–insoluble particles (and size distributions),
–organic acids,
–organo-halogens

Some of these species exist as gases and as particulates. These phases should be considered separately, and the problems with sampling (destruction on filters) should be considered. Three key locations suggested by the group where these experiments should be conducted are *(a)* the more easily accessible alpine areas (e.g., Switzerland, Alaska, Canada), *(b)* South Pole, and *(c)* Dye 3.

Finally, the group was unanimous in recommending that deep drilling should be accompanied by studies comparing C_{air} and C_{snow}.

HOW IS THE RECORD MODIFIED FROM THE SOURCE TO THE ATMOSPHERE ABOVE THE GLACIER?

In this step, we are primarily concerned with the transport of aerosols from the source regions on the surface of the Earth to the air above the glaciers. As in the previous step, much work remains to be done in order to ultimately link concentrations of particles found in glacial ice to source strengths and species distributions. The following serves as a brief synopsis of the problems to be addressed.

Though each water droplet in a cloud has condensed on an aerosol particle, not all aerosols are "activated" in the cloud formation process. In certain cases, fewer than one percent of the available particulates in suspension end up serving as cloud nuclei. In other cases virtually all aerosols are cloud condensing nuclei. The key simplifying parameter is the slope of the cloud condensation nucleus spectrum on a log–log plot; that is, the value K in the power law expression

$$N = \text{const} \times S^K,$$

where N is the number (per cubic centimeter) of nuclei activated at supersaturation S. Typically in natural cloud formations, S remains at all times less than a few tenths or even hundredths of one percent.

In the case $K<2$, the number of cloud droplets activated, Nc, is proportional to the concentration of aerosol nuclei. Most aerosol nuclei are sulfate particles a few hundredths of a micron in diameter. In this case, if one maintains cloud liquid water constant, the SO_4^{2-} concentration in the cloud water is proportional to Nb, where Nb is the aerosol concentration. In the condition being described (where $K << 2$ and therefore Nc is proportional to Nb), a finer drop size cloud would be less likely to precipitate so ice sheet accumulation, with all other factors being equal, would tend to decrease.

If the aerosols are distributed by size in such a way that $K >> 2$, the cloud droplet concentration is independent of Nb and dependent only on updraft velocity. In this case the cloud water and ice sulfate (and therefore precipitation) would be decoupled from each other.

We are ignorant of the value of the power law exponent K, and a question of high priority is to determine K at remote locations and at supersaturations relevant to cloud formation (e.g., $S < 0.1\%$).

The group identified several important questions which should be addressed:

1. How are the values of K distributed in the real atmosphere, especially in the proximity of ice sheets?

2. As pointed out by Shaw (this volume), large and small particles tend to be preferentially removed from the atmosphere, leaving as condensation nuclei predominantly sulfate nuclei in the size range of 0.01–0.1 microns. Can we develop a more comprehensive understanding of the evolution of these "removal resistant" nuclei?
3. Most of the condensation nuclei are thought to be biogenic sulfate nuclei produced from dimethyl sulfate (DMS) in the surface ocean. At the present time, anthropogenic production of sulfate is thought to be 2–5 times greater than the natural DMS production. Can evidence be found for or against the hypothesis that the increasing anthropogenic sulfur emissions increase the concentration of cloud droplets, and hence accumulation rates for polar ice sheets?
4. Almost all of what is known about condensation nuclei applies to "wet clouds," or clouds where a liquid phase is present. How important is ice nucleation in glacial clouds? How important is ice nucleation when a liquid phase is not present? Can we find a way of determining whether a liquid phase was ever present?
5. For particles to reach the ice sheets they must often pass through effective precipitation "scubbers," such as the belt of storms which often encircle Antarctica. How can we quantify the importance of these scubbers on the concentration and size distributions of particles reaching the ice sheets?
6. The delivery of particles to the ice sheets may be very episodic. How important are brief, high deposition periods (e.g. sodium storms) to the total delivery, and species and size distributions of particles?
7. What cooperative projects can be formed to tie the various disciplines in the ice core community to the cloud physicists and cloud chemists? Can stable isotopes be used to determine supersaturations in clouds (e.g., Jouzel and Merlivat 1984) and the possible existence of liquid water?

For the stable isotopes of snow (δD, $\delta^{18}O$, and the combination of these values, deuterium excess: $d = \delta D - 8 * \delta^{18}O$) the step from the vapor source region to snow falling over the glacier determines the nature of the climatic information recorded in the stable isotope values. Thus, a comprehensive understanding of this large step is critical to the interpretation of δD and $\delta^{18}O$ values in glaciers and ice sheets. Models based on Rayleigh distillation combined with kinetic effects during snow formation can now successfully predict key isotopic relationships (δD versus $\delta^{18}O$, and $\delta^{18}O$ versus deuterium excess) as well as key isotope–climate relationships ($\delta^{18}O$ and δD versus surface temperature gradients and deuterium excess seasonal cycles) for snow falling on the Antarctic (Jouzel and Merlivat 1984) and Greenland ice sheets (Johnsen et al. 1988). These models attempt to account for isotopic fractionations occuring during evaporation, vapor transport and rainout, and finally snow formation.

Less is known about the source to snow pathway and concurrent isotopic fractionations for low and midlatitude glaciers, although this can in part be blamed on postdepositional effects such as wind scouring and melting, which tend to obscure the isotopic record in surface snow. As more cores from these glaciers are recovered and the isotopic data base expanded, increased effort should be put into modeling the δD, $\delta^{18}O$, and deuterium excess values in these glaciers.

In addition to the above recommendation, the group identified two additional areas where research efforts should be focused:

1. *The Isotope Paleothermometer.* While there is no doubt that down-core variations in $\delta^{18}O$ and δD values in ice cores reflect predominantly changes in the temperature gradient from the vapor source region(s) to the ice core site, there is still some debate about the exact magnitude of the temperature changes so recorded. This debate centers on the problem of using spatially derived isotope-temperature slopes to interpret temporal isotopic changes, a problem shared with most other paleoclimatic indicators. Two steps can be taken to help improve this isotope thermometer. First, the collection of meteorological data, including vertical temperature soundings, on and near ice sheet and glaciers should remain a high priority. Snow and firn cores, particularly in areas with high deposition rates and little postdepositional alteration, contain the best records of temporal changes in stable isotopes in precipitation. As the length of the direct temperature observations increases, so will our confidence in comparing these temperature records with the isotope records in the time sense. Second, greater emphasis should be placed on testing models for their predictions of the isotope-temperature relationship, both in the temporal and spatial domains. The recent improvements in models for δD and $\delta^{18}O$ values in snow—in both the Antarctic and Greenland ice sheets—should be exploited to examine what these regional models predict for the down-core relationship between isotope values and temperature. Another promising approach is the incorporation of stable isotopes of water in general circulation models (GCM). The pioneering work of Joussaume et al. (1984) and Jouzel et al. (1988) should be expanded to other GCMs and these models tested using different boundary conditions for different temperatures over the ice caps.
2. *Deuterium Excess.* The deuterium excess of snow falling on the ice sheets has recently shown great promise as a tool for identifying vapor source regions over the ocean (Johnsen et al. 1988) as well as recording in ice cores the conditions of evaporation over the ocean during past

climates (Jouzel et al. 1982). Deuterium excess also provides a sensitive check on the models used to calculate isotopic fractionation during snow formation. Greater effort should be focused on measuring δD values which, with the exception of the French glaciology group, are not routinely done when measuring $\delta^{18}O$ values in ice.

REFERENCES

Davidson, D.I.; Harrington, J.R.; Stephenson, M.J.; Monaghan, M.C.; Pudykiewicz, J.; and Schnell, W.R. 1987. Radioactive cesium from the Chernobyl accident in the Greenland ice sheet. *Science* **237**: 633–634.

Davidson C.I., and Wu, Y.L. 1988. Dry deposition of particles and vapors. In: Acid Precipitation, vol. 2, ed. D.C. Adriano. New York: Springer-Verlag.

Enting, I.G., and Mansbridge, J.V. 1985. The effective age of bubbles of polar ice. *Pure Appl. Geophys.* **123**: 777–790.

Johnsen, S.J. 1977. Stable isotope homogenization of polar firn and ice. *IAHS–AISH Publ.* **118**: 210–219.

Johnsen, S.J.; Dansgaard, W.; and White, J.W.C. 1988. The origin of arctic precipitation under present and glacial conditions. *Tellus*, in press.

Joussaume, S.; Jouzel, J.; and Sadourny, R. 1984. A general circulation model of water isotope cycles in the atmosphere. *Nature* **311**: 24–29.

Jouzel, J., and Merlivat, L. 1984. Deuterium and oxygen-18 in precipitation: modeling of the isotopic effects during snow formation. *J. Geophys. Res.* **89**: 11 749–11 757.

Jouzel, J.; Merlivat, L.; and Lorius, C. 1982. Deuterium excess in an East Antarctic ice core suggests higher relative humidity at the oceanic surface during the last glacial maximum. *Nature* **299**: 688–691.

Jouzel, J.; Russell, G.L.; Suozzo, R.J.; Koster, R.D.; White, J.W.C.; and Broecker, W.S. 1988. Simulations of HDO and $H_2{}^{18}O$ atmospheric cycles using the NASA/GISS general circulation model: the seasonal cycle for present day conditions. *J. Geophys. Res.*, in press.

Mulvaney, R.; Wolff, E.W.; and Oates, K. 1988. Sulfuric acid at grain boundaries in antarctic ice. *Nature* **331**: 247–249.

Neftel, A.; Jacob, P.; and Klockow, D. 1986. Long-term record of H_2O_2 in polar ice cores. *Tellus* **38B**: 262–270.

Neftel, A.; Oeschger, H.; Schwander, J.; and Stauffer, B. 1983. Carbon dioxide concentration in bubbles of natural cold ice. *J. Physical Chem.* **87**: 4116–4125.

Petit, J.R.; Duval, P.; and Lorius, C. 1987. Long-term climatic changes indicated by crystal growth in polar ice. *Nature* **326**: 62–64.

Schwander, J; Stauffer, B.; and Sigg, A. 1988. Air mixing in firn and the age at pore close-off. *Ann. Glaciol.* **10**: 141–145.

Siegenthaler, U.; Friedli, H.; Loetscher, H.; Moor, E.; Neftel, A.; Oeschger, H.; and Stauffer, B. 1988. Stable-isotope ratios and concentration of CO_2 in air from polar ice cores. *Ann. Glaciol.* **10**: 151–156.

Suter, M.; Beer, J.; Bonani, B.; Hofmann, H.J.; Michel, D.; Oeschger, H.; Synal, H.A.; and Wolfli, W. 1988. ^{36}Cl studies at the ETH/SIN AMS Facility. *Nucl. Instr. Meth. Phys. Res.*, in press.

SUGGESTED READING

Bodhaene, B.A.; De Luisi, J.J.; Harris, J.M.; Houmere, P.; and Bauman, S. 1986. Aerosol measurements at the South Pole. *Tellus* **38B**: 1223–2351.

Brimblecombe, P.; Clegg, S.L.; Davies, T.D.; Shooter, D.; and Tranter, M. 1987. Observations of the preferential loss of major ions from melting snow and laboratory ice. *Water Res.* **21**: 1279–1286.

Cunningham, W.C., and Zoller, W.H. 1981. The chemical composition of remote area aerosols. *J. Aerosol Sci.* **17**: 367–384.

Davies, T.D.; Brimblecombe, P.; Tranter, M.; Tsiouris, S.; Vincent, C.E.; Abrahams, P.; and Blackwood, I.L. 1987. The removal of soluble ions from melting snowpacks. In: Seasonal Snowcovers: Physics, Chemistry, Hydrology, eds. H.G. Jones and W.J. Orville-Thomas, pp. 337–392. Dordrecht: Reidel.

Herron, M.M.; Herron, S.L.; and Langway, C.C. Jr. 1981. Climate signal of ice melt features in Southern Greenland. *Nature* **293**: 389–391.

Twomey, S. 1980. Cloud nucleation in the atmosphere and the influence of nucleus concentration levels in atmospheric physics. *J. Physical Chem.* **84**: 1459–1463.

The Environmental Record in Glaciers and Ice Sheets
eds. H. Oeschger and C.C. Langway, Jr., pp. 99–121
John Wiley & Sons Limited

Dating by Physical and Chemical Seasonal Variations and Reference Horizons

C.U. Hammer

Geophysical Institute,
University of Copenhagen,
2200 Copenhagen N, Denmark

Abstract. Stratigraphical dating methods have made high accuracy ice core dating possible, especially those methods based on the isotopic composition of the ice and the seasonal variation of atmospheric trace substances. Various methods based on $\delta^{18}O$, dust, ECM (acidity), and trace substances (e.g., NO_3^-, Cl^-, SO_4^{2-}, H_2O_2, etc.) are presented, as well as their advantages and limitations.

While Holocene ice can often be dated with a "tree-ring like" precision, the dating of pre-Holocene ice is still problematic. A potential exists for dating pre-Holocene ice by high resolution and continuous methods, as exemplified by the results from the Byrd core (ECM) and the Dye 3 core (dust).

The use of reference horizons, a subject of special importance for the dating of ice cores from low accumulation areas, will be discussed. A major problem in dating is related to the rapid climatic changes observed in Greenland ice cores during the past glacial period.

INTRODUCTION

Dating of ice sheets, ice caps, and glaciers can be approached in different ways and with different accuracy requirements. In some cases, a "tree-ring like" precision is needed while in other cases a rather rough dating may suffice.

Recent trends in ice core dating point to an increasing demand for more precise, detailed, and absolute dates; it is becoming increasingly important to compare data from various drill sites in an unambiguous way. In the past, data from various sites have often been compared without too much attention being paid to the dating accuracy and this, not surprisingly, has led to wrong conclusions. It is necessary to ensure that "yesterday is not tomorrow and tomorrow not yesterday."

In order to secure sufficiently accurate dating, some minimum requirements must be met whenever an important and relevant drilling operation takes place. In many smaller projects, such a requirement cannot be fulfilled and may not be necessary. In the case of intermediate or deep drilling, however, it should be a natural part of the project, even more so as such operations are rather costly. Special dating efforts would only require a little extra funding.

Ice sheets, ice caps, and glaciers are large-scale, "layered" bodies consisting of past frozen precipitations which have been compressed and strained by gravitational forcing, i.e., they flow. During the ice flow the annual layers are changed in various ways depending on the stress distribution in the ice body. In certain parts of the ablation zone and at certain regions close to an undulating bedrock, the layer thickness may increase, but generally the annual layers are thinned with increasing depth.

In this chapter only the dry facies of the ice sheets will be considered, both for practical reasons and due to the following considerations:

1. The central parts of the ice sheets of Antarctica and Greenland present the most "ideal" records in glaciers of large-scale environmental and climatic changes.
2. During ice flow, the thinning of annual ice layers can be described fairly well by modeling. Thus, favorable conditions exist allowing models to be refined using the information obtained from analysis of the core, e.g., from information on the flow properties of the ice and on general changes in the annual layer thickness.
3. In these regions the oldest, well-layered ice, as well as the oldest ice on Earth, can be found.

The latter point is of particular importance because large parts of the glacial history of the Quaternary may be inferred from ice cores obtained in these regions, and the potential for accurate dating is high.

As dating by ice flow models and radioactive isotopes will be treated in other chapters, I will concentrate on seasonal variations and reference horizons.

SEASONAL VARIATIONS

Any component which varies seasonally in the atmosphere over the ice sheets and which is removed by the precipitation falling on the ice surface may, in principle, be used as a seasonal dating parameter of ice cores, i.e., if the ice surface receives sufficient precipitation to store the seasonal signal and if the transfer from atmosphere to ice conserves the seasonal signal. This is, however, not always the case. In the central parts of East Antarctica, for example, the yearly snow accumulation is so small (Lorius et al. 1979) that the annual layering is disturbed by surface winds. The transfer may also obliterate the signal, as in the case of gases, where the time resolution is poor (ten to several thousand years) (Schwander and Stauffer 1984; see also Budd et al., this volume). In the following, the emphasis will therefore be on records of isotopic composition and aerosol-like trace substance concentrations of the ice.

Not all seasonally varying ice core constituents are useful dating parameters, either because the seasonal signal is weak, too noisy, or "unstable" or they may be technically difficult to measure in sufficient detail along the entire core. Below a number of ice core properties or constituents used for dating ice cores are presented. The criteria for parameter selection are either the ease with which they can be measured, the strength of the seasonal signal, or their usefulness in dating very old ice layers ($t > 10{,}000$ years).

$\delta^{18}O$

The method of dating glacial ice by seasonal changes in the isotopic composition of the deposited snow was introduced to glaciology nearly 30 years ago by W. Dansgaard (1954) and has since been widely used wherever the amount of annual snow accumulation has allowed it. This method is based on the relation between the temperature of condensation and the isotopic composition of the precipitation; accordingly, it also offers information on past climatic conditions.

In the broader scope of meteorology and climate, this method relies on some main features of the ocean-atmosphere water transport and condensation system. It is a method especially suited for the Arctic and Antarctic latitudes, where seasonal temperatures changes are large. Both the concentration of deuterium and oxygen 18 in the water may serve as parameters, as they are linearly related.

The isotopic composition is measured by mass spectrometry, and the isotopic composition is expressed by the δ function, i.e., for ^{18}O:

$$\delta^{18}O = \left(\frac{R_S - R_{SMOW}}{R_{SMOW}}\right) 1000\ ‰ = \left(\frac{R_S}{R_{SMOW}} - 1\right) 1000‰ \qquad (1)$$

where R_S is the isotopic ratio of oxygen 18 and 16 in the sample and R_{SMOW} the ratio in a standard (SMOW: standard mean ocean water). The relation between $\delta^{18}O$ and δD can be expressed as

$$\delta D = 8.0\,\delta^{18}O + 10‰ \qquad (2)$$

(Craig 1961; Dansgaard 1964). Small but important deviations from the latter relation exist, but they need not concern us here.

The dating precision can be quite high, as demonstrated in Fig. 1 a, b, and c; this figure is a detailed version of previously published results obtained on the central Greenland ice core Crête (Hammer 1977b). Due to the annual accumulation at Crête (25 cm water equivalent/yr) the original δ profile at the time of deposition has been smoothed by diffusive processes; dating, however, was achieved without complications and a high precision was obtained by using the dust profile as a cross-check. The record is from 1975 and many refinements have taken place in stratigraphic dating since then, as will be exemplified below. In high accumulation areas, such as the Dye 3 region in South Greenland, the δ method could be applied over ca. the last 8000 years, as illustrated in Fig. 2. In the case of older ice layers, other methods must usually be used, yet while no special "clean room" precautions have to be taken for the δ sampling, the conditions change when trace substances are used for dating.

Dust

Visible "dirt" horizons in glaciers have been recognized as a seasonal stratigraphic parameter for a long time; however, for the clean polar ice sheets this method had to await technical developments. The Coulter microparticle counter is one such development as it allowed a quantitative measurement of microparticles down to the 0.3 μm size range (radius).

In the polar ice sheets, the bulk of insoluble aerosols show radii between 0.1–2 μm, which is a prerequisite for the somewhat "surprising" fact that lightscattering methods can be used to infer the volume and hence the mass concentration in polar ice (Hammer 1977a; Hammer, Clausen, Dansgaard et al. 1985). The method is "rough," but it has the advantage of revealing dust concentrations in great detail continuously along the core. As a quantitative tool, this method is only applicable for the high elevations on ice sheets and larger ice caps because the average refractive index of the particles, as well as their upper size range, must be within certain limits. To some extent this is secured by the remoteness of the source regions and by a strong mixing in the mid-troposphere, as the size of the precipitated particles are generally kept within 0.1–2 μm (radius). Furthermore, the mixing, at these atmospheric levels, of the material from the various remote source regions ensures that the optical property of the assembly of particles changes only slightly.

As a stratigraphic dating method the advantages are great:

1. During the thinning of the annual ice layers, diffusive processes on microparticles are small.
2. The technique can be refined to offer a resolution of a few mm along

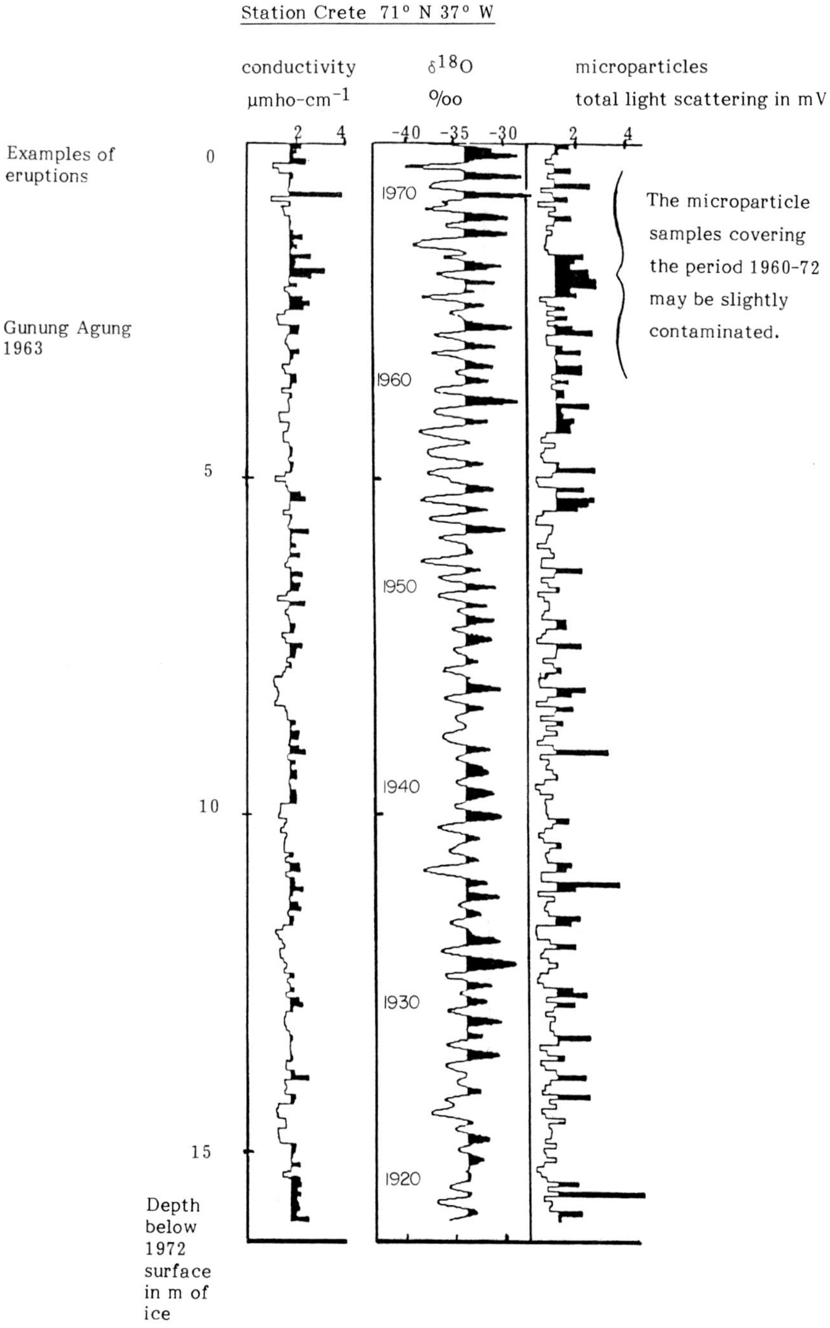

Fig. 1 *a*, *b*, *c*—Detailed profiles of liquid electrical conductivity (uncorrected), $\delta^{18}O$, and dust from Crête, Central Greenland. Note the seasonal variations and the volcanic influence on the conductivity profile. *(b & c overpage)*

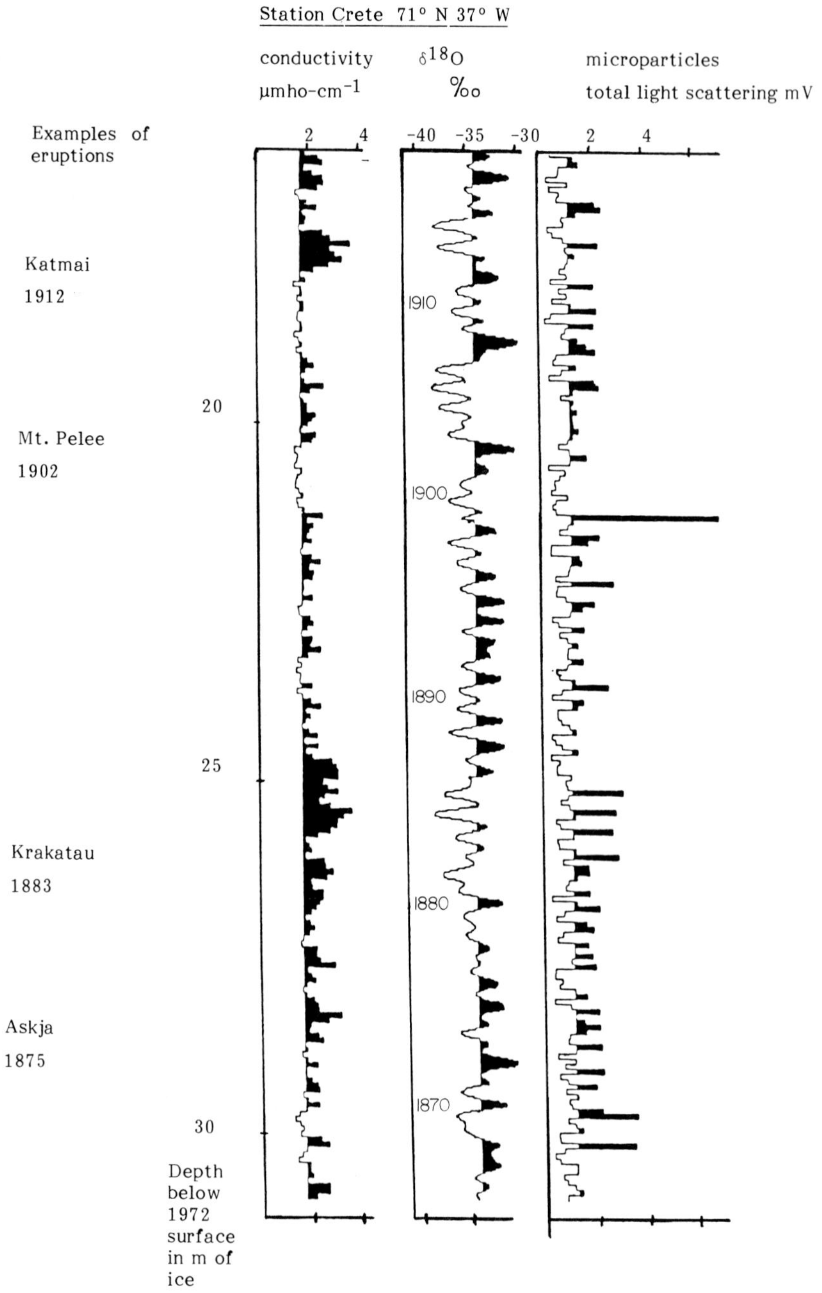

Fig. 1b—see legend on p. 103.

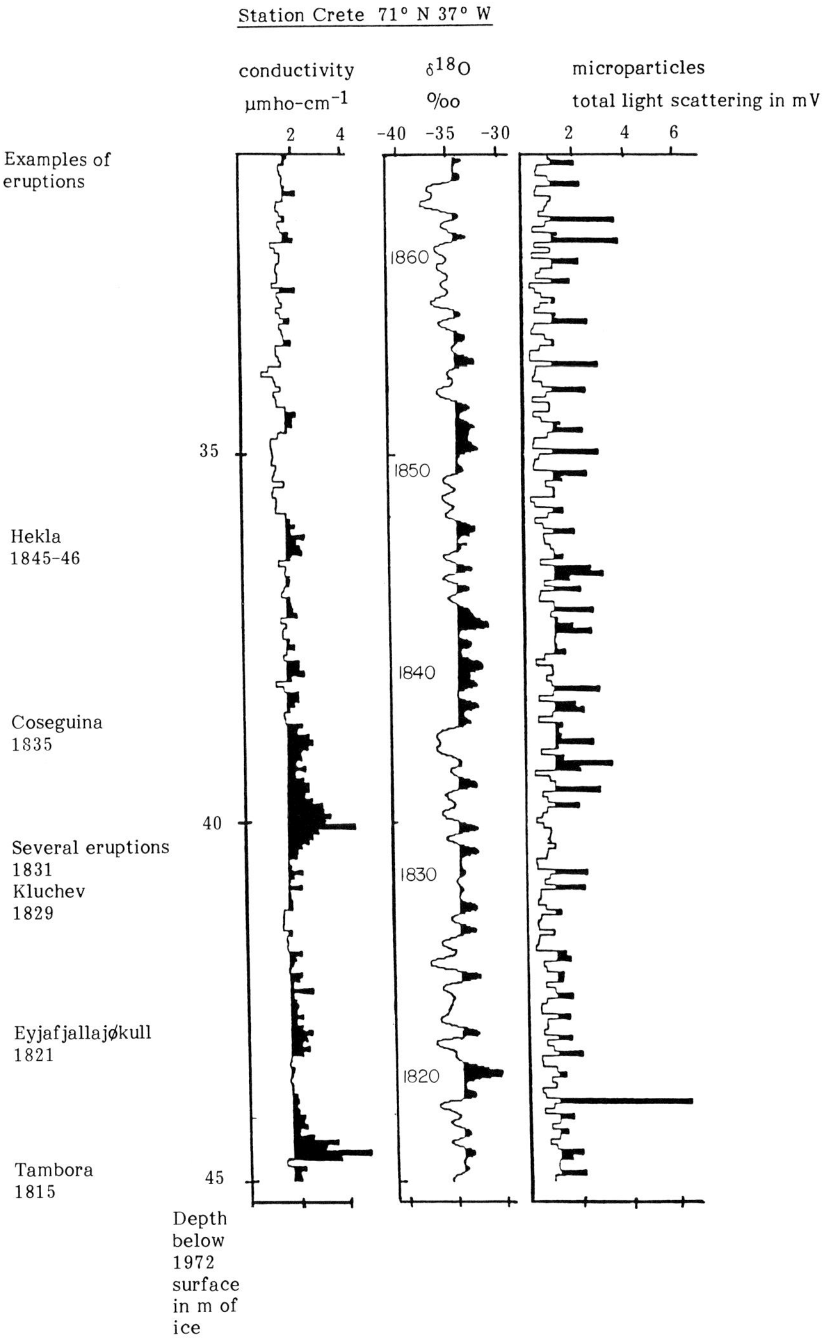

Fig. 1c—see legend on p. 103.

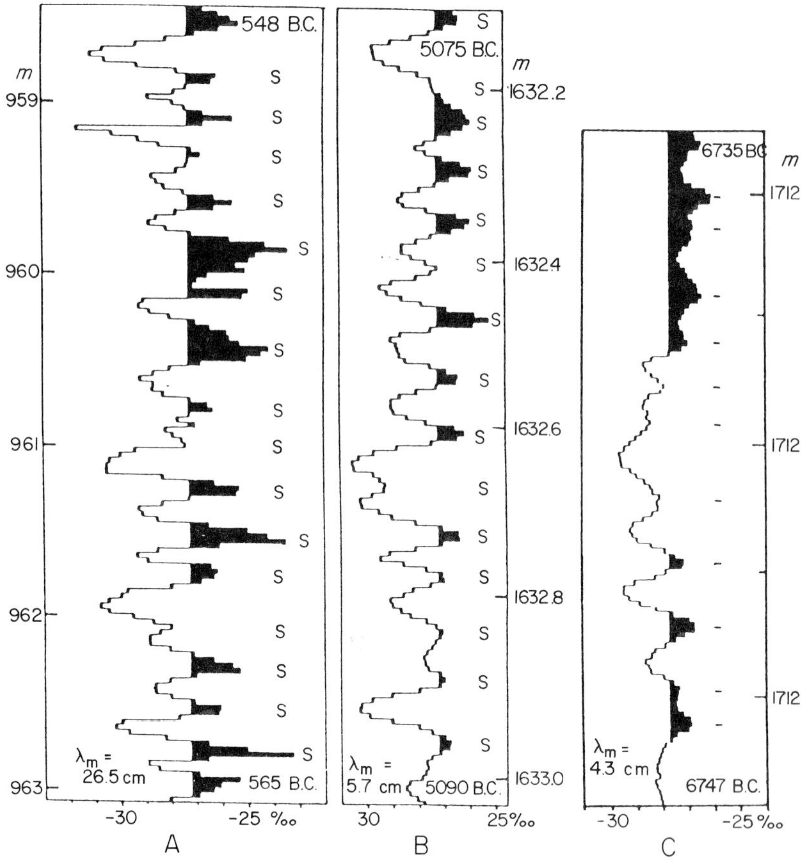

Fig. 2—Seasonal variations of $\delta^{18}O$ at various depths in the Dye 3 core from South Greenland. (Reprinted with permission from *Radiocarbon*, 1986, **28, No. 2A**: 284–291.)

the core, as compared to the present resolution of a few cm.

3. The technique can be used *in situ*, as was done during the Dye 3 deep drilling (see Figs. 3 and 4).

It has been suggested that boudinage may set a limit to the dating potential (Staffelbach et al. 1988), but it is also believed that the seasonal layering in favorable cases may be conserved down to a few tens of meters above the bedrock, e.g., in Central Greenland.

As the history of the Greenland ice sheet is not well known prior to some 10^3 yrs B.P., only a deep drilling at the central part of the ice sheet can demonstrate the full potential of the dust-dating technique. Dust-dating of Antarctic ice cores is limited to regions of sufficiently high snow accumulation

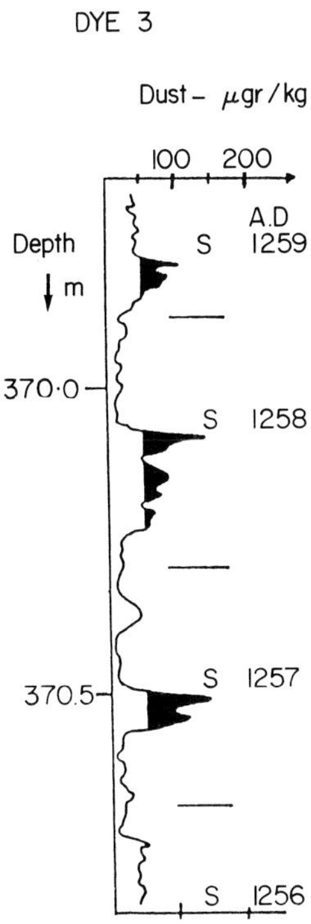

Fig. 3—Continuous dust measurements on the Dye 3 core; the profile shown is an example of "low" resolution application of the lightscattering technique.

(e.g., in West Antarctica) and has been successfully applied on the Holocene part of the Byrd core (Thompson 1977). The Byrd core has recently been dated by the continuous ECM technique (Hammer 1983; Hammer, Clausen, and Langway 1985), which yields a dating far into the last glacial. Cross-dating with the dust method, however, would increase the accuracy.

ECM Acidity

The ECM method (Hammer 1983) is a continuous, high resolution technique based on the electrical conductive characteristics of ice from the central parts of the polar ice sheets. This dating method presents us with great advantages, but also with some serious limitations. The advantages are quite clear as the ECM technique is easy to apply, is extremely fast, and delivers high resolution results down to a few mm along the cores, e.g., 2 m of core can be measured *in situ* in a few minutes.

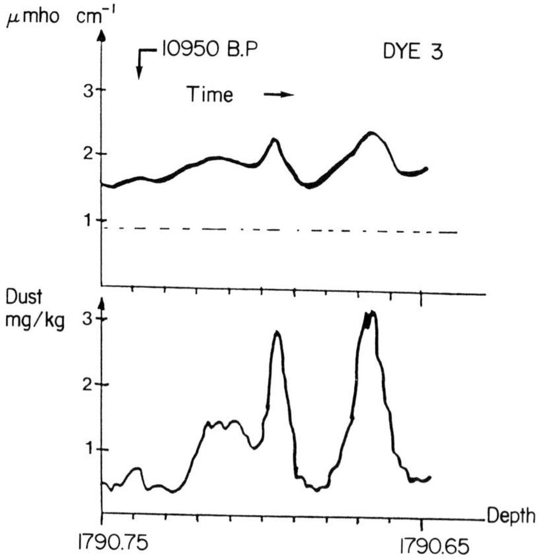

Fig. 4—High resolution dust measurements in the Younger Dryas part of the Dye 3 core; above the less resolved electrical conductivity (liquid). Four annual layers are seen.

As a quantitative estimate of acidity, the precision of the method has been questioned (Wolff and Peel 1985); however, for cores from the polar ice sheets there is little doubt that seasonal acidity variations, enhanced acidity due to volcanic eruptions or low acidity (or even alkaline ice), can be inferred from the electrical DC current measured. The ECM current is calibrated by pH measurements. Strong acidity can then be inferred for the special set-up used when the temperature of the ice is known (Hammer 1983).

The limitations of this method stem mainly from two important facts. First, the Greenland ice sheet was alkaline during glacial times, hence the interpretation of the low ECM currents representing glacial ice is not straightforward. Second, the concentration of strong acids is a composite of various acids which do not vary seasonally in the same way, and volcanic eruptions and other phenomena may add to the acidity.

Generally, HNO_3 and H_2SO_4 are the dominant acids, but in Greenland HNO_3 is responsible for some 80% of the acidity, and the seasonal signal is reasonably strong in most years. In the Antarctic Byrd core both acids are important; however, sulfuric acid shows the most clearcut seasonal variation which is perhaps related to regional oceanic production of sulfur-rich gases (Mulvaney and Peel 1988).

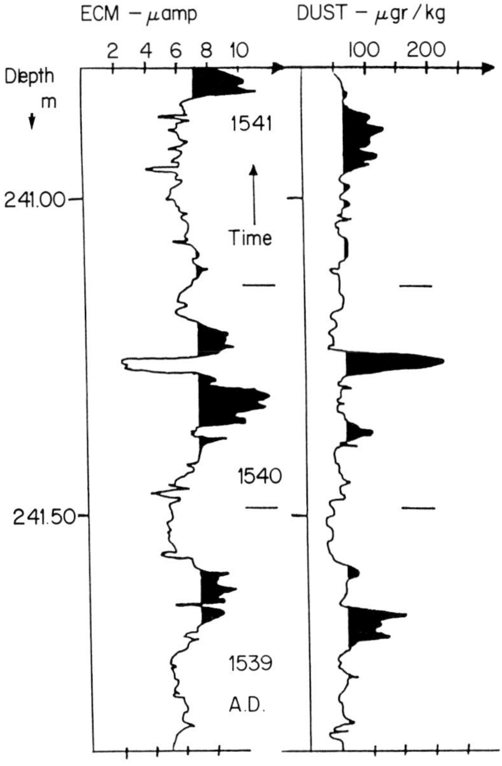

Fig. 5—ECM and moderate resolved dust measurements over the years A.D. 1536–1541 in the Dye 3 core, the ECM always has a high resolution. Note the relation between the dust and ECM results in late summer A.D. 1540.

The most successful applications of the ECM were on the Dye 3 deep core (Hammer et al. 1986) and the Antarctic Byrd core (Hammer 1982), as shown in Figs. 5 and 6, respectively. In the case of the Dye 3 core, cross-dating between $\delta^{18}O$, dust, and EMC profiles improved the dating accuracy (summarized in Table 1; see also Figs. 5 and 7).

The ECM method is most advantageous in the case of ice cores from Central Greenland or West Antarctica cores. This, however, does not exclude the fact that ECM also works well in other regions if the general conditions are favorable, e.g., the ice must generally be acidic, and this usually excludes sites where the impurity concentration in the ice is very high.

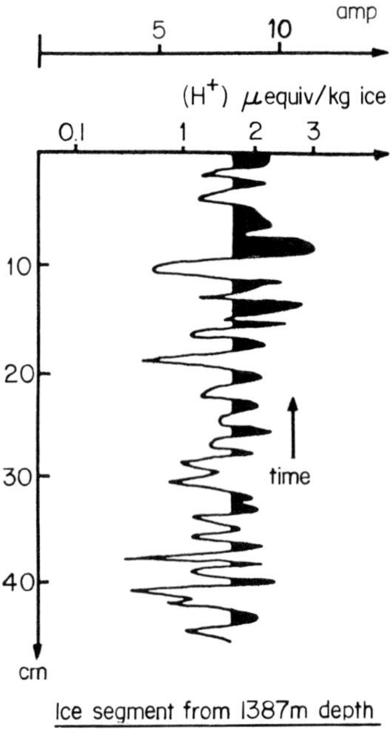

Fig. 6—Seasonal variations in (H^+) as measured by ECM on deep ice from the Antarctic Byrd core. Age approximately 18,500 years (from Hammer 1983). (Reprinted with permission of the American Chemical Society from *The Journal of Physical Chemistry*, 1983, **87**: 4099–4103.

Other Trace Substances: NO_3^-, Cl^-, SO_4, H_2O_2, etc.

Many trace substances, apart from the above mentioned, vary seasonally over the ice sheets and may, like H_2O_2 (Neftel et al. 1984), show great potential as dating methods. The major obstacle is sampling and analysis time. However, it is within our reach to develop continuous, moderate (high) resolution techniques based, e.g., on melting a groove along the cleaned ice core surface and subsequent chemical reactions and detection (see Figs. 8 and 9).

Seasonal variations on sequences of ice cores have been demonstrated for a great number of sites (e.g., Hammer et al. 1978; Dansgaard et al. 1977; Dansgaard et al. 1973; Langway et al. 1977; Koerner and Russell 1979; Cragin et al. 1977; Finkel et al. 1986), though the seasonal character may vary with site and time interval covered. Nitrate, Fig. 10 (Steffensen 1988), and hydrogen peroxide (only for the Holocene) seem good dating parameters for central Greenland ice cores, while sulfate and nitrate work well for the Byrd ice core (Fig. 11) and the Ross Ice Shelf (Herron 1982).

Less data exist on cations. As ionic analysis of ice cores becomes more and more frequent, however, we will probably soon see the number of data

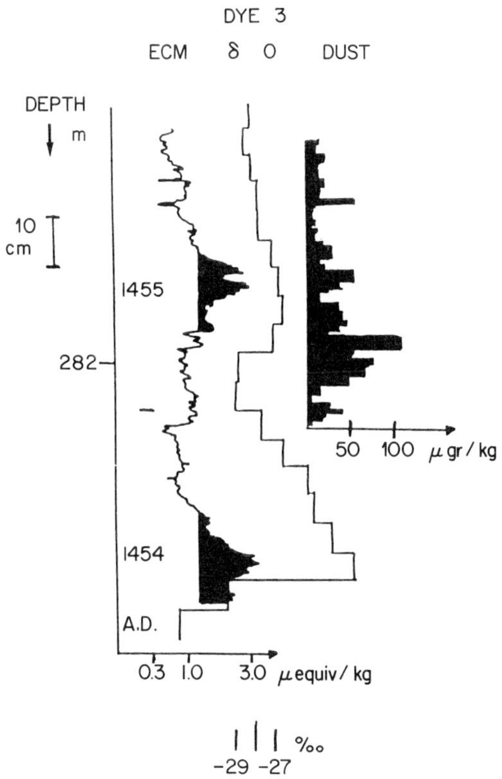

Fig. 7—ECM, $\delta^{18}O$, and dust profiles in the Dye 3 core over the years A.D. 1454 and 1455. Note the low resolution of the δ profile. The dust measurements were performed with extreme high resolution, but have been plotted here as an average value for each "individual" precipitation: the actual resolution is 20 times higher.

Table 1 Dye 3. Dating methods and dating precision in various depth intervals. (Reprinted with permission from *Radiocarbon*, 1986, **28, No. 2A**: 284–291)

Depth interval (m)	Time interval	Dating precision	Dating methods: Continuous	Dating methods: Selected segments
0–980	A.D. 1979–625 B.C.	625 B.C. ± 5 (estimated standard deviation)	Acidity $\delta^{18}O$	Dust
980–1540	626–3870 B.C.	3870 B.C. ± 10 (estimated standard deviation)	Acidity $\delta^{18}O$	
1540–1656	3871–5500 B.C.	5500 B.C. with est. error limit of 30 yr	Acidity	$\delta^{18}O$
1656–1785	5501–8770 B.C.	8770 B.C. with est. error limit of 150 yr	Acidity	$\delta^{18}O$ Dust

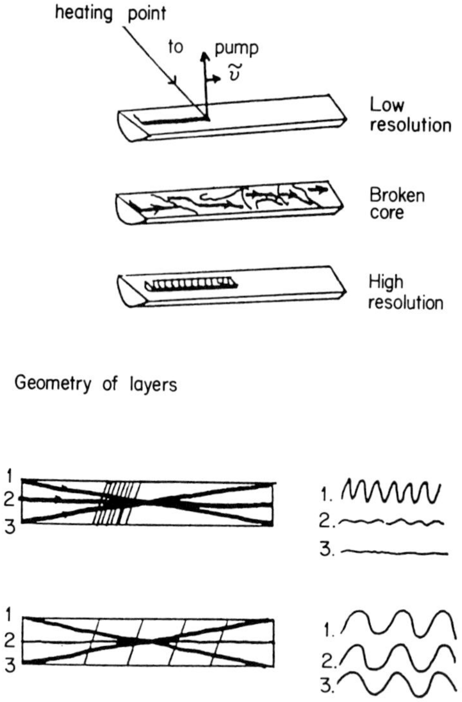

Fig. 8—In the upper part of the figure the melting profile in three cases is schematically shown: low resolution, broken core, and high resolution. In the lower part the obtained resolutions dependence on the actual melt-profile is illustrated; the resolution is indicated by the seasonal signals to the right.

strongly increase. In any case, as more and more seasonal dating techniques become available, the multi-element approach to dating should lead to a very high dating precision, e.g., the dependence on the distribution of seasonal precipitation will diminish as the signals vary in different ways during the year. In principle, we are already able to date ice cores from well chosen sites by a "tree-ring like" precision. A prerequisite, however, is that the ice core is continuous and that the annual ice layers have not been "misplaced", e.g., if the layers have flowed close to the bedrock.

REFERENCE HORIZONS

Atmospheric conditions, which mark precipitation falling on a large part of an ice sheet or even the entire hemisphere/global precipitation, may serve as reference horizons. Such horizons serve as "time-markers" along the core, although they may not always be absolutely dated. The general validity

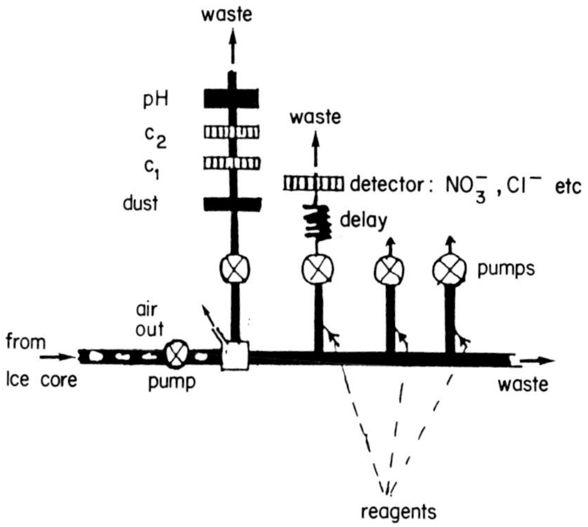

SCHEME OF MULTIFLOW SYSTEM

Fig. 9—Schematic set-up for continuous measurements of various kinds performed on the meltwater from the cleaned core surface. The air is removed from the meltwater before entering the system. The delay line is inserted in order to let the chemical reactions finish. In the future, this kind of system may replace the present standard analytic techniques.

of the reference horizons, of course, depends on the particular phenomena which cause the horizons. Since 1952, for example, nuclear bomb tests clearly marked the precipitation over the entire globe; the radioactive impurities in the precipitation, however, did not show the same detailed time-like pattern over the entire globe. The extensive data collected over this period, including ice core data, showed that it was necessary to distinguish between three geographical zones: equatorial and the remaining part of the Northern and Southern Hemisphere. Hence the detailed time profile of fallout deviated in Arctic and Antarctic ice cores. The features, however, were quite similar within each hemisphere and the radioactive data obtained on the ice cores were used as reference horizons, as exemplified in Fig. 12 (Hammer et al. 1980).

The spread of the radioactive debris was also used to infer the production of volcanic acids from single eruptions of considerable magnitude, by means of acid concentrations observed in the cores (Delmas et al. 1982; Legrand and Delmas 1987; Clausen and Hammer 1988; Langway et al. 1988). By nature of the phenomena, volcanic eruptions of considerable magnitude have proved to be useful "time-markers" via their acid gases SO_2, HCl, and HF.

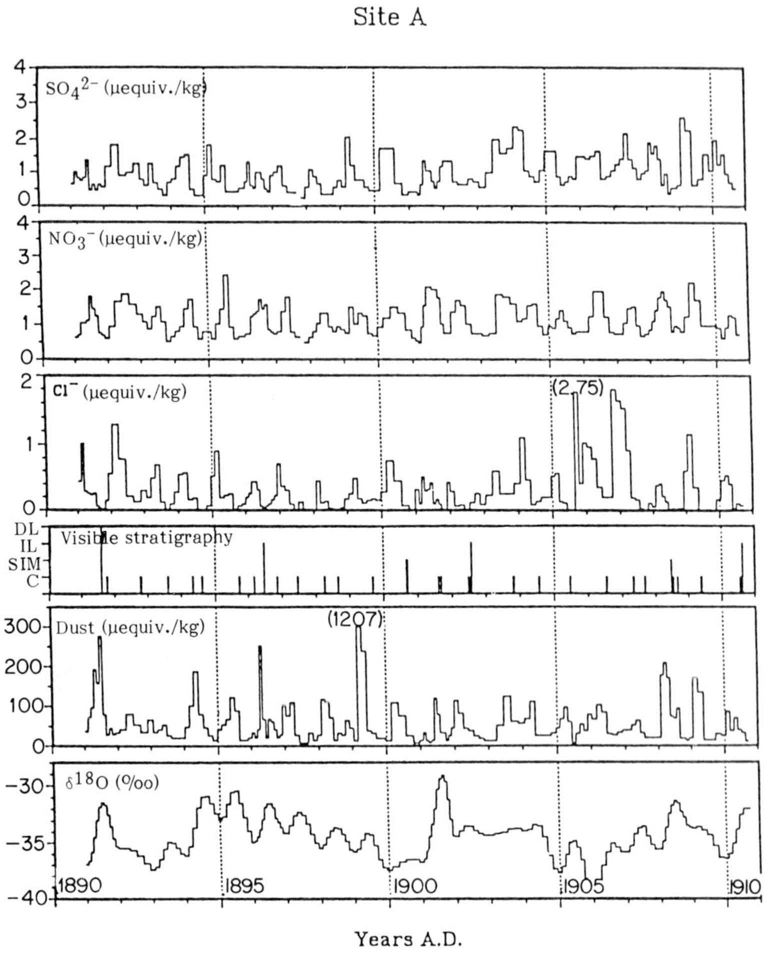

Fig. 10—Various ice core parameters which are often used for dating. Note that the site A is a site in Central Greenland, and the moderately high accumulation conserves the seasonal δ variation; diffusion, however, is clearly smoothing the record. Nitrate seems to be the most clear seasonally varying parameter (from Steffensen 1988). (Reproduced by courtesy of the International Glaciological Society from the *Annals of Glaciology*, 1988, **10**: 172, Fig. 2a.)

The use of volcanic reference horizons in ice cores, however, has not been widely used. The reason is twofold: First, before volcanic horizons could be used for dating purposes it was necessary to establish a time scale independent of any subjective interpretations of the volcanic signals (by seasonal variations). Second, the information on past volcanic eruptions is limited and the dating of the eruptions is not very precise, apart from certain well-documented historical eruptions. The dating of ice cores, however, has

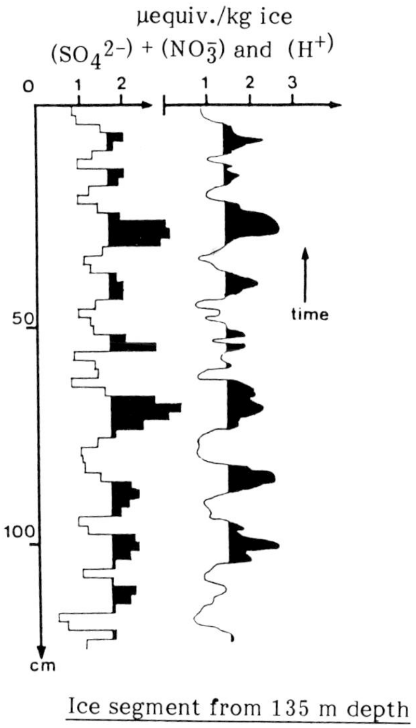

Fig. 11—Seasonal variations in $SO_4 + NO_3^-$ and H^+ as measured by ECM on deep ice from the Antarctic Byrd core. Approximate age is 1000 yrs (from Hammer 1983). (Reprinted with permission of the American Chemical Society from *The Journal of Physical Chemistry*, 1983, **87**: 4099–4103.)

now reached a state where it becomes legitimate to use volcanic eruptive ice core signals as reference horizons (see Fig. 1), even though care still has to be exercised.

The Icelandic Laki eruption in A.D. 1783 is a very strong reference horizon in Greenland and Canadian ice cores. This horizon was recently discussed in detail for 11 Greenland sites (Clausen and Hammer 1988). In the low accumulation area of central East Antarctica, both fission products and historical eruptions have served to date ice cores (Crozaz et al. 1966; Picciotto and Wilgain 1963) because seasonal stratigraphic methods were not applicable due to the low accumulation in this region.

These kinds of reference horizons are not the only ones existing in ice cores (e.g., regional melt features, unusual annual precipitations, etc. may also be useful). They are, however, fairly easy to deal with, which is not always the case, e.g., a large amount of additional information was required to use the end of the Younger Dryas climatic period as a reference horizon in Greenland ice cores (Hammer et al. 1986). This brings me to a

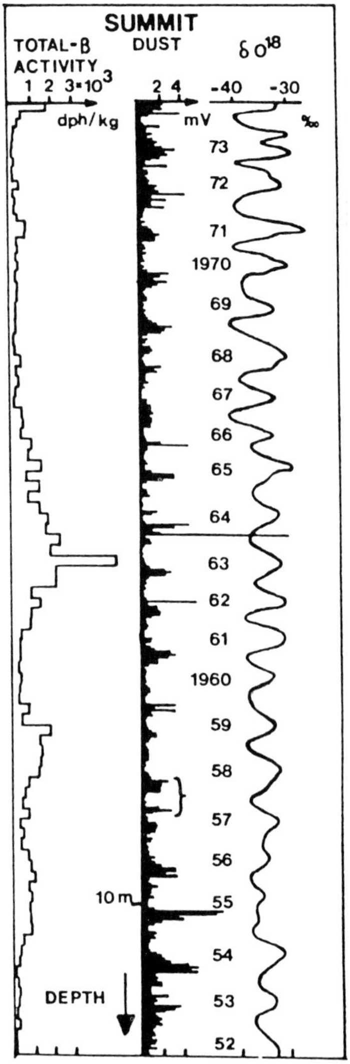

Fig. 12—Fission products, dust, and δ profiles in the upper part of the Summit core from Central Greenland (reprinted from Hammer 1977).

controversial issue: While there is general agreement on the main results from deep ice cores, we are left with some serious problems strongly related to the dating of pre-Holocene ice. This is very crucial as it is also related to some of the most intriguing features of the global climatic/environmental system, i.e., how will the climatic system react if it is pertubed, and how strong must pertubations be in order to cause *serious* changes? The word *serious* here is relative to the impact on Mr. and Mrs. Anthropus—not to mention the impact of anthropogenic activity on the climatic/environmental system of the globe.

THE PRE-HOLOCENE DATING PROBLEM

While dating of Holocene polar ice cores is in principle solved, even if a high dating accuracy is desired, considerable problems arise for pre-Holocene ice. The problem is not entirely dependent upon technical developments but is also related to the characteristics of the sites, where the few existing deep cores were obtained.

The Camp Century (Fig. 13, Hammer et al. 1978), Dye 3 (Fig. 4), and Byrd core (Fig. 6) have conserved what must be interpreted as seasonal variations within the layers representing the Wisconsin glacial period. However, as data are scarce (apart from the Byrd core), no well established time scales exist for the Camp Century and Dye 3 core. If the resolution of

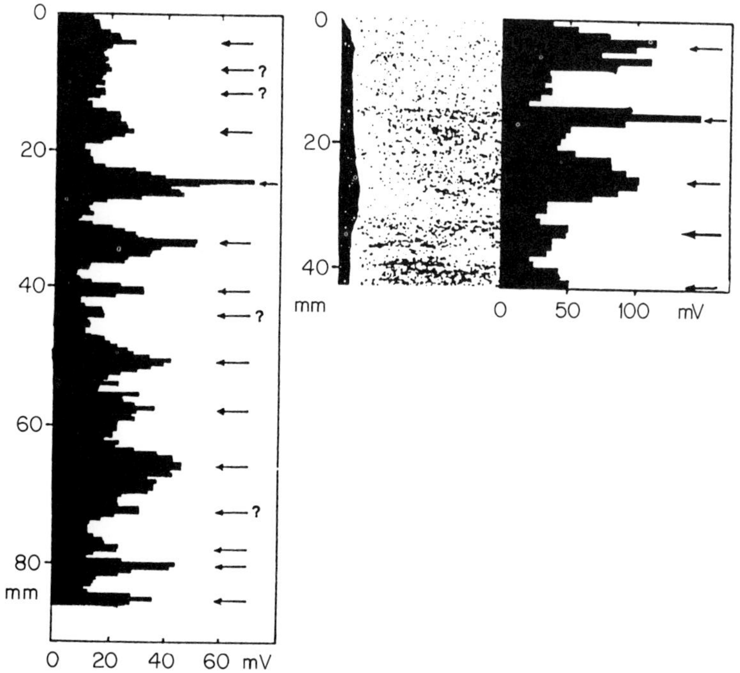

Fig. 13—High resolution dust measurements in the Younger Dryas part of the Camp Century deep core. The sampling was time consuming, as it was mechanically performed; today this technique is strongly improved by the "melt procedure" (reprinted from Hammer et al. 1978).

the measurements is too low, the seasonal signal may easily be lost, as exemplified in Fig. 14. Consequently, the present dating of pre-Holocene ice has been based on less strict considerations (Dansgaard et al. 1982; Lorius et al. 1985).

While there is little doubt as to the approximate dating of glacials and interglacial periods during the Quaternary, the situation is quite different when the approximate time scales are used to derive important relations between the various data obtained on the Quaternary period, i.e., from deep-sea records, peatbogs, ice cores, etc. Papers dealing with this subject are numerous; below I mention only a few important problems related to ice cores:

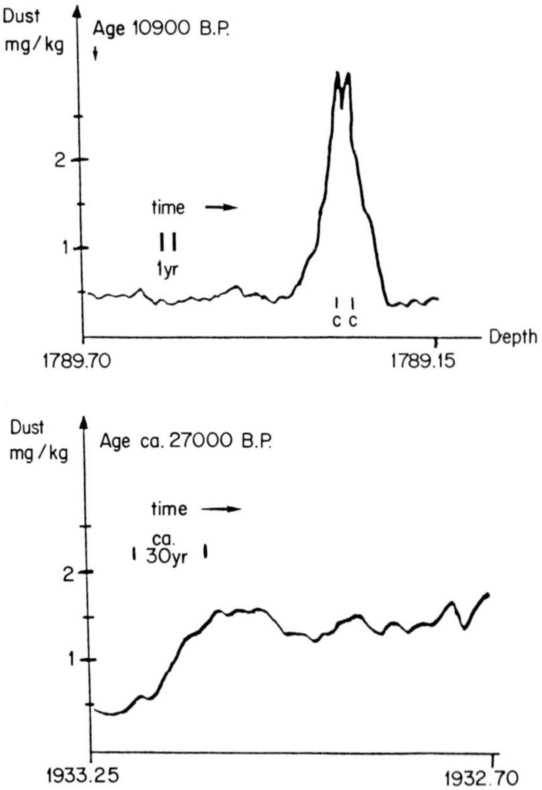

Fig. 14—Examples of "insufficient" resolution along the Dye 3 core; seasonal variations are not resolved. Only in the upper part of the figure are some "remains" of seasonal dust variations seen (cf. Fig. 4). The C's in the upper part indicate cloudy bands

1. How representative are the "seasonal" variations of dust and ECM in existing deep cores?
2. How far back in time can seasonal changes be traced before either the seasonal signal is "lost" or the continuous annual layering is broken?
3. Are the "rapid climatic changes" observed in Greenland ice cores artifacts or not?

The latter point is the most important, due to its serious impact on our understanding of the major features of the climatic/environmental system. It is a complex problem involving many different aspects of ice core analysis, but also invokes data from outside the glacier record (see also Budd et al. and Hecht et al., both this volume).

A FUTURE PERSPECTIVE

Technical developments have occurred for a multitude of dating techniques which can be applied on ice cores. I have mentioned only a few, but feel quite confident that within the next decade dating of ice cores will no longer be a major problem. Such a bold statement needs of course to be proven. If the present strong interest in ice core data continues, it must necessarily lead to a strong intent to get the time scale right!

We are well on the way and hopefully this workshop will set the standard for the following decades of ice core analysis.

Acknowledgments. The data presented is derived from many sites and projects but substantial funding was obtained from the Commission for Scientific Research in Greenland, the Danish Natural Science Research Council, and the U.S. National Science Foundation; Department of Polar Programs. I would also like to acknowledge the enthusiastic and friendly help and support from U.S. Polar Ice Coring Office; Lincoln, Nebraska U.S. Air Force 109 TAG; and the Royal Danish Air Force.

REFERENCES

Clausen, H.B., and Hammer, C.U. 1988. The Laki and Tambora eruptions as revealed in Greenland ice cores from 11 locations. *Ann. Glaciol.* **10**: 16–22.

Cragin, J.H.; Herron, M.M.; Langway, C.C., Jr.; and Klouda, G. 1977. Interhemispheric comparison of changes in the composition of atmospheric precipitation during the late Cenozoic era. In: Polar Oceans, ed. M.J. Dunbar, pp. 617–631. Calgary, Alberta: Arctic Institution of North America.

Craig, H. 1961. Isotope variations in meteoric waters. *Science* **133**: 1702–1703.

Crozaz, G.; Langway, C.C., Jr.; and Picciotto, E. 1966. Artificial radioactivity reference horizons in Greenland firn. *Earth Plan. Sci. Lett.* **1**: 42–48.

Dansgaard, W. 1954. Oxygen-18 abundance in fresh water. *Nature* **174**: 234.

Dansgaard, W. 1964. Stable isotopes in precipitation. *Tellus* **16**: 436–468.

Dansgaard, W.; Clausen, H.B.; Gundestrup, N.; Hammer, C.U.; Johnsen, S.J.; Kristinsdottir, P.M.; and Reeh, N. 1982. A new Greenland deep ice core. *Science* **218, No. 4579**: 1273–1277.

Dansgaard, W.; Johnsen, S.J.; Clausen, H.B.; and Gundestrup, N. 1973. Stable isotope glaciology. *Meddelelser om Grønland* **197, No. 2**: 1–53.

Dansgaard, W.; Johnsen, S.J.; Clausen, H.B.; Hammer, C.U.; and Langway, C.C., Jr. 1977. Stable isotope profile through the Ross Ice Shelf at Little America, Antarctica. *Meddelelser om Grønland* **197, No. 2**: 322–325.

Delmas, R.J.; Briat, M.; and Legrand, M. 1982. Chemistry of South Polar snow. *J. Geophys. Res.* **87, C6**: 4314–4318.

Finkel, R.C.; Langway, C.C., Jr.; and Clausen, H.B. 1986. Changes in precipitation chemistry at Dye 3, Greenland. *J. Geophys. Res.* **91, No. D9**: 9849–9855.

Hammer, C.U. 1977a. Dating of Greenland ice cores by micro-particle concentration analyses. In: Isotopes and Impurities in Snow and Ice, Proc. IUGG Symp., Grenoble, 1975. *IAHS-AISH Publ.* **118**: 297–301.

Hammer, C.U. 1977b. Past volcanism revealed by Greenland ice sheet impurity. *Nature* **270**: 482–486.

Hammer, C.U. 1982. The history of atmospheric composition as revealed in ice sheets. In: Atmospheric Chemistry, ed. E.D. Goldberg, pp. 119–134. Dahlem Konferenzen. Berlin, Heidelberg, New York: Springer.

Hammer, C.U. 1983. Initial direct current in the build-up of space charges and the acidity of ice cores. *J. Physical Chem.* **87**: 4099–4103.

Hammer, C.U.; Clausen, H.B.; and Dansgaard, W. 1980. Greenland ice sheet evidence of post-glacial volcanism and its climatic impact. *Nature* **288**: 230–235.

Hammer, C.U.; Clausen, H.B.; Dansgaard, W.; Gundestrup, N.; Johnsen, S.J.; and Reeh, N. 1978. Dating of Greenland ice cores by flow models, isotopes, volcanic debris, and continental dust. *J. Glaciol.* **20**: 3–26.

Hammer, C.U.; Clausen, H.B.; Dansgaard, W.; Neftel, A.; Kristinsdottir, P.; and Johnson, E. 1985. Continuous impurity analysis along the Dye 3 deep core. In: Greenland Ice Core: Geophysics, Geochemistry and Environment, eds. C.C. Langway, Jr., H. Oeschger, and W. Dansgaard. *Geophys. Monog.* **33**: 90–94. Washington, D.C.: Amer. Geophys. Union.

Hammer, C.U.; Clausen, H.B.; and Langway, C.C., Jr. 1985. The Byrd ice core: continuous acidity measurements and solid electrical conductivity measurements. *Ann. Glaciol.* **7**: 214.

Hammer, C.U.; Clausen, H.B.; and Tauber, H. 1986. Ice-core dating of the Pleistocene/Holocene boundary applied to a calibration of the ^{14}C time scale. *Radiocarbon* **28, No. 2A**: 284–291.

Herron, M.M. 1982. Glaciochemical dating techniques. In: Nuclear and Chemical Dating Techniques, Interpreting the Environmental Record, ed. L.A. Currie. *ACS Sympos. Ser.* **176**: 303–318.

Koerner, R., and Russell, R.D. 1979. $\delta^{18}O$ variations in snow on the Devon Island ice cap, Northwest Territories, Canada. *Can. J. Earth Sci.* **16(7)**: 1419–1427.

Langway, C.C., Jr.; Clausen, H.B.; and Hammer, C.U. 1988. An interhemispheric volcanic time-marker in ice cores from Greenland and Antarctica. *Ann. Glaciol.* **10**: 102–108.

Langway, C.C., Jr.; Klouda, G.A.; Herron, M.M.; and Cragin, J.H. 1977. Seasonal variations of chemical constituents in annual layers of Greenland deep ice deposits. In: Isotopes and Impurities in Snow and Ice, Proc. IUGG Symp., Grenoble, 1975. *IAHS-AISH Publ.* **118**: 302–306.

Legrand, M., and Delmas, R.J. 1987. A 220-year continuous record of volcanic H_2SO_4 in the Antarctic ice sheet. *Nature* **327**: 671–676.

Lorius, C.; Jouzel, J.; Ritz, C.; Merlivat, L.; Barkov, N.I.; Korotkevich, Y.S.; and Kotlyakov, V.M. 1985. A 150,000 year climatic record from Antarctic ice. *Nature* **316**: 591–596.

Lorius, C.; Merlivat, L.; Jouzel, J.; and Pourchet, M. 1979. A 30,000 yr isotope climatic record from Antarctic ice. *Nature* **280**: 644–648.

Mulvaney, R., and Peel, D.A. 1988. Anions and cations in ice cores from Dolleman island and the Palmer land plateau, Antarctic Peninsula. *Ann. Glaciol.* **10**: 121–125.

Neftel, A.; Jacob, P.; and Klockow, D. 1984. Measurements of hydrogen peroxide in polar ice samples. *Nature* **311(5981)**: 43–45.

Picciotto, E.E., and Wilgain, S.E. 1963. Fission products in Antarctic snow: a reference level for measuring accumulation. *J. Geophys. Res.* **68: No. 21**: 5965–5972.

Schwander, J., and Stauffer, B. 1984. Age differences between polar ice and air trapped in its bubbles. *Nature* **311**: 45–47.

Staffelbach, T.; Stauffer, B.; and Oeschger, H. 1988. A detailed analysis of the rapid changes in ice-core parameters during the last ice age. *Ann. Glaciol.* **10**: 167–170.

Steffensen, J.P. 1988. Analysis of the seasonal variation in dust, Cl^-, NO_3^-, and SO_4^{--} in two Central Greenland firn cores. *Ann. Glaciol.* **10**: 171–177.

Thompson, L.G. 1977. Variations in microparticle concentration, size distribution elemental composition found in Camp Century, Greenland, and Byrd Station, Antarctica, deep cores. In: Isotopes and Impurities in Snow and Ice, Proc. IUGG Symp., Grenoble, 1975. *IAHS–AISH Publ.* **118**: 351–364.

Wolff, E.W., and Peel, D.A. 1985. The record of global pollution in polar snow and ice. *Nature* **313**: 535–540.

The Environmental Record in Glaciers and Ice Sheets
eds. H. Oeschger and C.C. Langway, Jr., pp. 123–139
John Wiley & Sons Limited

Dating of Ice by Radioactive Isotopes

B. Stauffer

Physikalisches Institut
Universität Bern
3012 Bern, Switzerland

Abstract. Radioactive isotopes are enclosed in natural ice as part of the water substance (^{3}H), in form of aerosols (e.g., ^{32}Si, ^{10}Be), or in the air bubbles (e.g., ^{39}Ar, ^{14}C, ^{81}Kr). The suitability of radioactive isotopes for dating purposes depends mainly on the accuracy with which their initial concentration at the time of ice formation can be estimated. Applications of several isotopes are discussed in the paper. Generally, the radioactive isotopes of gases, reaching an equilibrium concentration in the atmosphere and getting enclosed in air bubbles, are especially well suited for dating. For many purposes other dating methods are available, but problems remain which most probably can only be solved by dating with radioactive isotopes.

INTRODUCTION

Natural ice contains at least two important archives related to atmospheric processes. The first is a record of precipitation from the past, including aerosols scavenged with the precipitation or deposited on the snow surface. The second is a record of air with atmospheric composition enclosed in bubbles if snow and firn transform at low temperature by a dry sintering process into solid ice. In both archives radioactive isotopes are included as well. Radioisotopes are produced in the atmosphere by cosmic radiation (e.g., ^{32}Si, ^{39}Ar, ^{14}C, ^{10}Be, ^{81}Kr); they can be emitted to the atmosphere as a product of a natural decay series (e.g., ^{210}Pb), by nuclear weapons tests (e.g., ^{3}H, ^{137}Cs, ^{90}Sr), or by the nuclear industry (^{85}Kr). To a certain extent radioisotopes may also be produced in the ice itself by cosmic radiation; this possibility will be discussed at the end of this paper.

Many isotopes are attached to aerosols. Generally they have a short residence time in the troposphere since they are deposited on the Earth's surface by precipitation or dry deposition. Other isotopes form gaseous compounds and are becoming part of the atmosphere (e.g., ^{39}Ar, $^{14}CO_2$).

They have a relatively long residence time in the troposphere and, assuming a constant production rate and a constant atmospheric concentration of the corresponding element, the ratio of radioisotopes to inactive isotopes for a given element will reach an equilibrium.

If radioisotopes are embedded in a snow layer or separated from the atmosphere by enclosure in air bubbles, their concentration levels start to decrease due to the radioactive decay. The method of dating assumes that the initial concentration of radioisotopes in a snow layer or in freshly formed bubbles can be accurately estimated. If the present concentration in the ice sample can be measured, the ratio of the two concentrations allows the calculation of the age:

$$t = (\ln 2)^{-1} T_{1/2} \ln[c_0/c(t)], \quad (1)$$

where t: age (time elapsed since precipitation or since separation from the atmosphere)
$T_{1/2}$: half-life of isotope
c_0: concentration of isotope at $t = 0$
$c(t)$: concentration of isotope at time t.

Generally, the half-life of a radioisotope is known. The problem of radioactive dating is therefore reduced to estimating the initial concentration c_0 and measuring the present concentration $c(t)$. These procedures are discussed in the next two sections, and then typical applications using different isotopes are given.

THE CONCENTRATION OF RADIOISOTOPES IN FRESH SNOW AND IN AIR BUBBLES

The most extensively measured radioisotope for dating earth materials is ^{14}C. When the ^{14}C method was first applied, it was assumed that the ^{14}C production rate by cosmic radiation and the concentration of atmospheric CO_2, as well as the amount of all living organic carbon and of dissolved carbonate in seawater, was constant. This assumption was principally confirmed by two results (Libby et al. 1949):

1. The specific activity calculated, based on the measured neutron flux, is in reasonable agreement with measured activities,
2. Measurements of c(t) on historically dated samples gave the expected value within the error limits.

Later, precise measurements of c(t) from dated tree-ring samples showed that there are changes with time in the ^{14}C concentration levels of atmospheric CO_2 (Stuiver et al. 1986). Such changes can either be due to changes of the production rate, changes of the size, or changes of exchange rates between the different carbon reservoirs.

^{10}Be becomes attached to aerosols after it is produced in the atmosphere. The ^{10}Be concentration levels along an ice core show variations, which may be caused by changes in the production rate, changes in the transport and local distribution, and changes in the snow accumulation rate. A comparison between the ^{14}C and ^{10}Be record must consider the geochemical behavior differences of the two isotopes (gas and aerosol). Due to the short residence time of ^{10}Be, compared with ^{14}C, changes in the production rates are directly reflected in the ^{10}Be record in ice without attenuation, whereas ^{14}C has a long atmospheric residence time and also exchanges with other carbon reservoirs, leading to an attenuation in the production rate variations recorded in the ice. ^{14}C production rate changes can be calculated based on a carbon cycle model, assuming that all observed ^{14}C variations in the tree-ring record are due to production rate changes (Beer et al. 1988). A good correlation between the main short-term variations of ^{10}Be based on ice cores and the ^{14}C record based on tree-rings, representing the same period of time, strongly suggests that these variations are due to changes of the production rate.

For ^{10}Be, however, not all changes can be attributed to a changing production rate; there are changes in the distribution pattern. Also, considering other climatic epochs such as the last glaciation, changes in the global precipitation pattern are responsible for large changes of c_0. In Table 1 the present concentration and annual deposition of ^{10}Be (Raisbeck and Yiou 1985) and ^{32}Si (Clausen 1973) are given for different locations. Based on the ^{10}Be results, Raisbeck and Yiou concluded that a constant deposition rate is a more appropriate assumption than a constant concentration. However, the local variations of both concentration and deposition rate are large.

Based on the results of ^{14}C, ^{10}Be, and ^{32}Si measurements, we can conclude the following:

1. Well mixed gaseous radioisotopes in the global atmosphere (like ^{14}C and all radioactive noble gases with a long enough half-life) reach an equilibrium concentration c_0, defined as the concentration of the radioactive isotope compared to the corresponding element.
2. These radioisotopes are stored as part of the atmospheric gases contained in the air bubbles of natural ice.
3. Concentration c_0 changes in the past occurred mainly due to changes in the production rate. According to ^{14}C measurements on tree-rings, c_0 deviated during the last 13,000 years by not more than about 10% from a mean value (Stuiver et al. 1986).

Table 1 Concentration of ^{10}Be and ^{32}Si in fresh snow and annual deposition rate of ^{10}Be and ^{32}Si at different locations

For ^{10}Be:

Location	Latitude	Longitude	Precipitation rate [m water equiv.]	^{10}Be concentration [atoms (kg ice)$^{-1}$]	^{10}Be deposition [atoms (m^2 year)$^{-1}$]
Vostok	78°28′S	106°48′E	0.023	7.9 10^7	1.8 10^9
Dome C	74°39′S	124°10′E	0.037	5.0 10^7	1.9 10^9
South Pole	90°00′S		0.085	3.0 10^7	2.5 10^9
Agassiz	81° N	73° W	0.17	1.5 10^7	2.6 10^9
Camp Century	77°11′N	61°09′W	0.35	0.75 10^7	2.6 10^9
Milecent	70°18′N	44°35′W	0.48	1.05 10^7	5.0 10^9
Dye 3	65°11′N	43°50′W	0.50	0.93 10^7	4.7 10^9

For ^{32}Si:

Location	Latitude	Longitude	Precipitation rate [m water equiv.]	^{32}Si concentration [atoms (kg ice)$^{-1}$]	^{32}Si deposition [atoms (m^2 year)$^{-1}$]
Camp Century	77°11′N	61°09′W	0.35	8.6 10^4	3 10^7
Dye 3	65°11′N	43°50′W	0.50	6.2 10^4	3 10^7
Inge Lehman	77°54′N	39°21′W	0.09	10.1 10^4	0.9 10^7
Jarl Joset	71°22′N	33°28′W	0.29	7.1 10^4	2 10^7
Byrd Station	79°59′S	120°01′W	0.16	9.3 10^4	1.5 10^7

4. Radioisotopes attached to aerosols do not reach an equilibrium state compared to the stable isotopes of the same element. The concentration of c_0 is defined as the concentration per mass of ice. From c_0 the annual flux of the radioisotope may be calculated if the annual snow precipitation at the location is known.
5. The initial concentration of radioisotopes attached to aerosols shows larger variations than the gaseous ones (by about a factor of two instead of 10%). The larger variations are caused by production rate changes that are not attenuated, by changes of the distribution pattern, and by changes on the annual rate of snow accumulation.

THE MEASUREMENT OF THE PRESENT CONCENTRATION

The present concentration c(t) is determined by measuring the specific radioactivity of a sample or by measuring directly the concentration with a mass spectrometer or with resonance-ionization spectroscopy. Until a few years ago, the most common method for measuring radioactive isotopes in ice was measuring the specific activity by decay counting.

To measure the activity, an adequate sample has to be extracted from the ice. The gaseous, liquid, or solid sample will cause a counting rate d(t) which is above background d′ in a counting system (proportional counter, liquid scintillation counter, or GeLi detector). If the efficiency of the counting system is known, the specific activity and the concentration c(t) (number of atoms/g) can be calculated as

$$c(t) = A(t)/\lambda m = (d(t) - d')/(\lambda m f), \tag{2}$$

where m: mass of sample (mass of ice for aerosol samples, mass of the same element for gaseous samples)
A(t): activity of the sample
d(t): gross counting rate
d′: background counting rate
f: counter efficiency (counting rate/effective decay rate)
λ: decay constant of radioisotope $\lambda = \ln 2/T_{1/2}$.

The ratio $c(t)/c_0$ can be determined without knowing the counter efficiency since

$$c(t)/c_0 = (d(t) - d')/(d_0 - d'), \tag{3}$$

where d_0 represents the gross counting rate of a recent sample of the same size as the investigated sample. The counting rates d(t), d_0, and d′ are subject to statistical errors. A characteristic figure describing the precision of a counting system is the figure of merit (FOM), defined as the ratio d_0^2/d'. Counters for ^{14}C, using samples of about 3 g carbon, have a FOM of about 2,000 cpm. The ratio $d(t)/d_0$ for a sample two half-lives old can be determined using such a system, with a precision of about 0.2%. A counter for ^{39}Ar, using samples of 2 l Ar, has a typical FOM of 0.27 cpm. The precision of $d(t)/d_0$ for a sample two half-lives old is accordingly only about 15%.

The accelerator mass spectrometry (AMS) permits a much smaller sample size to be used for determining c(t). This is of course beneficial for dating ice, since ice core samples are limited and smaller samples permit a more refined depth resolution. For instance, a ^{14}C measurement by AMS requires only 1 mg instead of 3 g, by traditional counting, to obtain a precision of about 0.5% for a 10,000 year old sample (Suter et al. 1984). The measuring time is reduced from one week to about three minutes. It does not make

Table 2 Data of most important radioisotopes for dating ice samples. The values for the concentrations and activities are rounded averages for fresh snow

Isotope	$T_{1/2}$ [years]	Origin	Located in	Concentration [atoms $(kg\ ice)^{-1}$]	Activity [Bq$(kg\ ice)^{-1}$]	Concentration	Activity
Fission products		Nuclear Weapons Tests	aerosol	$7\ 10^8 - 2\ 10^9$	$0.5 - 1.5$		
Tritium	12.26		ice	$5\ 10^{10}$	80	$^3H/H = 7\ 10^{-16}$	
^{85}Kr	10.76	Nuclear industry	bubbles	$4\ 10^3$	$8\ 10^{-6}$	$^{85}Kr/Kr = 1.4\ 10^{-12}$	750 Bq$(1\ Kr)^{-1}$
^{210}Pb	22.3	^{238}U decay family	aerosol	$2\ 10^7 - 10^8$	$0.02 - 0.1$		
^{32}Si	172	Produced in the atmosphere by cosmic radiation	aerosol	$4\ 10^4$	$5\ 10^{-6}$		
^{39}Ar	269	Produced in the atmosphere by cosmic radiation	bubbles	$2\ 10^4$	$1.7\ 10^{-6}$	$^{39}Ar/Ar = 8\ 10^{-6}$	$1.7\ 10^{-3}$ Bq$(1\ Ar)^{-1}$
^{14}C	5,730	Produced in the atmosphere by cosmic radiation	bubbles	10^6	$4\ 10^{-6}$	$^{14}C/C = 1.4\ 10^{-12}$	0.1 Bq$(1\ CO_2)^{-1}$
^{10}Be	1,500,000	Produced in the atmosphere by cosmic radiation	aerosol	$10^7 - 8\ 10^7$	$1.5\ 10^{-7} - 10^{-6}$		
^{36}Cl	301,000	Produced in the atmosphere by cosmic radiation	aerosol	10^6	$7\ 10^{-8}$		
^{81}Kr	213,000	Produced in the atmosphere by cosmic radiation	bubbles	$1.5\ 10^3$	$1.5\ 10^{-10}$	$^{81}Kr/Kr = 5\ 10^{-13}$	$1.5\ 10^{-3}$ Bq$(1\ Kr)^{-1}$

much sense to compare the two methods, AMS and decay counting, by their figure of merits because statistical errors are limiting the precision for decay counting, whereas in AMS the stability of the transmission of the accelerator and contaminations of the sample target are at least as important in limiting the precision. For glaciology, the AMS method has opened up many new possibilities. First of all there is a reduction by about three orders of magnitude in the sample size. Series of ^{10}Be and ^{36}Cl measurements have only become possible with the AMS method. For ^{14}C, dating the carbon of the CO_2 in about 12 kg ice is needed. This amount corresponds to about 1.2×10^7 ^{14}C atoms, corresponding to an activity of 5×10^{-5} Bq or one decay every 5 hours. For ^{10}Be measurements, about 2 kg of ice is needed, corresponding to about 3×10^7 ^{10}Be atoms having an activity of 4×10^{-7} Bq or one decay every month. The numerical example illustrates that AMS has special advantages for radioisotopes with a long half-life. In the future, however, the AMS will also be applied for isotopes with half-lives on the order of some 100 years, like ^{32}Si. Since AMS systems are based on tandem accelerators, which use negative ions at the accelerator inlet, they are not suitable to measure isotope ratios of noble gases.

A new technique has been recently developed to measure rare noble gas isotopes. The resonance ionization mass spectroscopy (RIMS) ionizes free atoms with laser pulses of appropriate wavelength and energy and separates the ions by a quadrupole or time-of-flight mass filter. The method is interesting for glaciology mainly with respect to ^{81}Kr, ^{85}Kr, and ^{39}Ar. The main interest concentrates on ^{81}Kr, which as yet cannot be measured by any other method. With the new method it is possible to measure the ^{81}Kr/Kr concentration in about 2×10^{-3} cm^3 Kr. This amount corresponds to only about 3×10^4 ^{81}Kr atoms, but still to about 20 kg of ice. Tests of the method have successfully been performed for ^{81}Kr and ^{85}Kr (Lehmann et al. 1985).

RESULTS OF RADIOISOTOPE MEASUREMENTS IN ICE

Table 2 presents a summary of radioisotope data for isotopes used for ice dating. In the sections below examples and descriptions are given for each of the radioisotopes listed in Table 2.

Fission Products, Tritium, and ^{85}Kr

The extensive tests of nuclear weapons have emitted large quantities of different radioisotopes into the atmosphere within short, well known time periods. The radioactivity of air and of precipitation samples have been recorded at many stations around the globe. The enhanced radioactivity in snow layers deposited after about 1954 is still measurable with relatively simple methods. The depth profile of an ice core can be compared with the record of measurements of precipitation samples during the last 30 years. The dating occurs by comparison of typical patterns and events and does not use the decay law for radioactive isotopes. The comparison allows a dating of certain layers within one year.

Fission products are scavenged as aerosols. Increases of the total β-activity in precipitation mainly occurred after the Castle test in 1954 and after the extensive tests in 1961–1962. In the Southern Hemisphere, the increase from the tests in the Northern Hemisphere was delayed by about two years. The specific activity of the layers representing the 1963 horizon in the Northern Hemisphere and the 1965 horizon in the Southern Hemisphere are on the order of 1.4 Bq (kg ice)$^{-1}$ in Central Greenland (Clausen and Hammer 1988) and on the order of 0.5 Bq (kg ice)$^{-1}$ at the South Pole (Lambert et al. 1977). The remaining β-activity of the fission products in these layers is mainly due to the isotopes ^{90}Sr ($T_{1/2}$ = 28.5 yrs) and ^{137}Cs ($T_{1/2}$ = 30.17 yr).

Tritium ($T_{1/2}$ = 12.32 yr) is mainly produced by tests of thermonuclear weapons. It has, as part of the water substance, a geochemical behavior which is different from that of the fission products. It is precipitated by co-condensation with water vapor. The main emission to the atmosphere occurred during the tests in 1961–1962. In the Northern Hemisphere, peak values of the tritium concentration due to these tests reached a maximum in the spring of 1963. At that time the ^{3}H/H concentration in precipitation was about 3.5×10^{-15}. The present concentration in ice layers formed during then is on the order of 7×10^{-16}, whereas the natural prebomb concentration is below 10^{-17} (Oerter and Rauert 1983).

The ^{85}Kr concentration in atmospheric air has been steadily increasing since 1955 due to emissions from the nuclear industry. The present concentration of ^{85}Kr/Kr is about 1.4×10^{-12}, corresponding to an activity of about 750 Bq (1 Krypton)$^{-1}$. ^{85}Kr could be used to investigate mixing processes of air in firn and the enclosure in bubbles. Up to now its concentration is measured by decay counting, which needs large samples. Therefore, it has not yet been applied very often in glaciology (Loosli 1983).

Dating by fission products and tritium has lost, in the past, a little of its importance in polar regions due to a competition of reference horizons formed after volcanic eruptions. The Tambora eruption in 1815, which can be identified in ice cores from Greenland and Antarctica (Clausen and Hammer 1988), is one such example. However, the fission product horizons of 1963 in the Northern Hemisphere and of 1965 in the Southern Hemisphere are so well defined that they are frequently used as the starting point for annual layer counting, instead of the actual snow surface. The dating by fission products and tritium is also important for midlatitude glaciers, where clear imprints of volcanic eruptions are not observed. In temperate glaciers the fission products can be redistributed by meltwater. An identification of the 1963 horizon by tritium is therefore more reliable in temperate glaciers.

^{210}Pb

^{210}Pb is a far decay product of ^{222}Rn which diffuses from the Earth's crust into the atmosphere. Both isotopes are members of the ^{238}U decay family. The half-life of ^{210}Pb is 22.3 years. The concentration in the atmosphere and in precipitation changes between different geographical sites and shows seasonal variations. A mean value for precipitation over Europe is about 0.07 Bq kg^{-1}. For South Greenland the value is about 0.05 Bq kg^{-1}, for the South Pole about 0.03 Bq kg^{-1}, and over the ocean on the order of 0.02 Bq kg^{-1} (Gäggeler et al. 1983). The ^{210}Pb activity can be determined by isolating the lead from a sample and measuring its β-activity. It can also be determined through its daughter product, ^{210}Po. The advantages of this second method are a higher counting efficiency and a lower background for the detection of the α-decay. Measurements with this method need precipitation samples of about 200 g each (Gäggeler et al. 1983).

On glaciers and ice sheets ^{210}Pb starts to decay after the snow deposition. A dating of buried snow and ice layers is possible under the following conditions:

1. The sample represents a mean value of the ^{210}Pb activity.
2. The mean value of the ^{210}Pb activity in precipitation has remained constant during the last two centuries (dating range).
3. An internal production of ^{210}Pb in ice due to ^{226}Ra in the embedded dust, or ^{222}Rn in the air filling the pore space, can be neglected.

Based on measurements of ice samples from Colle Gnifetti, a cold glacier in the Swiss Alps, Gäggeler et al. (1983) conclude that internal production can be neglected. Single adjacent samples, representing annual layers, show large fluctuations. Most of the fluctuations disappear for mean values representing an average of about six annual layers. Some minor variations, however, remain. These could be due to changes of the transport and origin of air masses with time.

With the ^{210}Pb method it is possible to determine the accumulation history during the last two centuries of cold glaciers and ice sheets. For Colle Gnifetti, Gäggeler et al. give a mean annual accumulation of 39±2 cm w.e. This figure is based on measurements of a series of 46 samples, distributed over the whole length of the ice core. It seems unlikely to get a reliable dating by just measuring the surface activity and one or a few samples from the layer of interest. The method will probably gain increasing importance, in spite of the uncertainty in respect to the initial concentration, for locations where dating by counting annual layers or by reference horizons are not possible. These are especially glaciers in mid latitudes.

Ice with a high content of volcanic dust, e.g., tephra-banded ice from Antarctica, can also be dated by dating the dust with other uranium-series methods (as described in the Budd et al., this volume).

^{32}Si

^{32}Si is produced by cosmic radiation in the lower stratosphere and in the troposphere by Ar spallation. Its half-life is still a matter of debate. Early estimates were in the range 60 to 700 years. Based on glaciological investigations, an apparent half-life of 295±25 yrs was obtained (Clausen 1973). First measurements of the half-life with the AMS method gave about 100 years but a recent and more reliable measurement, by counting the activity of a sample during 48 months, indicates a value of 172±4 years (Alburger et al. 1986). Further measurements will have to show whether this is the real half-life within the given error limits. The ^{32}Si concentration in freshly deposited snow is on the order of 4×10^{4} atoms (kg ice)$^{-1}$, corresponding to an activity of 5×10^{-6} Bq (kg ice)$^{-1}$. To measure the ^{32}Si concentration by decay counting, samples of hundreds of kilograms of ice are needed. Samples are extracted as silicic acid. The radioactivity measurements are carried out on the daughter isotope, ^{32}P. After two months of storage, the ^{32}P reaches a concentration above 95% of its equilibrium value. ^{32}P is easier to measure due to the higher β-energy of 1.7 MeV and the short half-life of 14.3 days, which is an additional help for identifying the isotope and for reducing the background (Clausen 1973). It should also be possible to measure the ^{32}Si concentration with the AMS method (see above). The ^{32}Si concentration is, however, two orders of magnitude lower than the ^{10}Be concentration. It is therefore estimated that a measurement with the AMS method would still need samples on the size of 20 to 100 kg of ice.

The ^{32}Si concentration in freshly deposited snow, like all radioisotopes attached to aerosols, shows large fluctuations with time and geographical location (Table 1). An important part of the fluctuations disappear if large samples, representing the snow of several years, are measured. Clausen (1973) concluded that dating by ^{32}Si is possible, based on measurements of independently dated ice samples from the Greenlandic stations Dye 3 and Camp Century. There is, however, the discrepancy between his "apparent" half-life of 295 years and the half-life of 172 years given by Alburger et al. (1986). A longer "apparent" half-life could be due to an increased initial concentration, c_0, in the past. The increased concentration can either be caused by a higher production rate or a change of the distribution pattern. I have calculated hypothetical concentrations based on the activity and the age of the samples given by Clausen, but with a half-life of 172 years. The calculated concentrations for the 9 younger samples are between 10% and 75% higher than today at the same location. This is not more than the differences between the present concentration for different sites in Greenland and it is of the same order as changes with time observed for the ^{10}Be concentration (Beer et al. 1988). The three oldest samples, however, show larger deviations from the present c_0 and the deviation is steadily increasing with increasing age. The 712 year old sample would need a c_0 which is 235% higher than today in order to explain the "apparent" half-life. There is the possibility that all activities given by Clausen are too high by a constant value, due to an underestimate of a possible contamination. This value could, however, not be more than about $10^{-7}-2.5\times10^{-7}$ Bq (kg ice)$^{-1}$ since Clausen measured the activity of several old ice samples and obtained values in this range (personal communication). If the higher value is subtracted from all measurements, the calculated c_0 for the 712 year old sample would still be 150% higher than today.

Due to the uncertainties discussed above, ^{32}Si can at present not be considered as a reliable dating method. It is questionable if the possibility to measure its concentration with the AMS method will change this situation.

^{39}Ar

^{39}Ar is produced mainly by the $^{40}Ar(2n,n)^{39}Ar$ reaction in the stratosphere. It has a half-life of 269 years. The produced ^{39}Ar mixes with atmospheric Ar and reaches an equilibrium concentration $^{39}Ar/Ar$ of about 8×10^{-6}, corresponding to an activity of 1.7×10^{-3} Bq per liter Ar or 1.7×10^{-6} Bq per kilogram ice. It is estimated that the atmospheric concentration did not deviate by more than 7% from this value during the past 1,000 years (Loosli 1983). The ^{39}Ar activity is measured in proportional gas counters. A sample of 2 l Ar has to be measured for one week, a sample of 0.4 to 1 l Ar for one month, in order to get statistically significant results. The Ar from about 1,000 kg of ice is therefore needed for one sample. A dating of a sample with an age of a few hundred years can be done with an accuracy of about 30 years (Loosli 1983).

In the future it should be possible to measure the ^{39}Ar concentration with the RIMS method. If the same number of ^{39}Ar atoms are needed as for ^{81}Kr measurements, the Ar extracted from 1 kg of ice should be sufficient. Under these circumstances ^{39}Ar would certainly become the most important and most reliable radioisotope for dating ice in the range of 100 to 1,000 years. This is a range which is especially important for midlatitude glaciers, but since Ar is enclosed in bubbles, the method is restricted to cold glaciers. It is important to remember that the ^{39}Ar age is the time since air bubble enclosure and not since snow deposition. The difference depends on temperature and accumulation rate and is for polar ice sheets typically in the range of 90 to 2,000 years (Schwander and Stauffer 1984).

^{14}C

Radiocarbon is produced by the reaction $^{14}N(n,p)^{14}C$ in the stratosphere and troposphere. ^{14}C reacts to CO_2 which mixes with atmospheric CO_2 and enters all carbon reservoirs which are in exchange with the atmospheric CO_2. The concentration in atmospheric CO_2 and living material is about $^{14}C/C = 1.4 \times 10^{-12}$ corresponding to an activity of about 0.225 Bq per g carbon. The half-life is 5,730 years. To calculate ^{14}C ages, the older value of 5,568 years is still used. Carbon is found in natural ice in different compounds. The largest amount is found in the CO_2 enclosed in air bubbles ($\approx$ 15 μg C per kg ice). There is also carbon dissolved in the ice mainly in the form of formaldehyde. ($\approx$ 1 μg C per kg ice) and embedded as solid organic material like pollen and tissue fragments ($\approx$ 0.1 μg C per kg ice) (Fredskild and Wagner 1974). In the dust, enclosed in polar ice, there is on the order of 3 μg of inorganic carbon per kilogram ice. In Greenland ice from the last glaciation this value is increased, due to the high dust concentration, to about 30 μg of inorganic C per kilogram ice. It is assumed that the carbon in the different forms does not interact with each other in cold ice. Attempts to date ice by ^{14}C have concentrated on the CO_2 enclosed in the air bubbles. Dating by measuring the radioactivity of extracted CO_2 needs ice on the order of 1,000 kg (Oeschger et al. 1967). A routine dating with the AMS method uses samples of 1 mg C, which still would correspond to 60 kg of ice. It is possible to use smaller samples but at the expense of the attainable precision. Andree et al. (1986) have dated several samples from an ice core from Dye 3, which have been dated with other, independent methods as well. The CO_2 from about 12 kg of ice, extracted by opening the bubbles mechanically to avoid any interaction of meltwater with carbonate, is needed. The carbon of the extracted CO_2 is deposited as amorphous carbon on a copper disk serving as target for the AMS analysis. Due to the reduced sample size and due to a changing contamination with modern carbon during the extraction of CO_2 from the ice and by preparing the target, the accuracy of the measured age is limited to about 5–8% (Andree et al. 1986). Taking into account that the enclosed CO_2 is younger than the surrounding ice, the obtained ages of six samples from an age range 7,000–11,000 yrs B.P. have all been too young by an average of about 600 yrs or 7%. This is within the error limits of a single measurement but indicates, nevertheless, a systematic trend of too young ages which could be caused by a contamination or by *in situ* production of ^{14}C, as will be discussed below. The goal is to improve the method in order to date ice samples of 12 kg size with the AMS, with an accuracy better than 5%. Several problems in glaciology could then be solved:

1. Time lags between the deglaciation at the end of the last ice age in the Northern and Southern Hemisphere could be determined.

2. Indications of fast climatic changes have been observed in ice cores from Camp Century and Dye 3. There is the possibility of an artifact due to an irregular stratigraphy. A ^{14}C dating of subsequent layers representing both climatic conditions could clarify the situation.
3. The dating of ice from coastal areas can give important answers concerning the present and past rheology of ice sheets.

In coastal areas, ice can be collected from the surface. For the other applications ice core samples have to be used. A sample of 10 kg corresponds to a whole core of 1.5 m length. This is still a lot of ice, considering that a mild period during the ice age is presented in the Dye 3 ice core by a core section of only about 2 m.

^{10}Be and ^{36}Cl

^{10}Be and ^{36}Cl are produced mainly by cosmic ray-induced spallation reactions in the atmosphere. The half-life of ^{10}Be is 1.5×10^6 years; that of ^{36}Cl is 301,000 years. The specific concentration of ^{10}Be in large snow samples covering the precipitation of several years varies from site to site from about 10^7 atoms (kg ice)$^{-1}$ to about 8×10^7 atoms (kg ice)$^{-1}$. The ^{36}Cl concentration is in the range of 10^6 atoms (kg ice)$^{-1}$. Both concentrations can be measured with the AMS method. Beryllium is extracted from the ice as BeO, chlorine as AgCl. It was clear from the beginning that a dating method could not only be based on just one of the isotopes, due to the large concentration variability in freshly deposited snow. Since ^{10}Be and ^{36}Cl are both attached to aerosols, however, there was the hope that they would respond the same way to changes in the distribution pattern and that therefore the c_0 ratio $^{10}Be/^{36}Cl$ would be constant in freshly deposited snow. The age of an ice sample could then be calculated as follows:

$$t = \ln[(c'(t)\, c_0)/(c(t)\, c_0')]/(\lambda-\lambda') \qquad (4)$$

where c_0: concentration of ^{36}Cl at $t = 0$
c_0': concentration of ^{10}Be at $t = 0$
$c(t)$: concentration of ^{36}Cl at time t
$c'(t)$: concentration of ^{10}Be at time t
λ: decay constant for ^{36}Cl
λ': decay constant for ^{10}Be.

The ratio $c(t)/c'(t)$ decreases with an apparent half-life of 370,000 years. The ^{36}Cl concentration is higher during the Maunder minimum, indicating that ^{36}Cl, like ^{10}Be, reflects solar modulation effects and that ^{36}Cl shows an increase at the boundary from the Holocene to the last glaciation similar to ^{10}Be, which is attributed mainly to a change in the precipitation rate. In spite of these similarities between ^{36}Cl and ^{10}Be regarding solar modulation and accumulation rates, the $^{10}Be/^{36}Cl$ ratio of measured samples varies by a factor of 2–5. Surprisingly, there does not seem to be much difference in this variability, regardless of whether weekly, annual, or decadal precipitation samples are analyzed (Suter et al. 1988). Taking into account that a 10% change of the concentration ratio corresponds to an age difference of about 57,000 years, a dating by the $^{36}Cl/^{10}Be$ ratio does not seem feasible (Suter et al. 1988). It is possible, as mentioned by Budd et al. (this volume), that the $^{26}Al/^{10}Be$ ratio, with an apparent half-life of 1.38 10^6 yr, is better suited for dating purposes of old ice since Al is expected to have a geochemical behavior similar to Be.

^{81}Kr

^{81}Kr is produced in the atmosphere by cosmic ray-induced spallation reactions with stable Kr atoms and by neutron capture of ^{80}Kr. ^{81}Kr has a half-life of 213,000 years. After its production it mixes with atmospheric Kr. Its concentration in atmospheric Kr is expected to have been constant within a few percent during the last few hundred thousand years since small changes of the production rate caused by variations of the solar activity and the geomagnetic field are averaged out due to the long half-life. The present concentration $^{81}Kr/Kr$ is about 5×10^{-13}, the activity about 1.5×10^{-3} Bq $(1\ Kr)^{-1}$. Due to the low concentration and low activity it was impossible to apply this isotope using decay counting. With the RIMS technique it is now possible (see discussion on p. 129) to measure the ^{81}Kr concentration in 2×10^{-3} cm^3 Kr. This amount can be extracted from about 20 kg of ice. The measurement is a multistep process. After the Kr separation, several steps of isotopic enrichment are required to reduce the abundance of interfering adjacent Kr isotopes to acceptable levels, before counting with the RIMS technique. In three subsequent enrichment steps, the ratio $^{81}Kr/Kr$ is increased by about a factor of 10^3 in each step. About 10% of the total number of ^{81}Kr atoms are recovered in the final enriched sample. This sample consists of about 10^3 ^{81}Kr atoms and 10^8 other Kr atoms. The exact ratio is measured by RIMS counting. Tests of the method (including extraction of Kr), enrichment, and final detection by RIMS have successfully been performed. The overall accuracy of the concentration measurement is at present about 30%, but will be increased to 10% or better (Thonnard et al. 1987).

The dating range of ^{81}Kr is between 50,000 and 10^6 years. An accuracy of 10% corresponds to an age difference of about 30,000 years. The application for ^{81}Kr dating in glaciology is therefore restricted to very old ice. The question whether ice at the border of the Antarctic ice sheet is 20,000 or 200,000 years old is a typical question which can be answered by a ^{81}Kr dating.

IN SITU PRODUCTION OF RADIOACTIVE ISOTOPES IN ICE

A number of radioisotopes are produced *in situ* in the ice by spallation reactions of the cosmic radiation, mainly with oxygen atoms in the ice (Lal 1988). Lal has estimated that the annual production rate for an ice surface in an altitude of 3,000 m a.s.l. is about 1.5×10^5 ^{14}C atoms (kg ice)$^{-1}$ and 7.3×10^4 ^{10}Be atoms (kg ice)$^{-1}$. If we take a hypothetical site on a polar ice sheet at 3,000 m.a.s.l., with an accumulation rate of 5 cm w.e. per year, the total *in situ* production rate of radioisotopes can be estimated. According to Lal, this amounts to 10^6 ^{10}Be atoms (kg ice)$^{-1}$ which is about 2–10% of the total number enclosed with the aerosols in the ice. Therefore, the *in situ* production of ^{10}Be isotopes can probably be neglected, considering the large variation in the initial concentration levels. The total ^{14}C production is estimated to be 2×10^6 ^{14}C atoms (kg ice)$^{-1}$. This is twice the amount enclosed with atmospheric air in bubbles in freshly formed ice. Even with an amount ten times less, due to higher accumulation rates, the *in situ* produced ^{14}C would influence dating measurements. However, ^{14}C dates on ice have been made, as described above, and give the expected ages within 10%. This is explained by the fact that the *in situ* production of ^{14}C occurs mainly in the top 3 m. There is a metamorphosis of snow below this depth, exemplified by strong crystal growth. We know from measurements that the CO_2 adhering to the snow grains escapes during the metamorphic process (Stauffer et al. 1981). If *in situ* produced ^{14}C is present as CO or CO_2 there is a great chance that most of it escapes and mixes with the atmosphere. The small remaining part could, however, be responsible for the too young ages of the ice samples from Dye 3.

The question of *in situ* produced ^{14}C is by no means resolved. More research is required. Due to the problems with the metamorphosis of snow in accumulation areas, investigations in ablation areas with a small annual ablation could be especially promising.

CONCLUSIONS

The development of new detection methods for radioactive isotopes in very low concentrations has made enormous progress in the past few years, and a similar progress is expected for the coming years. Radioactive isotopes in the enclosed gases in air bubbles have better reliability for ice dating than do isotopes attached to aerosols. Therefore, the improvement of the ^{14}C dating of very small samples and the development of the ^{81}Kr dating technique will be most important for glaciology and ice core research. With an improved ^{14}C dating technique it will be possible to show if there is an undisturbed stratigraphy in ice relatively close to the bottom and to solve problems concerning the synchronism of events during and at the end of the last glaciation. With ^{81}Kr it will be possible to date very old ice at the margins of ice sheets.

REFERENCES

Alburger, D.E.; Harbottle, G.; and Norton, E.F. 1986. Half-life of ^{32}Si. *Earth Plan. Sci. Lett.* **78**: 168–176.

Andrée, M.; Beer, J.; Loetscher, H.P.; Moor, E.; Oeschger, H.; Bonani, G.; Hoffman, H.J.; Morenzoni, E.; Nessi, M.; Suter, M.; and Wölfli, W. 1986. Dating polar ice by ^{14}C accelerator mass spectrometry. *Radiocarbon* **28(2a)**: 417–423.

Beer, J.; Siegenthaler, U.; Bonani, G.; Finkel, R.C.; Oeschger, H.; Suter, M.; and Wölfli, W. 1988. ^{10}Be in the Camp Century ice core: information on past solar activity and geomagnetism. *Nature* **331**: 675–679.

Clausen, H.B. 1973. Dating of polar ice by ^{32}Si. *J. Glaciol.* **12(66)**: 411–416.

Clausen, H.B., and Hammer, C.U. 1988. The Laki and Tambora eruptions as revealed in Greenland ice cores from 11 locations. *Ann. Glaciol.* **10**: 16–22.

Fredskild, B., and Wagner, P. 1974. Pollen and fragments of plant tissue in core samples from the Greenland ice cap. *Boreas* **3**: 105–108.

Gäggeler, H.; von Gunten, H.R.; Rössler, E.; Oeschger, H.; and Schotterer, U. 1983. ^{210}Pb dating of cold alpine firn/ice cores from Colle Gnifetti, Switzerland. *J. Glaciol.* **29(101)**: 165–177.

Lal, D. 1988. *In situ* cosmogenic ^{14}C and ^{10}Be for determining the hydrodynamic flow characteristics of ice sheets, in press.

Lambert, G.; Ardouin, B.; Sanak, J.; Lorius, C.; and Pourchet, M. 1977. Accumulation of snow and radioactive debris in Antarctica: a possible refined radiochronology beyond reference levels. *IAHS-AISH Publ.* **118**: 146–158.

Lehmann, B.E.; Oeschger, H.; Loosli, H.H.; Hurst, G.S.; Allman, S.L.; Chen, C.H.; Kramer, S.D.; Payne, M.G.; Phillips, R.C.; Willis, R.D.; and Thonnard, N. 1985. Counting ^{81}Kr Atoms for analysis of groundwater. *J. Geophys. Res.* **90(B13)**: 11, 547–11, 551.

Libby, W.F.; Anderson, E.C.; and Arnold, J.R. 1949. Age Determination by Radiocarbon Content. *World-Wide Assay of Natural Radiocarbon Science* **109(2827)**: 227–228.

Loosli, H.H. 1983. A dating method with ^{39}Ar. *Earth Plan. Sci. Lett.* **63**: 51–62.

Oerter, H., and Rauert, W. 1983. Core drilling on Vernagtferner (Oetztal Alps, Austria) in 1979: tritium contents. *Z. Gletscher. Glazial.* **18(1)**: 13–22.

Oeschger, H.; Alder, B.; and Langway, C.C. 1967. An *in situ* gas extraction system to radiocarbon date glacier ice. *J. Glaciol.* **6**: 939–942.

Raisbeck, G.M., and Yiou, F. 1985. ^{10}Be in polar ice and atmospheres. *Ann. Glaciol.* **7**: 138–140.

Schwander, J., and Stauffer, B. 1984. Age difference between polar ice and the air trapped in its bubbles. *Nature* **311(5981)**: 45–47.

Stauffer, B.; Berner, W.; Oeschger, H.; and Schwander, J. 1981. Atmospheric CO_2 history from ice core studies. *Z. Gletscher Glazial.* **17(1)**: 1–15.

Stuiver, M.; Kromer, B.; Becker, B.; and Ferguson, C.W. 1986. Radiocarbon age calibration back to 13,300 years B.P. and the ^{14}C age matching of the German oak and U.S. bristlecone pine chronologies. *Radiocarbon* **28(2B)**: 969–979.

Suter, M.; Balzer, R.; Bonani, G.; Hofmann, H.; Morenzoni, E.; Nessi, M.; Wölfli,

W.; Andrée, M.; Beer, J.; and Oeschger, H. 1984. Precision measurements of ^{14}C in AMS – some results and prospects. *Nucl. Instr. Meth. Phys. Res.* **B5**: 117–122.

Suter, M.; Beer, J.; Bonani, G.; Hofmann, H.J.; Michel, D.; Oeschger, H.; Synal, H.A.; and Wölfli, W. 1988. ^{36}Cl Studies at the ETH/SIN–AMS facility. *Nucl. Inst. Meth. Phys. Res.* **B29**: 211–215.

Thonnard, N.; Willis, R.D.; Wright, M.C.; Davis, W.A.; and Lehmann, B.E. 1987. Resonance ionization spectroscopy and the detection of ^{81}Kr. *Nucl. Instr. Meth. Phys. Res.* **B29**: 398–406.

The Environmental Record in Glaciers and Ice Sheets
eds. H. Oeschger and C.C. Langway, Jr., pp. 141–159
John Wiley & Sons Limited

Dating by Ice Flow Modeling: A Useful Tool or an Exercise in Applied Mathematics?

N. Reeh

Alfred-Wegener-Institut Für Polar- und Meeresforschung
2850 Bremerhaven, F.R. Germany

Abstract. Different purposes of dating by ice flow modeling and the corresponding accuracy requirements are mentioned. Various flow models are described and their ability to meet these requirements is discussed. It is concluded that simple models requiring a minimum of input data can provide preliminary age-depth profiles over a period of a few thousand years, accurate enough, for example, to choose the frequency of sampling along ice cores. As regards precise dating of environmental records from glaciers and ice sheets over longer time spans, dating solely by ice flow modeling is not advisable. This does not mean, however, that flow model dating is useless, since a comparison of calculated and experimentally determined ages will throw light on the climatic and dynamic history of the ice masses, and thus will contribute to the interpretation of the environmental records obtained from the glaciers and ice sheets.

INTRODUCTION

If this paper ends by concluding that of course dating by ice flow modeling is useful, it will probably come as no surprise. Nevertheless, it may be worthwhile a priori not to take this for granted, but rather to look into such questions as can dating by ice flow modeling provide ages so reliable that dating by other methods becomes unnecessary? Or, is there a different purpose with flow model dating other than just to establish the age-depth relationship along, e.g., an ice core? Also, it might be useful to discuss to what extent the development towards more sophisticated ice flow models has actually improved the theoretical dating of ice in glaciers and ice sheets.

A sensible way to approach these problems seems to be to discuss the expectations one has to dating by ice flow modeling, in terms of which purpose the dating is supposed to serve and the accuracy required to meet these purposes. Three different applications for flow model dating occur immediately:

1. to obtain a general overview of the age distribution within large ice masses (for example, the Greenland or Antarctic ice sheets);
2. to establish a preliminary age profile or time scale (i.e., the age-versus-depth or age-versus-distance relationship) at a specific drilling or surface sampling site for the purpose of determining the sampling frequency for some paleoenvironmental variable; and
3. to date accurately the ice along a sampling profile (whether it be a vertical section of an ice core or a near-horizontal section at the surface near an ice sheet margin) with the purpose of establishing a dated environmental record.

The required dating accuracy is different for the different applications. Whereas a 10% or even 20% accuracy may be acceptable for the first two applications, the accuracy requirements for the third may range from exact dating on an annual or even seasonal basis to dating with error limits of up to several thousands of years, depending on the record and the time range in question.

In this chapter the different applications of flow model dating will be discussed. Moreover, an application that goes beyond the purpose of simply dating the ice will be discussed, i.e., combining theoretical and experimental dating methods to derive information on the past dynamic and climatic history of the ice masses. With the recent improvements in experimental ice dating techniques (see Hammer and Stauffer, both this volume) this may show up to be a more fruitful application of ice dynamic models than just establishing time scales.

SETTING THE PROBLEM

Flow Pattern

Ice sheets and glaciers can be considered sedimentary deposits consisting of sequences of layers, generally deposited annually in the accumulation zone (i.e., the region of positive mass input) as snow accumulation. The snow layers sink into the ice mass subject to continuous thinning, initially as a result of densification, by which the snow is transformed into ice, but then mainly due to flow-induced vertical compressive strain. In this process the layers are stretched horizontally until they are advected by the ice motion into the ablation zone, where the ice is removed by melting or calving of icebergs. If ablation by melting is predominant, the layers move upwards relative to the surface where they eventually melt away. The ice flow pattern, illustrated in Fig. 1, shows particle paths in a cross section of an ice sheet. It appears from the figure that the higher the ice is originally deposited in the accumulation area, the deeper into the ice mass it will sink, and the nearer to the ice sheet margin it will reappear at the surface in the ablation zone. Consequently, the oldest ice is found near the base and along the margin of the ice sheet. In principle, a complete sequence of all the deposited layers can be obtained either by deep core drilling from the surface to the base in the accumulation zone or by surface sampling from the equilibrium line (the line separating the accumulation and ablation zones) to the ice margin. In contrast, a core drilled from the surface to the base somewhere in the ablation area will miss the most recent part of the record—the more so, the closer to the ice margin the core is drilled.

If the temperature reaches the melting point at the ice sheet base, the oldest layer sequences may also have been removed by basal melting. This is certainly the case for temperate glaciers, but also the Greenland and Antarctic ice sheets most likely have extended areas with basal melting. As shown in a later section, basal melting may drastically reduce the time range of the layer sequences left behind in the ice mass.

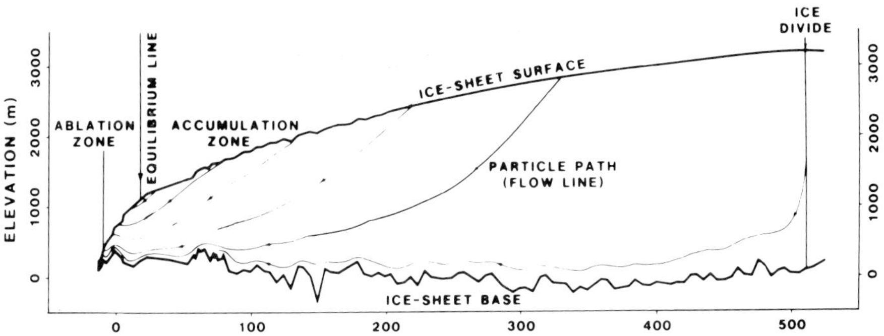

Fig. 1—Flow in a cross section of an ice sheet.

Age and Annual Layer-thickness Profiles

The age profile t(z) along a vertical line through an ice mass and the vertically measured annual layer-thickness profile λ(z) are related by the equation

$$t(z) = \tau \int_z^H dz/\lambda(z) \tag{1}$$

where z is a vertical coordinate, τ is 1 year, and H is surface elevation. Therefore, if either t or λ is known, the other can be calculated. In a paper on dating, the age profile, of course, is of primary interest. However, it is sometimes more convenient first to calculate the λ-profile, and then use Eq. 1 to determine the age profile. Moreover, the λ-profile can often be determined experimentally by stratigrafic analysis of ice cores, and comparison of the experimental and theoretical λ-profiles is an obvious way to check the theoretical dating.

The age of an "ice particle" at a given location in an ice mass is calculated by first determining the path of the particle from the site of deposition on the surface (see Fig. 1) and then calculating the time used by the particle to travel along this path from the surface to its present position. This requires modeling of the dynamic and thermal history of the ice mass during the entire period elapsed since the ice particle was deposited at the surface. This again presupposes past upstream histories of accumulation rate/ablation rate, ice thickness, ice temperature, and ice-flow-law parameters to be known or calculated, and it becomes an increasingly difficult and uncertain task to perform the further back in time the dating is extended. Also, past changes in other boundary conditions, e.g., sea level in the case of an ice sheet terminating in the sea, must be considered.

On an ice sheet dome or on a horizontal crest there is no horizontal ice motion and consequently no need for corrections for advective transport, at least not for as long as the dome or crest position has remained constant. This is an argument for choosing drill sites on domes or slightly sloping crests, since at least the more recent parts of ice core records from such locations should be more easy to date and interpret. Still, however, accumulation rate, ice thickness, internal ice temperature, and ice-flow-law parameters most likely have changed in time, even in case the dome position has not. This causes temporal changes in the depth distribution of the vertical rate of straining of the layers, an effect which must be considered when calculating the time scale.

Often past variations in accumulation rate, ice thickness, ice temperature, etc. are disregarded in the models used to date theoretically the ice (steady state models). This is done in order to simplify the calculations, but also because our knowledge about past changes of ice sheet climate and dynamics is seldom good enough to justify application of complicated time-dependent models for dating the ice. Moreover, even the present upstream distributions of the relevant glaciological parameters may not be known, thus imposing the application of even more simplified models. With the improved understanding of climate and ice sheet dynamics and their interaction it may, however, in the future be profitable to use elaborate time-dependent models for the purpose of ice dating.

STEADY STATE MODELS

Dating the Near-surface Layers

The minimum information needed to make an estimate of the age-depth profile in the accumulation region of a glacier or ice sheet is the amount of annual snow accumulation. In the near-surface layers thinning is mainly due to densification, by which the snow/firn is transformed to ice. According to Paterson (1981, pp. 13–16) the depth to the firn-ice transition ranges from about 10 m for relatively warm sites, where refreezing meltwater contributes to the densification process, to about 100 m for the interior cold areas (the dry snow zones) of the Greenland and Antarctic ice sheets. The corresponding time range is from about 5 years to several hundred years in central Greenland and even several thousand years for the low accumulation areas of interior Antarctica.

Formulae relating density to depth and age in the near-surface layers of interior ice sheet locations have been derived, e.g., by Herron and Langway (1980). They have also compared model-estimated, age-depth profiles to observations at several Greenland and Antarctic ice sheet locations. Over a depth range of 60 m and an age range of up to 250 years, observed and predicted ages agreed to within 5 years.

Constant Vertical Strain Rate Model

With increasing depth (for thin glaciers or ice caps even before the transformation of firn to ice is completed), flow-induced thinning of the layers becomes important. The simplest possible model to account for this effect is due to Nye (1963) and Haefeli (1963). Based only on two assumptions, (*a*) that there is negligible bottom melting and (*b*) that the vertical strain rate along any vertical line in the ice mass is uniform at any given instant, Nye (1963) deduced the thickness of an annual layer at distance z above the base as

$$\lambda(z) = \lambda_0 z / H_0, \tag{2}$$

where λ_0 and H_0 are surface layer thickness and ice thickness, respectively, at the time and location when the layer was originally deposited.

It is worthwhile to note that this simple expression does not presuppose that the ice mass is in steady state. However, it is generally not as useful as one should immediately think, first because the assumption of constant vertical strain rate is often a poor approximation except in the upper part of an ice mass and second because in order to use Eq. 2 for dating purposes, λ_0 and H_0 must be known as far upstream and as far back in time as the dating is desired. Haefeli (1963) went around this problem by simply assuming that λ_0 and H_0 are constant in time and space, and he deduced the following logarithmic time scale, often referred to as the Nye time scale,

$$t = (H/a_S)\ln(H/z), \tag{3}$$

where a_S = accumulation rate. This formula has been used intensively for ice dating purposes. However, except for making rough estimates, its use can not be recommended. The map in Fig. 2, showing calculated ages at 90% relative depth in the Greenland ice sheet, is produced by means of a steady state, uniform strain-rate model (column flow model), albeit, with due account of the nonuniform distributions of accumulation rate and ice thickness over the ice sheet area (Radok et al. 1982). Due to the application of such a simple model, the calculated ages may well be wrong by a factor of two or even more. The map, however, illustrates the general trends of the age distribution in the deep layers of the Greenland ice sheet and thus serves a useful purpose.

Sandwich Models

Based on the assumption of constant accumulation rate and ice thickness, a whole class of models has been developed that is simple and yet much improved as compared to the Nye-Haefeli model. The assumptions imply that the thickness of the annual layers does not vary horizontally. To reflect this, the models will be denoted as "sandwich" models. The general expressions for the horizontal and vertical velocity components u and w are

$$u = u_m\phi(z/H) \quad \text{and} \quad w = -a\psi(z/H) + w_B, \tag{4 a,b}$$

where u_m is the depth-averaged horizontal velocity, and

$$a = a_S + a_B, \tag{5}$$

where a_S is the accumulation rate at the surface and a_B is the rate of melting/refreezing at the base. ϕ and

$$\psi = \int_0^{z/H} \phi\,(z/H)d(z/H) \tag{6}$$

are shape functions for the horizontal and vertical velocity profiles while w_B is the vertical velocity at the bottom of the ice mass. If the ice mass is frozen to the bottom, $w_B = 0$. For an ice mass in steady state, $w_B = a_B$ if

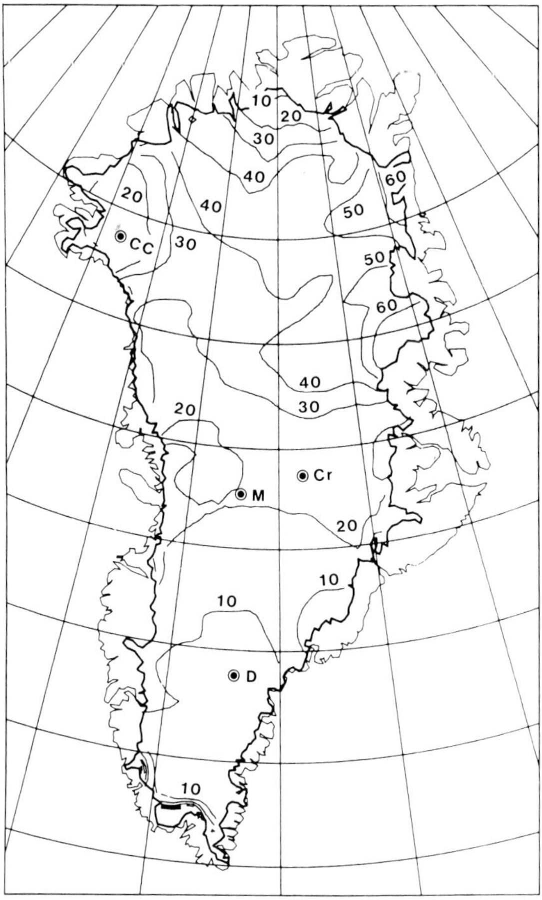

Fig. 2—Age (ka) of ice at 90% depth in the Greenland ice sheet. Location of important drill sites are also shown. CC = Camp Century, Cr = Crête, M = Milcent, D = Dye 3 (modified from Radok et al. 1982).

the base is horizontal. Moreover, in this case the annual layer thickness is distributed in the same manner as the vertical velocity. Hence, by means of Eq. 1, the time scale becomes

$$t(z) = (H/a)\int_{z/H}^{1} (\psi(z/H) - a_B/a)^{-1}\, d(z/H). \qquad (7)$$

In order to apply this expression, the ice thickness, the accumulation rate, and the rate of basal melting/refreezing must be known. Furthermore, the profile function ϕ (or ψ) must be specified. The Nye-Haefeli time scale given by Eq. 3 is a special case of Eq. 7, emerging for $a_B = 0$, $\phi(z/H) = 1$, and $\psi(z/H) = z/H$. (Incidentally, the vertical strain rate varies with depth in the same manner as ϕ. Therefore, $\phi = 1$ corresponds to the case of uniform vertical strain rate.) As previously mentioned, the uniform distribution is a poor approximation to the real strain-rate variation below a certain depth in the ice sheet (approximately halfway down). Generally, the strain rates decrease towards the glacier bed where they approach zero if the bed is horizontal and the ice mass is frozen to it. To account for this fact, Dansgaard and Johnsen (1969) introduced a model with ϕ constant down to some distance h above the bed and from there decreasing linearly to zero at the bed (see Fig. 3). This model has proved to be very useful for establishing preliminary time scales along ice cores. The success of the model is illustrated by comparing the predicted ages at the bottom of three ca. 400 m drill holes on the Greenland ice sheet with the ages later obtained by stratigraphic dating methods. For the Dye 3, Milcent, and Crête locations (see map in Fig. 2), predicted and "observed" dates were respectively A.D. 1196, 1176, and 554 (S. J. Johnsen, personal communication) and A.D. 1244, 1177, and 553 (Hammer et al. 1978). This precision is fully satisfactory, e.g., for deciding how to sample a core to ensure that seasonal variations in some ice constituent will not escape detection.

The Dansgaard-Johnsen time scale depends on the value of h, which is

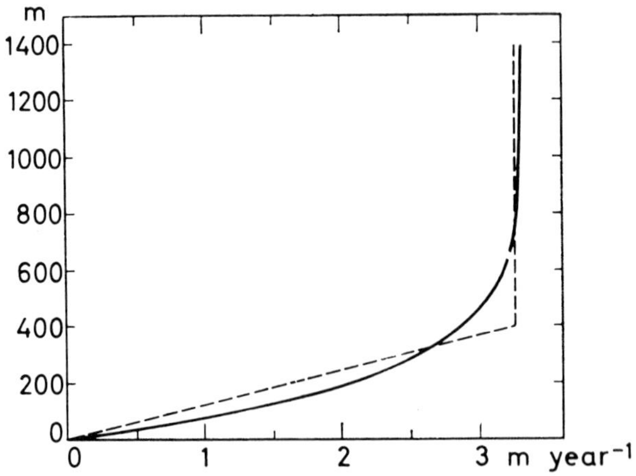

Fig. 3—Horizontal velocity profiles for Camp Century. Full curve: Weertman's (1968) calculation; dashed curve: approximation introduced by Dansgaard and Johnsen (1969).

usually estimated from the way in which the horizontal velocity component varies with depth. Unless measured in a borehole, this variation must be calculated by integrating the shear strain rates from the base to the surface of the ice sheet. In this integration it has been common practice to neglect the effect of normal stresses. Dansgaard and Johnsen (1969), for example, used such a velocity profile calculated by Weertman (1968) for the Camp Century drill site to estimate h; they then calculated the time scale by means of Eq. 7, applying present values of a and H. a_B was taken as zero because the basal temperature at Camp Century is −13°C. The model successfully predicted the age at the Holocene/Wisconsinan transition, revealed by a large shift in $\delta^{18}0$ (see Dansgaard and Oeschger, this volume) to be ca. 10,000 yr. Also, the calculated annual layer thickness profile is in close agreement with a least-squares fit to the observed annual layer thicknesses (Hammer et al. 1978).

Philberth and Federer (1971), Paterson and Waddington (1984), and Wolff and Doake (1986) also calculated time scales for the Camp Century ice core by means of sandwich models, however with different choices for the profile functions ϕ and ψ. They all compared their model time scales to the observed time scale and took the close agreement back to ca. 10,000 B.P. as support for their models. For example, Wolff and Doake (1986) applied a value of 1 for the exponent in Glen's law, instead of the commonly accepted value of 3, and advocated that their modeling of the Camp Century time scale supported this choice. However, such a conclusion cannot be drawn because their model neither accounts for variations in the upstream distributions of ice thickness and accumulation rate, for the influence of normal stress on the shear strain rates, for enhanced flow of Wisconsinan ice as compared to Holocene ice, nor for time variations of these quantities. In fact, as long as the above mentioned features are not accounted for, there is no reason to try to "improve" the Dansgaard-Johnsen model, which has the advantage of being very simple. However, in spite of the fact that it predicts the time scale for the Camp Century core rather accurately back to ca. 10,000 B.P., the Dansgaard-Johnsen model, no more than the Philberth-Federer, the Paterson-Waddington, and the Wolff-Doake models, explains the age and layer thickness profiles in terms of ice physics or ice sheet dynamics. Flow line models, e.g. the one used by Budd and Young (1983) for the Camp Century location, and the model to be described in the following section attempt to do this. The Dansgaard-Johnsen model should be considered a simple way to obtain an approximate time scale at a location where upstream accumulation rate and ice thickness variations are moderate. Also, the interval over which the model can be expected to work well is limited by the assumption of steady state. For the thick Greenland and Antarctic ice sheets this is at most 10,000 years, i.e., back to the last ice age; for thin ice caps which react faster and more violently to climatic change, the interval may be much shorter (see section below on *Dynamic History of the Devon Island Ice Cap*).

Flow-line Models

In an attempt to establish an accurate steady state reference for the age and layer thickness profiles along the Dye 3 deep ice core (for location, see map in Fig. 2), a rather elaborate flow model was developed (Reeh et al. 1985). The ice thickness at Dye 3 is about 2,000 m, but there are bedrock undulations with amplitudes of 200 m and wavelengths of a few kilometers along the upstream flow line. These produce 20 m surface undulations which in turn cause spatial variations in the accumulation rate superimposed on the smoothed distribution that decreases in the upstream direction. Consequently, a sandwich model with constant ice thickness (H) and accumulation rate (a) would not be realistic. Instead, a model was designed as a first approximation to consider the "basic" flow between smoothed surface and subsurface boundaries with an ice flux determined by the trend of the accumulation rate distribution. A second order approximation model included the flow perturbations caused by the deviations of H and a from their trend lines. The flow models also accounted for varying temperature and normal stress with depth, and enhanced flow of ice of Wisconsinan origin.

Figure 4 displays calculated and observed λ-profiles. The step curve shows measured mean annual layer thicknesses with ages added to the right of the curve. The full, smooth curve and the dashed, oscillating curve shows the results produced by the first- and second-order models, respectively. The fit of the dashed curve to the measured step curve back to ca. 2,500 yr B.P. shows that most of the oscillations in the corresponding depth range are due to upstream flow and accumulation effects and do not reflect temporal variations as one would be tempted to conclude if the calculations had stopped with the first-order model. Since detailed data are not available farther upstream along the flow line, the perturbation modeling cannot be extended further back in time. Therefore, it cannot yet be decided whether the relatively thick, ca. 2700–6000 yr old annual layers can also be ascribed to steady state upstream effects or whether they indicate increased accumulation rates in the period in question. The Dye 3 study illustrates the importance of using accurate flow models to interpret layer thickness records in terms of past accumulation rates. As far as dating is concerned, the second-order model improves the precision from ca. 30 yr to a few years for the period back to ca. 1000 yr B.P. Back to ca. 1500 B.P. the improvement is from 70 yr to 30 yr. For the older ice the main improvement arises from introducing enhanced flow of the bottom ice of Wisconsinan origin, which changes the theoretical date for the Holocene/Wisconsinan transition from ca.12,300 yr B.P. to ca. 10,100 yr B.P.

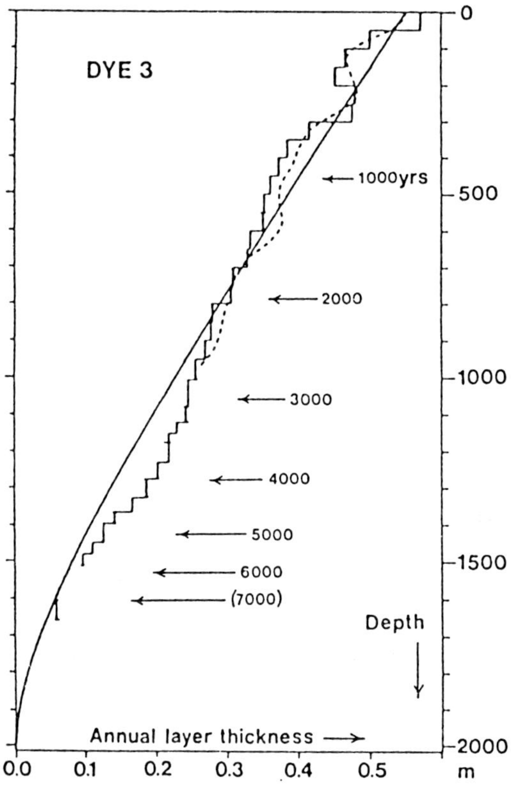

Fig. 4—Dye 3 deep core. Full curve: annual layer thickness profile predicted by a first-order approximation model; dashed curve: prediction by a second-order approximation model; step curve: measured annual layer thickness profile; numbers: absolute ages (from Dansgaard et al. 1985).

Models for Dome and Crest Regions

As mentioned earlier, domes and near-horizontal crests on ice sheets are favorable locations for ice core drilling. However, at these locations special conditions prevail since the internal deformations are dominated by normal strain rates, in contrast to farther down the slope where shear strain rates are at least as important and are in fact generally domineering in the near-bottom part of the ice column. Due to this difference, the vertical strain rate, and therefore also the age and layer thickness profiles have special distributions in dome/crest regions. Raymond (1983) studied this problem by means of a finite element model and found an approximately linear depth distribution of the vertical strain rate on a symmetrical ice divide. This corresponds to a Dansgaard-Johnsen distribution for the case of h =

H. Raymond also showed that a few ice thicknesses away from the crest, the strain rate distribution had already changed to that characteristic for an ice sheet slope, so that the special crest distribution is very local. Reeh (1988) derived the strain rate distribution at a symmetrical dome by an analytical model. Reeh's distribution deviates from the linear approximation suggested by Raymond, being concave-up near the surface, concave-down near the base, and having an inflection about halfway down the ice sheet. The corresponding time scale is plotted in Fig. 5 together with the Raymond (1983) time scale and the classical, logarithmic Nye-Haefeli time scale (Eq. 3). The three time scales give order-of-magnitude different ages for the near-bottom ice layers. For example, using the present-day Central Greenland accumulation rate (ca. 0.3 m/yr) and ice thickness (ca. 3,000 m), the age of the ice at 90% depth is predicted by the three models to be ca. 1,000,000, 100,000 and 20,000 yr. Neither of these ages have much significance from the point of view of dating, if only for the reason that they are obtained by steady state models. They merely illustrate the large range of ages produced by different models and also that ice of a given age is found in less depth on a dome than on a slope, everything else being equal. This is further illustrated by the depth to the 10,000 yr age horizon, which is predicted to be ca. 1280, 1450, and 1850 m, respectively, by the three models.

Dating the Surface Ice in the Ablation Zone

As previously mentioned, a complete sequence of the layers originally deposited in the central accumulation region is found along the ice sheet surface in the ablation zone, except maybe for the oldermost layers that may have been removed by basal melting. Generally the age of the layers increases from the equilibrium line to the ice margin. Various studies have substantiated that paleoenvironmental information can be obtained by sampling the easily accessible surface ice (e.g., Lorius and Merlivat 1977; Reeh et al. 1987). This has stressed the need for developing flow models that can provide age profiles along the ice surface. In principle there is no difference between flow model dating of the deep ice in the central ice sheet areas and the dating of the ice along the ice sheet margin. However, since the ice at the margin has in general travelled a long distance before it resurfaces, accurate flow line modeling is a must for dating this ice. For dating over longer time spans, the relatively fast and large amplitude response of the ice margin to climatic change must be considered, even though a study by Reeh et al. (1987) indicated that at least back to ca. 6,500 yr B.P. a steady state model gave reasonable ages for a West Greenland ice margin location. With Hammer's (personal communication) demonstration that datable acid ice layers of volcanic origin can be detected on the surface

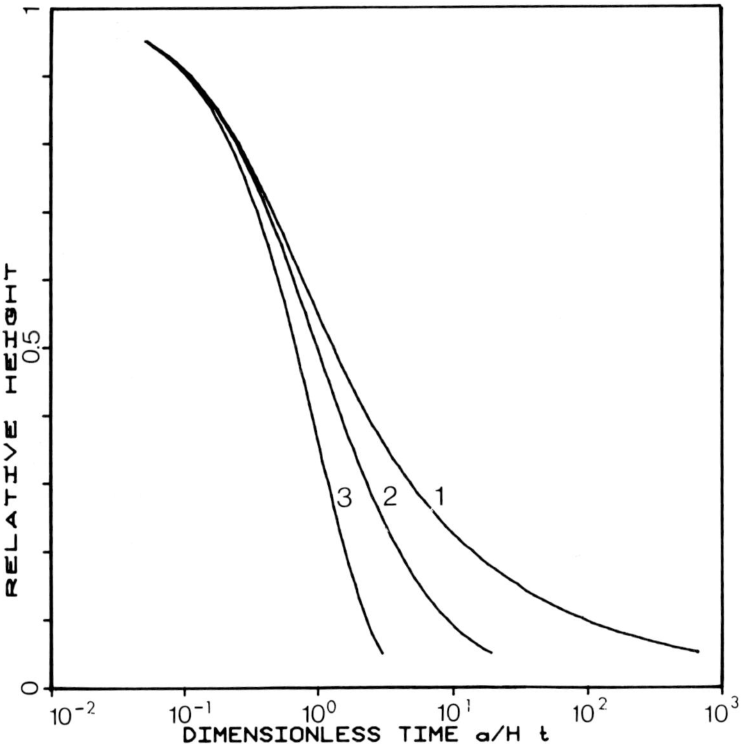

Fig. 5—Age profile at the dome of a symmetrical, isothermal, steady state ice sheet (1). The age profiles resulting from Raymond's (1983) linear approximation to the strain-rate distribution (2) and Nye's (1963) uniform strain-rate distribution are also shown (3).

of the Greenland ice sheet margin, and with the possibility, yet to be demonstrated, of dating the ice near the surface by means of accelerator based ^{14}C techniques, it may soon be possible to compare flow model and experimental dates also for the marginal ice. As will later be discussed, the main benefit of such a comparison will probably be information about the past dynamic history of the ice mass. It should be pointed out, however, that close to the ice margin the layer chronology is likely to be disturbed due to the presence of shear zones. Models that can handle this phenomenon at the quantitative level are not yet available.

Basal Melting

Ice masses are generally cooled from the surface by low air temperatures and heated from below by geothermal and deformational heat. If more heat is supplied to the base than can be conducted or advected away, the surplus heat will cause melting of the basal ice. Restricting ourselves to considering land-based ice masses, the amount of basal melting is typically a few mm/yr but may reach a value of a few cm/yr. Hammer et al. (1978) used a modified Nye-Haefeli model to illustrate the effect of basal melting on the time scale and showed that basal melting may considerably reduce the age of the basal ice. Figure 6 presents some results produced by means of Eq. 7 with the Dansgaard-Johnsen strain rate distribution, which is a more realistic approximation to the strain rate variation in an ice sheet than the Nye-Haefeli distribution. The figure shows the age of the basal ice plotted versus the ratio of basal melting to surface accumulation, parameterized for various values of h/H. By way of example, take $a_S = 0.15$ m/yr, $a_B = -0.002$ m/yr, $H = 3000$ m, and $h/H = 1$ (Raymonds approximation for a crest). Then, by means of Fig. 6, the age of the bottom ice is found to be ca. 250,000 yr. For comparison, curve 2 in Fig. 5 shows that 250,000 yr old ice is found about 220 m above the base if there is no bottom melting and that, in this case, the theoretical age of the ice at the very bottom is infinite.

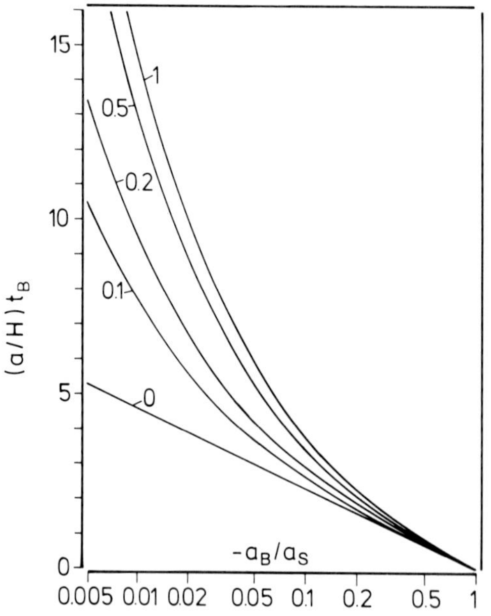

Fig. 6—Age of basal ice versus the ratio of basal melting to surface accumulation rate, parameterized for various values of h/H.

TIME-DEPENDENT MODELS

As already mentioned, there is an age limit beyond which dating by steady state ice flow models breaks down, because substantial changes in flow pattern, accumulation rate, and/or ice thickness have occurred prior to that time. These changes have influenced the past deformation history of the ice mass, as have changes in ice temperature and ice rheology. Very little work has been done to study how past changes in the dynamical, thermal, and rheological conditions influence the present age and layer thickness distributions in an ice mass. Also, there are very few examples of nonsteady state models used to establish a time scale along an ice core. One example is the ca. 160,000 yr chronology for the Vostok ice core from East Antarctica (78°28′S and 106°48′E) established by Lorius et al. (1985).

Time Scale for the Vostok Core

The Vostok model accounted for changes in past accumulation rate estimated from the $\delta^{18}O$ profile along the core by means of a model linking δ to the temperature of formation of precipitation. Ice thickness changes along the flow line due to bedrock topography were also considered. The calculated age profile agreed well with the established chronology from deep-sea foraminifera records back to ca. 110,000 yr B.P. For the oldest part of the record there were discrepancies of up to 10,000 yr. This may be due to inaccuracies in the dating of the deep-sea records, as argued by Lorius et al. (1985), but might as well be explained by shortcomings of the model used to predict the accumulation rate, or by the fact that temporal changes in ice thickness and temperature were neglected. During the glacial-interglacial cycles even interior East Antarctica may have experienced changes in ice thickness, and ice temperatures definitely have varied in response to temperature changes at the surface, causing changes in the thinning history of the internal layers, and thus influencing the time scale.

Dynamic History of the Devon Island Ice Cap

Paterson and Waddington (1984) compared a steady state λ-profile calculated by means of a finite element model to observed layer thicknesses along a 300 m ice core from Devon Island ice cap, Arctic Canada. Down to a depth corresponding to an age of ca. 1300 yr the profiles agreed well, thus confirming the steady state assumption. However, between 1300 and 4800 yr B.P. Paterson and Waddington found an increasing discrepancy between calculated and observed layer thicknesses, and concluded that accumulation rate increased steadily with age. The steady state profile, shown as the dashed curve in the upper diagrams of Fig. 7, is determined by a rather elaborate flow line model (Reeh and Paterson 1988) and shows similar deviations from the measured layer thicknesses. Two attempts to interpret these deviations by means of a simple time-dependent model are also illustrated in the figure. The model is essentially a time-dependent version of the "sandwich" model, using a shape function for the vertical velocity determined by the steady state profile. If the accumulation rate (respectively the ice thickness) is assumed to be known, an ice thickness history (respectively an accumulation rate history), which is compatible with the observed layer thickness profile, can be determined by means of the model. For example, for constant accumulation rate the ice thickness would have had to increase steadily from ca. 120 m about 6000 yr ago to the present-day thickness of ca. 300 m (see lower left diagram in Fig. 7). On the other hand, as shown in the lower right diagram of Fig. 7, the assumption of

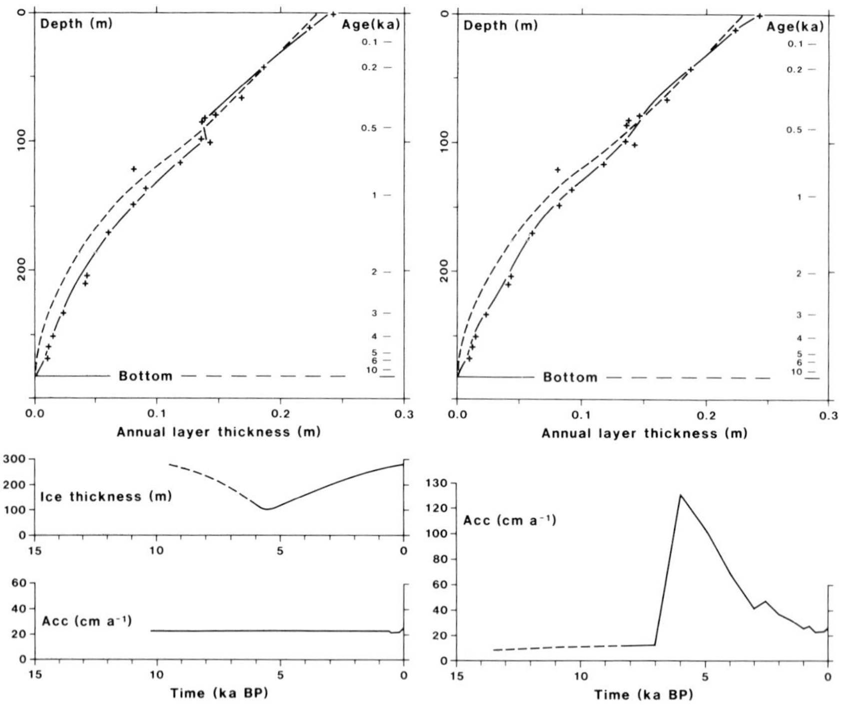

Fig. 7—Devon Island ice cap. Two examples of nonsteady state model calculations. Left side: derived ice thickness history assuming constant accumulation rate. Right side: derived accumulation rate history assuming constant ice thickness. Upper part: observed layer thicknesses shown by cross marks. The layer thickness profile calculated by a steady state flow model is shown by the dashed curve (Reeh and Paterson 1988). The layer thickness profiles calculated by the nonsteady state model are shown by full lines. Numbers are absolute ages.

constant ice thickness requires a sixfold decrease in accumulation rate since ca. 6000 yr B.P. Such a large change in accumulation rate is unlikely since the nearby Greenland Camp Century record indicates only 10–15% higher accumulation rate in the climatic optimum (Paterson and Waddington 1984). Also a threefold increase in ice thickness since 6000 B.P. is an unexpectedly large change which, however, would explain the abnormally large change found in the δ record from Devon ice cap between the climatic optimum and the present time (Paterson et al. 1977). One should, of course, be cautious not to overinterpret results obtained by such a simple model. Strain rate variations caused by varying ice temperature and ice rheological properties, e.g., due to the evolution of ice fabrics and downward migration of the transition between hard Holocene ice and soft Wisconsinan ice since the last ice age (Reeh 1985), may also have influenced the λ-profile. The importance of these effects can also be studied by means of the time-dependent sandwich model.

CONCLUSION

Dating by ice flow modeling serves several useful purposes related to the extraction of environmental records from glaciers and ice sheets. Simple models, requiring a minimum of input data, can provide preliminary time scales over some thousands of years, accurate enough to choose sampling frequency along an ice core. For accurate dating, elaborate flow models must be used. However, their use requires information about upstream distributions of ice thickness, accumulation rate, etc., and also about temperature and ice rheology within the ice mass. This means that their application is dependent on the accomplishment of extensive glaciological programs. For the large Antarctic and Greenland ice sheets, the precision of such flow model dating can be better than 10% over the time interval back to the termination of the last ice age, ca. 10,000 yr. ago. For smaller ice caps, the age limit for theoretical dating with a 10% accuracy seems to be more like 1000 to 1500 yr. Dating beyond these age limits requires consideration of temporal variations in accumulation rate, flow pattern, ice thickness, temperature, and ice rheology. For the large, relatively stable Antarctic and Greenland ice sheets, the accumulation rate is the most important time variable to consider. This is fortunate because it seems that past accumulation rate variations can be predicted with some precision from $\delta^{18}0$ records, at least as far as Antarctica is concerned (see above), but may be also in the case of Greenland as mentioned by Dansgaard and Oeschger (this volume). Also, ^{10}Be measurements can be used to estimate past variations in accumulation rate. For smaller ice caps, which react faster and more violently to climatic change, ice thickness variations are also important. Generally, the larger the ice mass and the less the accumulation rate, the further back in time theoretical dating is likely to be successful. The reason for this is that the larger the ice mass, the less sensitive the ice thickness in the interior regions is to climate variability (disregarding the possibility of a surge, i.e., a rapid disintegration of the ice mass caused by an instability), and the less the accumulation rate, the farther away from the base with its complicated strain conditions old ice will be found. Even under favorable conditions, dating of environmental records from glaciers and ice sheets solely by ice flow modeling will seldom, if ever, be advisable. However, calculated age and layer thickness profiles, when compared with observations, will throw light on the climatic and dynamic history of the ice masses, and thus are important links in the interpretation of the environmental records from glaciers and ice sheets. In particular, it seems promising to combine mutually the results of such comparisons from several ice sheet and ice cap locations, with information obtained from records of other indicators of past climate and ice dynamics, e.g., $\delta^{18}O$, ^{10}Be, and total gas content.

Acknowledgments. This work was sponsored by the Danish Commission for Scientific Research in Greenland. The Geological Survey of Greenland is thanked for use of its facilities.

REFERENCES

Budd, W. F., and Young, N. W. 1983. Application of modeling techniques to measured profiles of temperature and isotopes. In: The Climatic Record in Polar Ice Sheets, ed. G. de Q. Robin, pp. 150–179. Cambridge: Cambridge Univ. Press.

Dansgaard, W.; Clausen, H. B.; Dahl-Jensen, D.; Gundestrup, N.; and Hammer, C. U. 1985. Climatic history from ice core studies in Greenland. Data correction procedures. In: Current Issues in Climatic Research, Proc. European Economic Community Climatology Programme, eds. A. Ghazi and R. Fantechi, pp. 45–60, Symposium, Sophia Antipolis, France, 1984.

Dansgaard, W., and Johnsen, S. J. 1969. A flow model and a time scale for the ice core from Camp Century, Greenland. *J. Glaciol.* **8 (53)**: 215–223.

Haefeli, R. 1963. A numerical and experimental method for determining ice motion in the central parts of ice sheets. *IAHS Publ.* **61**: 253–260.

Hammer, C. U.; Clausen, H. B.; Dansgaard, W.; Gundestrup, N.; Johnsen, S. J.; and Reeh, N. 1978. Dating of Greenland ice cores by flow models, isotopes, volcanic debris, and continental dust. *J. Glaciol.* **20 (82)**: 3–26.

Herron, M., and Langway, C. C., Jr. 1980. Firn densification: an empirical model. *J. Glaciol.* **25 (93)**: 373–385.

Lorius, C.; Jouzel, J.; Ritz, C.; Merlivat, L.; Barkov, N. I.; Korotkevich, Y.S.; and Kotlyakov, V. M. 1985. A 150,000-year climatic record from Antarctic ice. *Nature* **316**: 591–596.

Lorius, C.; Jouzel, J.; Ritz, C.; Merlivat, L.; Barkov, N. I.; Korotkevich, Y. S.; in East Antarctica: observed changes with depth in the coastal area. *IAHS Publ.* **118**: 127–137.

Nye, J. F. 1963. Correction factor for accumulation measured by the thickness of the annual layers in an ice sheet. *J. Glaciol.* **4 (36)**: 785–788.

Paterson, W. S. B. 1981. The Physics of Glaciers. Oxford: Pergamon.

Paterson, W. S. B.; Koerner, R. M.; Fisher, D.; Johnsen, S. J.; Clausen, H. B.; Dansgaard, W.; Bucher, P.; and Oeschger, H. 1977. An oxygen-isotope climatic record from the Devon Island ice cap, arctic Canada. *Nature* **266 (5602)**: 508–511.

Paterson, W. S. B., and Waddington, E. D. 1984. Past precipitation rates derived from ice core measurements: methods and data analysis. *Rev. Geophys. Space Phys.* **22 (2)**: 123–130.

Philberth, K., and Federer, B. 1971. On the temperature profile and the age profile in the central part of cold ice sheets. *J. Glaciol.* **10 (58)**: 3–14.

Radok, U.; Barry, R. G.; Jenssen, D.; Keen, R. A.; Kiladis, G.N.; and McInnes, B. 1982. Climatic and Physical Characteristics of the Greenland Ice Sheet. Cooperative Institute for Research in Environmental Sciences, Univ. of Colorado, Boulder.

Raymond, C. F. 1983. Deformation in the vicinity of ice divides. *J. Glaciol.* **29 (103)**: 357–373.

Reeh, N. 1985. Was the Greenland ice sheet thinner in the late Wisconsinan than now? *Nature* **317 (6040)**: 797–799.

Reeh, N. 1988. A flow-line model for calculating the surface profile and the velocity, strain-rate, and stress fields in an ice sheet. *J. Glaciol.* **34 (116)**: 34–54.

Reeh, N.; Hammer, C. U.; Thomsen, H. H.; and Fisher, D. A. 1987. Use of trace constituents to test flow models for ice sheets and ice caps. *IAHS Publ.* **170**: 299–310.

Reeh, N.; Johnsen, S. J.; and Dahl-Jensen, D. 1985. Dating the Dye 3 deep ice core by flow model calculations. *Geophys. Monog.* **33**: 57–63. Washington, D.C.: Amer. Geophys. Union.

Reeh, N., and Paterson, W. S. B. 1988. Application of a flow model to the ice-divide region of Devon Island ice cap, Canada. *J. Glaciol.* **34 (116)**: 55–63.

Weertman, J. 1968. Comparison between measured and theoretical temperature profiles of the Camp Century, Greenland borehole. *J. Geophys. Res.* **73 (8)**: 2691–2700.

Wolff, E. W., and Doake, C. S. M. 1986. Implications of the form of the flow law for vertical velocity and age-depth profiles in polar ice. *J. Glaciol.* **32 (112)**: 366–370.

The Environmental Record in Glaciers and Ice Sheets
eds. H. Oeschger and C.C. Langway, Jr., pp. 161–175
John Wiley & Sons Limited

Physical Property Reference Horizons

H. Shoji* and C.C. Langway, Jr.

Ice Core Laboratory, Department of Geology
State University of New York at Buffalo
Amherst, NY 14226, U.S.A.

Abstract. Physical property variations are used to delimit the boundaries of past snow accumulation cycles. Density and visible stratigraphy are the main repetitive indicators which denote annual layers to at least the 100 m depth in north Greenland. Melt features are observed in a south Greenland deep ice core to a depth of over 1.5 km, representing nearly 5500 years of accumulation. Density, grain size, crystal size, melt features, c-axes orientation, air volume, and light transmissivity are all among the property changes with depth that serve as systematic reference indicators for accumulation and climate or describe characteristic changes in the physical nature of a glacier. At Dye 3, Greenland an abrupt shift exists in many of the physical and mechanical properties at about 1786 m and this reflects the Holocene/Wisconsin climatic boundary. This pronounced data set shift occurs in unison with different geochemical and climate signals in the Dye 3 core. Similar signs are recorded in other ice cores from Greenland and Antarctica. Wisconsin age ice exhibits different physical and mechanical characteristics than the ice above it. Near bottom, basal ice contains alternating layers of clear ice and high debris content bands.

INTRODUCTION

A polar ice sheet is a large glacier mass consisting of a physical assemblage of snow, firn, ice, enclosed atmospheric gases, and included solids. There are also various chemical constituents. Firn is deposited snow that is over one year old. The firn profile extends from near the glacier surface to a depth at which the formerly intercommunicating pore spaces close off from atmospheric circulation and form individual bubbles. By definition, below this firn/ice transition (zero permeability) and extending to the subglacier interface is glacier ice. The maximum thickness of an inland polar ice sheet

*Joint appointment with the Snow and Ice Laboratory, Department of Earth Sciences, Toyama University, Toyama 930, Japan.

is between 3000 m and 4000 m and may contain up to 1 to 3×10^6 yr of past snow accumulation and climate history.

Both firn and glacier ice incorporate a record of physically identified index features produced by climatic or dynamic processes, as explained later in this paper. These indices may be cyclical, periodic, or nonuniformly repetitive. In some cases several physical property indicators may identify a rapid change in a climatic condition and serve as a major stratigraphic reference horizon. The Holocene/Wisconsin age boundary illustrates such a case.

APPLICATION

Physical property measurements are used to determine the boundaries of annual accumulation layers and to disclose other meteorological or climatic events which are registered in ice sheets. They are also used to determine the physical parameters needed in ice dynamic and climate change models. For purposes of this paper, we identify stratigraphic reference horizons in terms of abrupt or subtle, but well-defined, physical property changes as revealed in ice cores. These features and their periodicities are best defined by continuous multiple-parameter physical property measurements over long profiles. A complete ice core study would involve cross-correlations of physical and chemical data sets from laterally adjacent samples to enhance the reliability and maximize the return of the resulting environmental data.

In this paper we first review the main physical stratigraphic criteria used to identify layered accumulation sequences from the surface to about 100 m. Then we discuss the significance of other stratigraphic indicators and physical parameters found deeper in the glacier. Most examples used are primarily from studies made on ice cores obtained from the Greenland ice sheet, but the principles discussed apply equally as well to the Antarctic ice sheet. A number of the methods and techniques presented were taken from earlier reports (Langway 1967; Langway et al. 1985; Shoji and Langway 1985).

THE ENVIRONMENT

The physical characteristics of deposited snow are the result of atmospheric, depositional, and postdepositional processes. Snow falls throughout the year on a high polar ice sheet, although it is generally assumed that more accumulates during the relatively warmer summer periods. The environmental conditions and geographical location of a site strongly influence the amount of net annual accumulation and physical nature of the deposited snow. With time, new snowfall buries the previous accumulation deeper, the density of the underlying firn increases by compaction, and with depth the firn gradually transforms into glacier ice.

An exposed vertical cut at the ice sheet surface will display well-defined strata and physical property changes. These megascopic and, in some cases, microscopic features become less distinct with depth, but are recognizable by careful examination and measurement. Important physical properties, useful for determining past annual accumulation, usually have a cyclical recurrence. The cycles are related to original variations in the precipitation supply, changes in the rate of deposition, fluctuations in the surface temperature, and other meteorological variables such as wind velocity.

Temperature is an important factor affecting polar strata. Diurnal and seasonal temperature waves penetrate and markedly alter the texture and structure of the upper stratigraphic layers. The alterations are a result of reversing temperature gradients which assist mass vapor transfer in the upper 10 m of porous firn. Longer-term temperature waves seep downward into the deeper firn, and ice and geothermal heat flow ascends upward into the basal and near-bottom ice.

Glacier motion can have a disruptive effect on the original chronological sequences of a glacier. The vertical strain rate causes a progressive thinning of the layered time units and near-bottom turbulent flow often obliterates any vestige of systematic bedding. On the other hand, deformational processes register their own imprint in the glacier mass. Deformation features represent the past and present dynamic history of the glacier. Knowledge of the regional ice flow patterns are also required for ice core dating and other modeling purposes.

POLAR STRATIGRAPHY

An annual snow deposit has certain indicative components. The summer layer usually has one or more melt features located in a porous and coarser-grained matrix with low bulk density. In contrast, a winter layer displays finer-grained, hard and homogeneous wind-packed layers of relatively high bulk density. Small dimension (1–2 mm thick) wind crusts are prominent in both summer and winter deposits in near-surface firn, but solar radiation crusts are present only in summer accumulation (Langway 1967). It is difficult to discriminate between wind and radiation crusts in the deeper firn.

The stratigraphic criteria used to interpret seasonal layers on a high polar glacier are listed in Table 1. The main features used to differentiate annual layers are density differences and melt phenomena (crusts, layers, wedges, or glands depending upon the severity and duration of the weather). Of secondary importance, but often useful in validating a questionable boundary, are radiation crusts or sublimation features (depth hoar, hoar frost), wind features (crusts and slabs), and diagenetic changes (grain size, hardness). In conjunction with other features, light transmissivity is also a useful guide for determining annual boundaries.

Table 1 Physical indicators of seasonal deposits in polar firn

	Seasonal Deposits		
Property or Feature	Summer	Fall	Winter
Density	lower		higher
Melt phenomena	prominent		
Wind crusts	present		present
Radiation crusts	present		none
Grain size	coarse		fine
Hardness	lower		higher
Depth hoar		present	
Total air content	high		low
Light transmissivity	darker and multilayered		lighter and homogeneous

Various stratigraphic features "observed" are difficult to categorize explicitly. The peculiarities of weather add to the difficulty by making it nearly impossible to rely on the cyclical recurrences of certain stratigraphic indicators. The indicators or features listed in Table 1 are generally diagnostic and useful only when considered in relationship to the whole. To a great extent, interpreting physical stratigraphy is an art dependent upon experience and foreknowledge of the contemporary environmental conditions at the site.

Table 2 is a flow diagram depicting the environmental and physical conditions associated with the development of some of the stratigraphic indicators.

Macroscopic Observations

Macroscopic stratigraphy refers to those structural features observed in firn/ice cores that are visible to the unaided eye, either by direct observation or by examination in transmitted light. When a light table examination is cross-correlated with direct physical measurements (see Table 1) a complete physical description of an ice core is possible. The main features observed in a firn core on a light table are melt phenomena, wind crusts, depth hoar structures, grain size, and variable light transmissivity. However, with an ice core it is only possible to observe melt features, variable light transmissivity, and a bulk measure of the presence of occluded air bubbles.

Density Variations

Density is a fundamental physical property of snow and ice and is an established seasonal stratigraphic indicator. Cyclical density variations are

Table 2 Flow diagram indicating conditions associated with formation of physical stratigraphic indicators

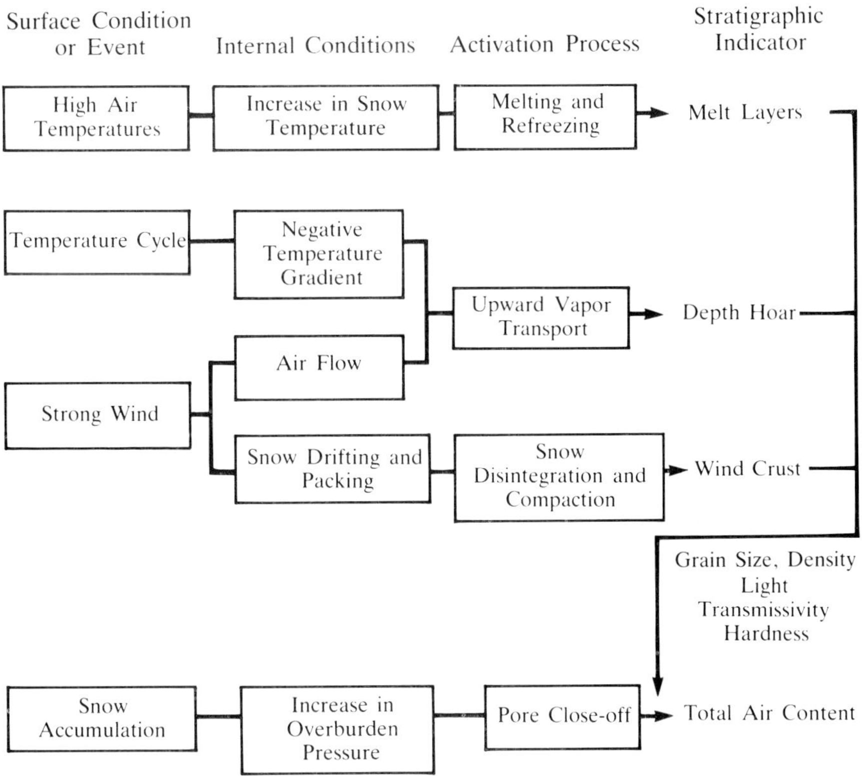

usually annual and are measurable to depths of 100 m and more. Less dense layers are associated with summer snowfalls and, conversely, more dense layers depict winter precipitation. Summer deposits are more often multilayered and composed of coarser-grained crystals with a looser packing than are the fine-grained, more homogeneous and hard-packed winter deposits. Seasonal variations in density diminish considerably below the pore close-off zone but still exist. Figure 1 presents a plot of the stratigraphy and density data from a 4 m long ice core interval at about the 100 m depth in Greenland. The summer (S) and winter (W) ticks to the right of the plotted density column are seasonal peaks and troughs determined by continuous measurements of the $^{18}O/^{16}O$ ratios. Note in Fig. 1 the good agreement in less dense layers with melt features and the isotopically determined summer markers. Similar agreement exists for winter strata.

In addition to the stratigraphic usefulness of density, these measurements provide information on other physical processes. A continuous smoothed

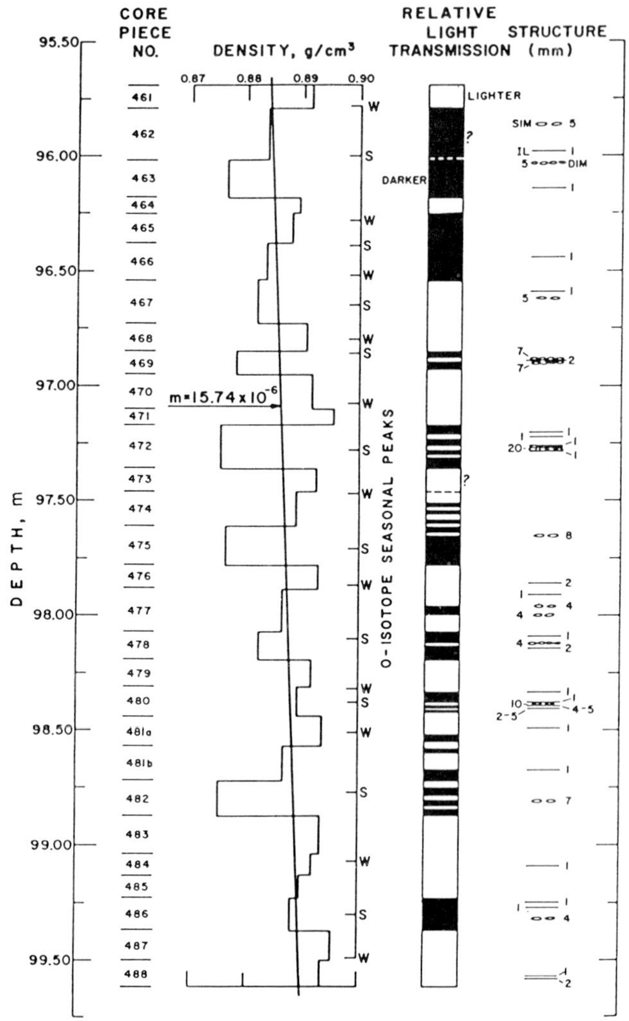

Fig. 1—Seasonal distribution of physical stratigraphic indicators compared with oxygen isotope measurements. The ice core is from northwest Greenland. The seasonal distribution of isotope peaks (summer) and troughs (winter) with depth compare favorably with the physical property measurements. Structure features SIM and DIM indicate slight indication of melt and definite indication of melt; horizontal lines represent melt layers (from Langway 1967).

density curve from the surface downward has several deflection points. Polar snows have an average density of about 0.3 g/cm^3 at the surface. In the upper layers load pressure causes densification by settling. The closer packing of grains results in a reduction in the amount of air-filled pore spaces. The first major change in slope occurs at about the 10 m depth at a density of about 0.55 g/cm^3. From there the overall form of the curve shows a nonlinear increase in density with depth at a decreasing rate, asymptotically approaching the density of pure bubble-free polycrystalline ice (0.92002 g/cm^3 at −25°C). The difference between the density of glacier ice and pure polycrystalline ice is the total air content (porosity) of the glacier sample. At the approximate depth, where the average density is 0.83 g/cm^3 (Langway 1967), the former intercommunicating pore spaces become closed off from atmospheric circulation and form individual air bubbles preserving specimens of air with the atmospheric composition at the time of ice formation. The transformation of firn to glacier ice is an important phenomena that has great bearing on atmospheric gas inclusions (see Schwander, this volume). The depth of zero permeability is a function of accumulation and temperature and varies with inland location.

Melt Features

Melt phenomena are postdepositional features and play a dominant role in polar stratigraphy. Thin ice layers (< 2 mm) form only on or close to the surface by solar radiation. Higher summer surface temperatures create more severe melt conditions and produce thicker ice layers, ice wedges, and glands, which always form below the surface, sometimes after percolating downward as much as 2 m (Benson 1959). Depth hoar formation develops most favorably when a warmer snow surface is quickly blanketed by a colder deposit. The new snow deposit restricts upward heat transfer and vapor migration causing the vapor to condense at the interface. Under ideal conditions a single distinctive depth hoar layer separates the summer/fall boundary, although these features may develop any time a steep temperature gradient exists (Schytt 1958).

In the Dye 3 deep ice core, distinct melt features and layers with fewer air bubbles have been observed and measured continuously on a light table to a depth of about 1500 m corresponding to an age of 5500 years B.P. A correlation of these summer temperature features (Herron, M. et al. 1981; Langway and Shoji, in preparation) with the mean annual temperature data from the same core determined by the stable isotope technique (Dansgaard et al. 1982, 1985) shows good coherence. A melt layer study has the advantage of being a rapid method to obtain independent proxy data on past summer climate, and it requires fewer assumptions to interpret results than does the stable isotope method.

WISCONSIN AGE ICE

There are natural discontinuities and variations in the physical properties and other changes in the mechanical behavior of glacier ice below the pore close-off zone. They are particularly noticeable in the Wisconsin age and basal ice. In general, if the ice core is of good physical quality (i.e., unfractured and continuous from the surface) all the physical property studies made on the firn core may be continued in the deeper ice core to the extent that the properties are discernable. In addition to the stratigraphic indicators, other physical and mechanical investigations are made, some for their intrinsic value. Ice deformation rates and the temperature profile for the Dye 3 borehole are discussed by Dansgaard and Oeschger (this volume).

The Holocene/Wisconsin boundary for the Dye 3 deep ice core was first identified at the 1786 m depth by its lower conductivity and high dust content during initial field measurements (Hammer et al. 1985). This boundary or transition was subsequently verified by stable isotope analyses (Dansgaard et al. 1982, 1985) and by chemical stratigraphy (Herron, M. and Langway 1985) and was dated as 10400 years B.P. Concurrent multiparameter measurements of crystal size, ice fabrics, ultrasonic wave velocity (Herron, S. et al. 1985), and material strength behavior (Shoji and Langway 1985, 1987a) also established that pronounced physical changes occurred at about the same depth level as the chemical signals.

Following the field measurements, several detailed laboratory studies on the Dye 3 core further confirmed other strong cross-correlations in the physical, chemical, stratigraphic, and climate records.

Crystal Size

Grain and crystal size tends to gradually increase with depth. There are, however, notable fluctuations in crystal size. A sudden major decrease in crystal size from 5 to 2 mm (Herron, S. et al. 1985) occurs at the Holocene/Wisconsin boundary. Figure 2 shows crystal size measurements made on the Dye 3 ice core from below 1786 m (Holocene/Wisconsin boundary) to the 2037 m depth, or 1 m above the glacier/sub-ice interface. The main characteristics of the crystal size curve appear to be a base level minimum size of about 0.4 mm for the Wisconsin age ice. However, size fluctuations range erratically within the 1812–1970 m zone and reach a maximum of around 2.0 mm. For the 1970–2008 m zone the measurements show an average crystal size of about 2 mm. Below 2008 m crystal size decreases to a minimum of about 0.4 mm.

Note the strong inverse correlation of the impurity (SO_4^{2-}) concentration levels with crystal size shown in Fig. 2. There is also a clear and strong correlation of general crystal size with other solid and ionic constituents and stable isotope curves in the Wisconsin age ice (Hammer et al. 1985; Herron, M. and Langway 1985; Herron, S. et al. 1985; Langway et al. 1988).

Ice crystal growth is a result of metamorphic processes. Time, temperature, and stress all play important roles in this process. It follows that crystal size

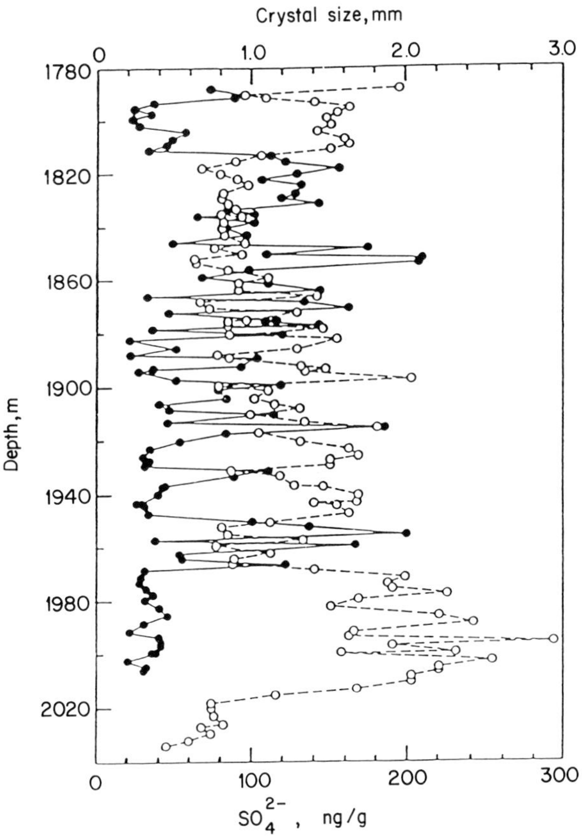

Fig. 2—Crystal size and SO_4^{2-} concentration levels in Wisconsin age ice. Crystal size is shown by open circles and SO_4^{2-} by solid circles (from Langway et al. 1987).

can be expressed as a function of time using values established for initial grain size, the nucleation rate of recrystallization, and the crystal growth rate. Both nucleation and crystal growth rates may be influenced by impurity content as also revealed in metallurgy. A recent paper (Petit et al. 1987) suggests that crystal size may be used to reconstruct paleotemperature records.

Ultrasonic Wave Velocity

Ultrasonic wave velocity is a function of the dynamic elastic constants of a material and, as is the case for ice, depends upon crystal orientation. Since glacier ice is a polycrystalline substance (with air bubbles, impurities, and structure), measurements of wave propagation through an ice core specimen provide an estimate of the average c-axis crystal orientation. Each specimen is taken as representative of the depth or location from which it was obtained. Several thousand individual ice crystals are usually contained in a single specimen. Results of ultrasonic wave velocity measurements on the Dye 3 core are shown in Fig. 3 (Herron, S. et al. 1985). These curves reveal gradual increases in wave velocity in Holocene ice and sharp increases in velocity at the Holocene/Wisconsin age boundary. These data indicate that a strong, near-vertical single maximum fabric pattern exists in all the Wisconsin age ice. This condition has been verified by thin section studies using a universal stage on both Dye 3 and Camp Century ice cores (Herron, S. et al. 1985; Langway et al. 1988). Although the cause of the strong vertically oriented c-axis patterns in Wisconsin age ice are not completely understood, they most probably involve selective growth, crystal rotation, and recrystallization, all influenced by deformational conditions and impurity content.

Strain Rate Enhancement Factor

The predominant mechanisms for the plastic deformation of glacier ice varies under certain sets of temperature and stress conditions. In addition, the flow velocity of polycrystalline ice can be altered by the effects of c-axis orientations or impurity content. The field conditions are complicated and difficult to simulate in laboratory tests. A comparison of the Dye 3 field borehole tilt measurements with the results of simplified laboratory tests and experiments on deep core samples (Shoji and Langway 1987a, 1988; Dahl-Jensen and Gundestrup 1987) are shown in Fig. 4. They indicate a very high enhancement factor value for the horizontal shear deformation of the Wisconsin age ice (1786 m to 2008 m) compared to the Holocene ice (250 m to 1786 m). High enhancement factors for the Wisconsin age ice reflect a softening of the material. This agrees favorably with the results of borehole tilt measurements made on the Agassiz Ice Cap Elsmere Island (Fisher and Koerner 1986) but varies from the analysis of the tilt measurements from Camp Century, Greenland, and Byrd Station, Antarctica (Paterson 1983).

Total Air Volume

The volume of air entrapped in polar glacier ice at pore close-off is related to the atmospheric pressure (elevation) and surface temperature at the location. If the temperature history is known, past elevation changes in the glacier can be inferred by measuring the total air content in core samples. To obtain elevation change data from total air volume measurements requires knowledge of today's surface elevation for existing flow conditions,

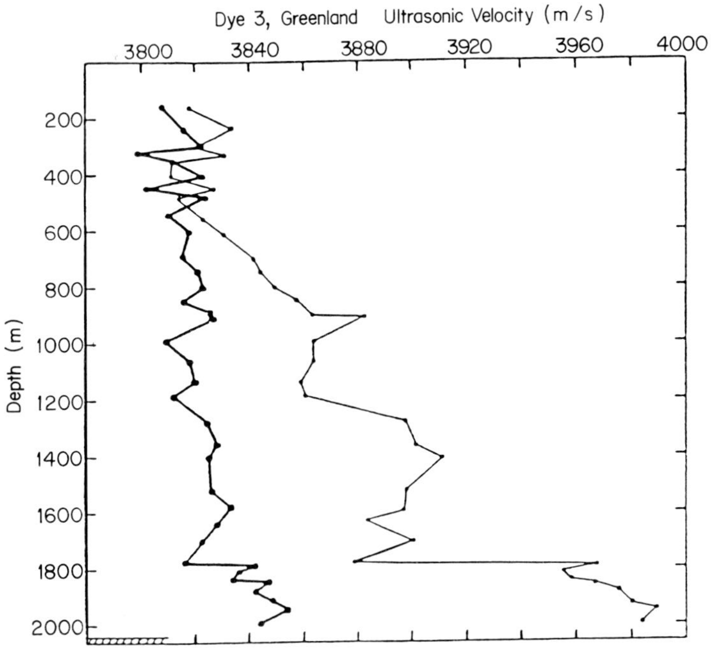

Fig. 3—Ultrasonic velocity profile. Thick line indicates horizontal velocities; thin line represents vertical velocities (from Herron, S. et al. 1985).

past changes in the flow line, today's surface temperature conditions, and past changes in the temperature regime.

Ice core studies on the Camp Century ice core (Raynaud and Lebel 1979) showed that the ice sheet elevation was 800 m to 900 m higher during the Wisconsin age than at present. A later study on the same ice core (Herron, S. and Langway 1987) considered seasonal variability in the analysis of the total air volume. This study revealed elevation difference of only 400 m to 500 m for the same time periods and reported that no elevation changes were detected for the Dye 3 area from measurements made on that core. A plot of air volume versus depth for the Camp Century ice core is presented in Fig. 5.

Basal Ice

The stratigraphy of near-bottom ice is complicated by turbulent flow and the inclusion of debris bands. Common to the three ice cores that have up to now reached bedrock in Greenland and Antarctica is a thick band of debris laden ice. At Camp Century and Dye 3, Greenland the multilayered bands are about 15 m and 30 m thick, respectively. At Byrd Station,

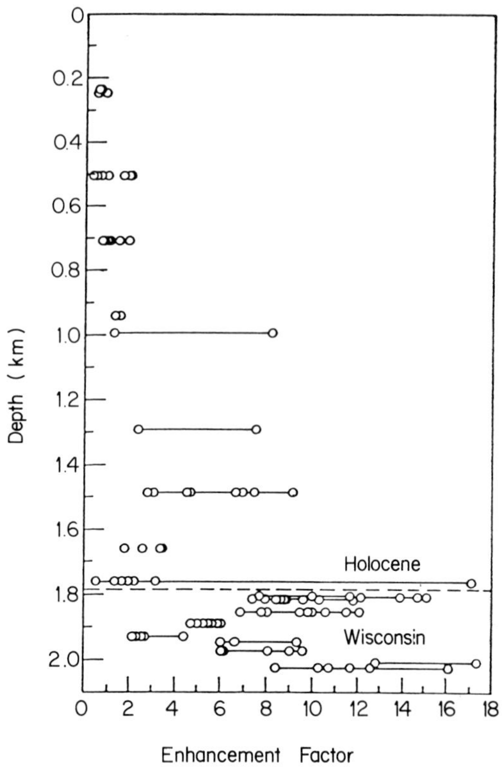

Fig. 4—Enhancement factor profile for Dye 3 ice core obtained from uniaxial compression tests (from Shoji and Langway 1987a).

Antarctica the band is about 5 m thick. The debris is abraded from the sub-ice bedrock by glacier movement and incorporated in the basal ice by a freeze-on process, usually as alternating clear and heavy debris content layers. At Camp Century and Dye 3 most of this material is very fine-grained rock flour but millimeter-sized fragments are often present (Herron, S. 1982).

From the surface to the 751 m depth at Dye 3 the ice core was absent of any visible or dark-appearing foreign substances. At 751 m a subtle, faintly discolored 2–3 mm thick horizontal layer was observed. Below this the ice once again became visibly free of any dark foreign materials until a single isolated 1 cm diameter pebble was encountered at the 1950 m depth at an age of 70 ka B.P. (Dansgaard et al. 1982; Herron, M. and Langway 1985). The core between 1950 m and 2008 m contained numerous granitic dust particles (Hammer et al. 1985) below which the multilayered debris and clear ice bands appeared sharply.

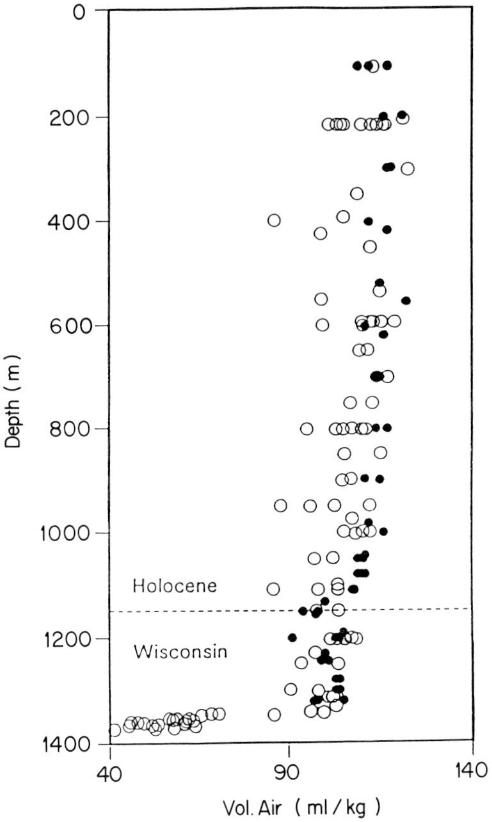

Fig. 5—Camp Century ice core total air volume versus depth (from Herron, S. and Langway 1987). Open circles from Herron, S. and Langway (1987); closed circles from Raynaud and Lorius (1973) and Raynaud (1976). The data sets show the variability introduced when sample selection considerations are given to the short-term cyclical effects of seasonal density variations and summer melt phenomena (open circles).

Air Hydrate-Bubble Transformation

Air hydrate inclusions were first observed in the Dye 3 core during an optical microscope examination conducted in the field trench laboratory (Shoji and Langway 1985). Later laboratory studies revealed that a significant number of air hydrate inclusions still existed in the deep ice core samples obtained at Camp Century in 1966 and at Byrd Station in 1968 (Shoji and Langway 1984, 1987b). The shallowest depths at which air hydrate inclusions are observed in the Dye 3, Camp Century, and Byrd Station cores are at 1092 m, 1099 m, and 727 m depths, respectively. The observed depths at which air hydrates appear in the Dye 3 and Camp Century ice cores agree with Miller's calculation for the phase transition (Miller 1969). However, the observed depth at which air hydrates appear is about 100 m higher in the Byrd Station core than in Miller's calculation. This apparent difference at Byrd Station may be attributed to the general upward flow trajectory of the ice at that location which begins about 5 km upstream. From laboratory studies it appears that once an air hydrate inclusion is created from an air bubble, the air hydrate remains in a semi stable state. Stability persists even at unstable pressure/temperature conditions until the surrounding ice matrix deforms and activates the nucleation process required for the transformation.

SUMMARY AND CONCLUSIONS

Ice sheets have evolved as a primary data source for paleoenvironmental information. This has been achieved, in part, by the development of drills capable of augering core samples to great depths from the inland ice sheets. The cores are studied by various physical and chemical methods to extract the paleodata contained in the ice layers. Physical property measurements are an integral part of a complete ice core study. They are used to identify past annual accumulation layers, meteorological events, climate trends, and to supplement and validate other geophysical and geochemical information. Physical measurements are also necessary for data input to calculate ice dynamic and climate models.

Acknowledgments. The authors thank the U.S. National Science Foundation, Division of Polar Programs, for their long-term support of the Greenland and Antarctic ice core research discussed in this paper. This paper was prepared as part of NSF/DPP Grant No. 8520911.

REFERENCES

Benson, C.S. 1959. Physical investigations on the snow and firn of northwest Greenland, 1952, 1953, 1954. U.S. Army SIPRE, Research Report 26.

Dahl-Jensen, D., and Gundestrup, N.S. 1987. A constitutive law for ice determined from the inclinometer survey at Dye 3, Greenland. In: The Physical Basis of Ice Sheet Modelling, Proc. 19th IUGG General Assembly, Vancouver, B.C. Canada, 1987. *IAHS Publ.* **170**: 31–43.

Dansgaard, W.; Clausen, H.B.; Gundestrup, N.; Hammer, C.U.; Johnsen, S.F.; Kristinsdottir, P.M.; and Reeh, N. 1982. A new Greenland deep ice core. *Science* **218**: 1273–1277.

Dansgaard, W.; Clausen, H.B.; Gundestrup, N.; Johnsen, S.J.; and Rygner, C. 1985. Dating and climatic interpretation of two deep greenland ice cores. *Geophys. Monog.* **33**: 71–76. Washington, D.C.: Amer. Geophys. Union.

Fisher, D.A., and Koerner, R.M. 1986. On the special rheological properties of ancient microparticle-laden Northern Hemisphere ice as derived from bore-hole and core measurements. *J. Glaciol.* **32**: 501–510.

Hammer, C.U.; Clausen, H.B.; Dansgaard, W.; Neftel, A.; Kristindottir, P.; and Johnsen, E. 1985. Continuous impurity analysis along the Dye 3 Deep Core. *Geophys. Monog.* **33**: 90–94. Washington, D.C.: Amer. Geophys. Union.

Herron, M.M.; Herron, S.L.; and Langway, C.C., Jr. 1981. Climatic signal of ice melt features in southern Greenland. *Nature* **293**: 389–391.
Herron, M.M., and Langway, C.C., Jr. 1985. Chloride, nitrate, and sulfate in the Dye 3 and Camp Century, Greenland ice cores. *Geophys. Monog.* **33**: 77–84. Washington, D.C.: Amer. Geophys. Union.
Herron, S.L. 1982. Physical properties of the deep ice core from Camp Century, Greenland. Ph.D. diss, Dept. of Geological Sciences, State Univ. of New York at Buffalo.
Herron, S.L., and Langway, C.C., Jr. 1987. Derivation of paleoelevations from total air content of two deep Greenland ice cores. In: The Physical Basis of Ice Sheet Modelling, Proc. 19th IUGG General Assembly, Vancouver, B.C., Canada, 1987, *IAHS Publ.* **170**: 283–295.
Herron, S.L.; Langway, C.C., Jr.; and Brugger, K.A. 1985. Ultrasonic velocities and crystalline anisotropy in the ice core from Dye 3, Greenland. *Geophys. Monog.* **33**: 23–32. Washington D.C.: Amer. Geophys. Union.
Langway, C.C., Jr. 1967. Stratigraphic analysis of a deep ice core from Greenland. USA CRREL Research Report 77.
Langway, C.C., Jr.; Oeschger, H.; and Dansgaard, W., eds. 1985. Greenland Ice Core: Geophysics, Geochemistry, and the Environment. *Geophys. Monog.* **33**: Washington, D.C.: Amer. Geophys. Union.
Langway, C.C., Jr.; Shoji, H.; and Azuma, N. 1988. Crystal size and orientation patterns in the Wisconsin age ice from Dye 3, Greenland. Symposium on Ice-Core Analysis, Bern, Switzerland, 1987. *Ann. Glaciol.* **10**: 109–115.
Miller, S.L. 1969. Clathrate hydrates on air in Antarctic ice. *Science* **165**: 489–490.
Paterson, W.B.S. 1983. Deformation within polar ice sheets: an analysis of the Byrd Station and Camp Century borehole-tilting measurements. *Cold Reg. Sci. Tech.* **8**: 165–179.
Petit, J.R.; Duval, P.; and Lorius, C. 1987. Long-term climatic changes indicated by crystal growth in polar ice. *Nature* **326**: 62–64.
Raynaud, D. 1976. Les inclusions gazeuses dans la glace de glacier, leur utilisation comme indicateur du site de formation de la glace polaire applications climatiques et rheologiques. These de Doctorate d'Etat, Universite Scientifique et Medicale de Grenoble, France.
Raynaud, D., and Lebel, B. 1979. Total gas content and surface elevation of polar ice sheets. *Nature* **281**: 289–291.
Raynaud, D., and Lorius, C. 1973. Climatic implications of total gas content in ice at Camp Century. *Nature* **243**: 283–284.
Schytt, V. 1958. Glaciology II: Scientific results of the Norwegian-British-Swedish Antarctic Expedition, 1949–1952 (Oslo, Norsk Polarinstitutt). **4**: 7–148.
Shoji, H., and Langway, C.C., Jr. 1984. Flow behavior of basal ice as related to modeling considerations. *Ann. Glaciol.* **5**: 141–148.
Shoji, H., and Langway, C.C., Jr. 1985. Mechanical properties of fresh ice core from Dye 3, Greenland. *Geophys. Monog.* **33**: 39–48. Washington, D.C.: Amer. Geophys. Union.
Shoji, H., and Langway, C.C., Jr. 1987a. Flow velocity profiles and accumulation rates from mechanical tests on ice core samples. In: The Physical Basis of Ice Sheet Modelling, Proc. 19th IUGG General Assembly, Vancouver, B.C., Canada, 1987, *IAHS Publ.* **170**: 67–77.
Shoji, H., and Langway, C.C., Jr. 1987b. Microscopic observations of the air hydrate-bubble transformation process in glacier ice. *J. Physique* **Cl**: 551–558.
Shoji, H., and Langway, C.C., Jr. 1988. Flow law parameters of the Dye 3, Greenland deep ice core. Symposium on Ice-Core Analysis, Bern, Switzerland, 1987. *Ann. Glaciol.* **10**: 146–150.

Standing, left to right:
Dominique Raynaud, Claus Hammer, Ed Fireman, Bernhard Stauffer, Jean Jouzel, Wolfgang Graf, Niels Reeh

Seated, left to right:
Robert Finkel, Hitoshi Shoji, John Andrews, Johannes Weertman, Bill Budd

The Environmental Record in Glaciers and Ice Sheets
eds. H. Oeschger and C.C. Langway, Jr., pp. 177–192
John Wiley & Sons Limited

Group Report
How Can an Ice Core Chronology Be Established?

W.F. Budd, Rapporteur
J.T. Andrews
R.C. Finkel
E.L. Fireman
W. Graf
C.U. Hammer
J. Jouzel
D.P. Raynaud
N. Reeh
H. Shoji
B.R. Stauffer
J. Weertman

ICE CORE CHRONOLOGY

OBJECTIVES

The establishment of reliable chronologies for ice cores is a primary requirement for their interpretation as environmental records.

This paper presents the outcome of group discussions addressing the question, "How can an ice core chronology be established?" Although ice cores with approximate chronologies provide valuable information on the extent of past environmental changes, it is important to stress the need for high accuracies for the chronologies.

The major objective for ice core chronology is to establish highly accurate continuous chronologies through the ice to determine sufficiently precise timing of major environmental changes and the time leads and lags between events to clarify the processes of global change.

The basic theme of this paper is to recommend that because of the need for high accuracy in the ice core chronologies, and because all of the dating techniques are subject to some error, multiple dating techniques should be used to provide sound control of the derived time scales.

A general introduction to the various types of dating involved in the multiple dating techniques will be given first. Then a number of the most important issues concerned with ice chronology are discussed in more detail. These issues are:

1. To what extent can the comparison with other records be used as a tool for dating ice core records?
2. How can we synthesize the chronologies vertically through the ice and along the surface in the ablation zone at the margin of ice sheets?
3. How can we improve our knowledge concerning the chronology of apparently rapid changes observed in ice core records?
4. Can changes in annual cycles or seasonality be determined from measurements of multiple parameters along the ice cores?
5. What are the capabilities and limitations of dating by use of radioactive isotopes?
6. What new dating techniques can be expected from the application of new analytical techniques to ice cores and from the availability of large ice samples?

Finally, a summary is given for a set of the most important types of research programs required to advance the objectives of improving ice core chronologies.

MULTIPLE DATING TECHNIQUES FOR ICE CORES

Summary of Major Ice Core Dating Techniques

Numerical modeling of the age distribution through an ice sheet by ice dynamics. The properties of the present ice sheet can be used to compute an age distribution through an ice sheet or glacier as described by Reeh (this volume). If there is information available on the past variations of the ice sheet or of the various input parameters, then that can also be used in the modeling. For this work it is important to have adequate coverage of data over the ice sheet particularly upstream from the site of the ice core drilling. Preliminary modeling for the age distribution should be carried out before the drilling and the core analysis, to give an initial indication of the age depth relation, which can later be improved with the additional ice core data. Subsequently, using the new data derived from the ice core concerning the past changes of the features of the ice sheet (such as accumulation rate, surface temperature, elevation, etc.), the age distribution can then be computed much more precisely.

In brief, the modeling requires data along the flowline, upstream from the borehole, beyond the distance from which the deep ice to be dated has originated. The data required includes surface elevation, ice thickness, surface accumulation rate, ice velocity, surface isotopes, and chemical concentrations. These data are used to model particle paths and age structures through the ice flow section over the time period of the accumulation and flow (cf. Budd and Young 1983). The data determined from the ice core, such as the past accumulation rates, can also be used to control the modeling. For locations near ice dome summits, upstream data collection is not so important, but information on gradients of variables across the summit and possible shifts of the summit are important. For flowlines a number of boreholes along the flow section can also help control the modeling for the past changes of the ice velocities and strain rates. The most important variables controlling the age depth profile are the past history of the accumulation and vertical strain rate.

The accuracy of the method depends on a number of factors. For the upper parts of ice sheets, which have not been subject to large changes, the accuracy of the choronology can be high. For regions near the bed, on the scale of the bed topography variations, the complications of the ice flow can make the derived chronology unreliable. If the region of the ice sheet has undergone large changes over the time period, the errors of the modeling can also be high. The inland regions of the thick ice sheets therefore have the best prospects for reliable long-term chronologies. These factors need to be considered in the selection of future sites for deep core drilling.

For the case of the Vostok ice core reaching a depth of 2,083 m where the ice thickness is about 3,700 m, the deformation and strain rate profile can be expected to be relatively simple (cf. Lorius et al. 1985). The main contribution to the errors in the dating come from errors in the measurements of the present accumulation rate and its past variations over the period (cf. Lorius et al., this volume). The measurements of the present accumulation rate from a variety of techniques, together with the past variations from the ^{10}Be and temperature changes, suggest the present accuracy of the dating from modeling is better than $\pm$ 10%, or 15 ka in 150 ka near the bottom of the core.

Annual layers counted from seasonal variations of various parameters. Closely spaced measurements along an ice core often reveal clear annual variations, for example, stable isotopes ($\delta^{18}O$, δD), ice electrical conductivity, dust, microparticles, chemical composition, crystals, physical properties, and stratigraphy (cf. Hammer, this volume). Where these annual layers are clear and distinct they provide the most accurate means of establishing the ice chronology over the depth of the core. At large depths the annual layers become more closely spaced and the annual variations of the parameters can become indiscernible. This often means that the old ice-age ice in the deeper layers cannot be dated by this method. A major aim of future research is to determine how annual layers can be traced back further in time through the ice sheets to reach well into the ice age period, when conditions were generally quite different from those of the present time. The sites for which the annual layers may be expected to be traced furthest back in time tend to have the following general features: thick ice, sufficiently high accumulation rate, large annual variations at the surface, ice thickening along the flowline, and clear annual layers extending far upstream.

Dating using radioactive isotopes. Radioactive isotopes provide a powerful means of obtaining an independent absolute age estimate for the ice cores as described by Stauffer (this volume). The techniques involve a number of problems which are addressed further in the section on *Prospects for Possible New Dating Techniques*. These problems include uncertainty in the changes of the initial concentrations, errors in the detection techniques for very small samples, and effects of contamination. The precision and reliability of the dating can be improved by using a suite of different isotopes and appropriate ratios of isotopes as will be indicated in this later section.

While the detection errors and contamination risks can be lowered by more sophisticated techniques, the uncertainty in the initial concentration is a basic problem of radioactive dating.

Reference horizons. In many ice cores, prominent features are found in the continuous records of some parameters measured along the ice. These features are often associated with large-scale changes in the surface deposition of the region and so they can be used as reference horizons for a common age. In some cases these reference horizons may be global in character and can be used to establish synchronous reference points in different cores, even when the absolute ages are not well known. Such reference horizons may also occur in other records, such as those for deep-sea sediments, loess, lake sediments, bogs, etc. Examples of reference horizons include volcanic ash (tephra), dust, chemicals, isotopes, and for ice cores, of particular importance are the entrapped gases which have long atmospheric residence times.

Comparison with other records. A large number of records of a wide range of different environmental features varying over time are available for comparison with data from ice cores. These include sea sediments, land or lake sediments, till, raised beaches, and sea level records, etc. Many of these records have been independently dated by several different techniques, such as the use of a number of different radioactive isotopes. When clear features of an ice core record can be identified to correspond to dated events in other records, then these dates can be used to help control the ice core chronology.

Astronomical time scales and the Earth's orbital changes to the radiation regime. The variations in the distribution of solar radiation (outside the atmosphere) over the Earth due to orbital changes can be computed accurately back in time to over half a million years. (Berger 1978, 1984). The expected response of the Earth's climate (and other consequential effects) to these radiation changes can be computed to various degrees of confidence using complex atmosphere-ocean-ice sheet climate models. These computed responses, which can include changes in the ice sheets, sea level, surface temperature, seasonal variations of the sea ice, etc., can be used to compare with the results of measurements on the ice cores. The objective of ice core chronology should be to date accurately the ice cores independently, by a combination of direct techniques, like ice dynamics, annual layers, and radioactive dating. This then allows the comparison with other records, and the astronomical forced responses, to provide more insight into the complex climatic mechanisms.

In all cases for the ice core records, the primary reference scale should be the depth along the core. The use of any time scale should include a clear indication of the means to convert the record back to the original depth scale. Any users of time scales should be aware of the basis for the time scale and the potential range of errors.

Prospects for New Techniques in Determining Ice Core Chronologies

Neutron activation of tephra in ice. Banded tephra in ablation zones of ice sheets provide a promising new technique for dating old ice. Uranium series dating of the ice with tephra has been described by Fireman (1986) from a site in the Antarctic. Tephra-banded ice samples in the main Allan Hills meteorite site are dated by a uranium-series method. The ages at four locations are $(66 \pm 7) \times 10^3$, $(98 \pm 8) \times 10^3$, $(95 \pm 20) \times 10^3$, and $(295 \pm 60) \times 10^3$ yrs. The locations of the dated ice samples are consistent with the horizontal stratigraphy expected from the ice flow, with the older ice being closer to the margin of the land barrier. The tephra particles in the bands are mainly (95%) fine volcanic glass shards; their chemical compositions indicate that the shards originated from at least two different volcanic sources.

The tephra may become useful reference horizons if they can be clearly identified. The technique of instrumental neutron activation analysis (INAA) has also been used by Fireman. High-resolution, high-sensitivity germanium detectors with computerized readouts need to be utilized more extensively in neutron-activation studies of ice cores. With 5 to 30 minute irradiations in the M.I.T. Nuclear Reactor and 30 minute counting times, the following elements were determined in 10 g samples of Allan Hills tephra-banded ice: Na, Mg, Al, Cl, K, Ca, Sc, Ti, V, Cr, Mn, Fe, Co, Zn, Se, Br, Rb, Sr, Ba, La, Ce, Nd, Sm, Eu, Gd, Yb, Lu, Hf, Hg, and Th. With similar irradiation and counting times, the following elements were determined in

20 g samples of Allan Hills clear ice: Na, Al, Cl, Sc, Cr, Mn, Fe, Zn, Br, La, and Sm. Approximately 20 such samples could be analyzed in one day.

Thermoluminescence (T.L.). The application of thermoluminescence techniques in general to dating has been well described by Aitken (1985). The application to ice core dating must still be established. The technique requires quartz or feldspar dust particles in the ice. Ice samples are needed of a size to contain 1 g of dust, but the samples need to have been extracted without exposure to light. For other sources studied so far, age estimates in the 70 000 yr range, have been obtained with estimated errors of ± 10% (cf. Andrews et al. 1983). Dust up to a few microns becomes T.L. age-zeroed by sunlight in between 3–8 hrs of exposure. The use of T.L. is recommended as an additional new method for dating the glacial age part of the Greenland ice cores. This may be one of the most ideal cases of using T.L.: fine surface dust, sufficient time in the atmosphere, composition, and size distribution change only little during the glacial period, rapid covering by snow, moving with ice in pristine surroundings. There are still some problems related to the suspension and distribution of the particles in the ice which need to be studied. The technique needs thorough testing in relation to other dating, and the effects of factors such as particle type and concentration need to be systematically investigated.

Further prospective new techniques including radioactive dating are discussed below in the section on *Prospects for Possible New Dating Techniques*.

COMPARISON WITH OTHER RECORDS

In the comparison of different records it is important to emphasize that caution is needed in associating variations which appear similar but for which complex phase differences may occur. Clear, well established reference horizons are needed to fix the relative time scales of the different records. Sea sediment records are now being examined for variations of indicators of sea surface and bottom temperatures as well as indicators of seawater isotope changes. These various records can then be compared with the ice core records and there are prospects that common reference horizons (such as dust and ^{10}Be) may be able to be used to fix the relative time scales.

Comparison of ice core records from Antarctica and Greenland provide a powerful technique for examining hemispheric differences in past environmental changes. Such comparisons need accurate relative time scales. The $\delta^{18}O$ of the oxygen gas in the air bubbles of the ice differs from the $\delta^{18}O$ of the ice. The atmospheric oxygen can be expected to be globally well mixed and uniformly distributed over the globe through time. The $\delta^{18}O$ of the air in the bubbles can therefore be used as a common stratigraphic marker for the Greenland and Antarctic ice core records. For locations where the atmospheric trace gases are well preserved, a similar feature applies to other atmospheric gases such as carbon dioxide (CO_2) and methane (CH_4). Detailed measurements of these gases in the ice may therefore also be used to control the relative chronologies of the different ice cores around the world.

The relatively rapid interaction between the atmospheric $\delta^{18}O$ in O_2 and the global oceans may also provide an additional means for control of the relative ice core and sea sediment record chronologies from the $\delta^{18}O$ of the oxygen in the entrapped air in the ice core bubbles.

CONNECTION BETWEEN ICE DEPTH CHRONOLOGY AND THE SURFACE CHRONOLOGY OF ICE MARGIN ABLATION ZONES

In the ablation zones near the margins of ice sheets, deep ice from the interior comes to the surface. The easy access to this old surface ice provides an effective means of obtaining large ice samples for more accurate dating, provided that the quality of the ice is appropriate and well preserved. If the continuous connection of the age horizons can be made to the deep interior ice, it may be possible to establish a more precise interior depth chronology as well. An example of a site where this type of connection has been examined is shown in Fig. 1 from a study in Terre Adelie, Antarctica, given by Lorius (1968).

In the ablation zone a multiple dating approach is recommended. The use of reference horizons should be combined with annual layer detection, either by coring or by direct measurement on the surface of features such as conductivity, dust, chemicals, or isotopes.

Common reference horizons may be found on the surface and at depth in the cores. These could include features of the glacial-interglacial transitions (e.g., $\delta^{18}O$), tephra, or dust layers dated by counting the annual variations. Other common reference horizons could include the atmospheric gases, ^{10}Be, and clearly identified prominent features of the various continuous records.

An ablation zone region is needed with a simple flow regime free from complications of folds or discontinuities. The common features exposed in the ablation zone can then be more readily dated, using radioactive isotope techniques, by taking large ice samples from near the surface. Once a chronology has been established along the ablation surface, and corresponding common features have been identified along a core at depth inland, it is possible to use numerical modeling of the ice dynamics to tie surface and depth stratigraphy together with a common chronology.

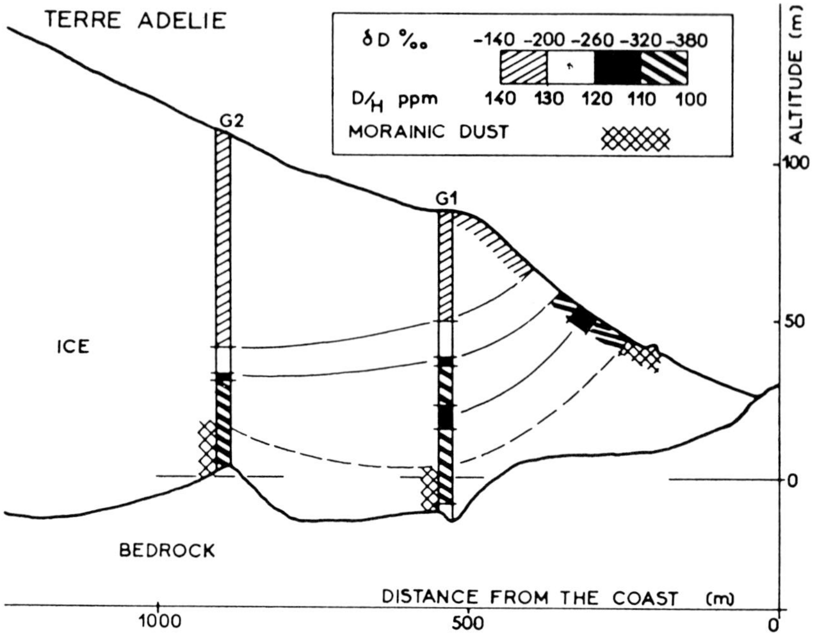

Fig. 1—The connection between the stratigraphies (from deuterium and dust) of the ice cores through the ice sheet and the surface in the ablation zone is illustrated by this study with ice core drilling (at G1 and G2) and sampling in the ablation zone near the coast of Terre Adelie, Antarctica, (from Lorius 1968).

If the bedrock in the region is rough, or if the ice flow is complicated, then the common features found in the surface and depth records can be used to help control the ice flow modeling. If the tephra is exposed at the surface, neutron activation (NA) or K/A dating can be used as described above. For very old ice (of 200 ka or more), the radioactive isotopes ratio $^{26}Al/^{10}Be$ (apparent half-life 1.38 Ma) may be used with sampling from large volumes of near surface ice. Surface and depth profiles of acidity may reveal common volcanic horizons, which could also be radioactively dated using large samples of surface ice.

Finally, there is a need to locate special sites in coastal ablation zones with large exposures of old ice, which provide optimum value for surface dating by using large volume samples to provide more accurate dates in order to control the deep interior ice stratigraphic chronology.

Radio echo sounding, for the detection and tracing of internal layer echos from deep within the ice, offers the potential for extending the depth age chronology throughout ice sheets. In many locations strong subsurface echos have been observed deep and widespread through the ice sheet (Millar

1981). The evidence suggests that these subsurface echos represent constant time horizons (or isochrones) within the ice sheet (Robin 1983).

In many cases these internal echos appear to be able to be traced over large horizontal scales. This means that once the time scales have been established at some locations, from deep ice cores, it may be possible to extend the chronology widely through the ice sheet by means of tracing the internal radio echos. Such layers may also be traced to the ice edge, where the ablation zone dates may be able to be used to control the internal layer dating. This clearly represents a valuable combination of techniques deserving further application.

RAPID CHANGES IN RECORDS OF ICE CORE FEATURES

One of the most important applications of a precise ice core chronology is to examine the possible rates of change of environmental variables from the record of their indicators in the ice cores. As an example, the Greenland deep ice cores from Camp Century and Dye 3 show large relatively rapid changes in various features through the record, especially within the part of the core of ice age origin, e.g., the $\delta^{18}O$ varies by about 3.5 to 5‰, which is approximately one-third of the total Wisconsinan-Holocene transition change (about 7 to 11‰) over a period of a few thousand years (Stauffer et al. 1984). Similar relatively large changes occur in the records for the CO_2 and other features, including dust, acidity, and chemical composition.

The observed rapid variations in the Greenland ice cores pose a dilemma, especially in respect to CO_2. The observed transitions appear to be unusually large and fast, especially when effects of diffusion are also taken into account. These large CO_2 changes have also not yet been observed in any of the Antarctic ice cores.

Atmospheric carbon dioxide should be relatively well mixed around the world, with long-term average concentrations only varying by a few ppm by volume. Therefore, general global changes in atmospheric concentrations of CO_2 should be detectable in ice cores from different locations around the world. This illustrates the importance of collecting and studying ice cores from a range of different locations over both hemispheres, to provide clear confirmation of past environmental changes of global significance.

In this way it may also be possible to use measurements of CH_4 to clarify the changes observed in the CO_2 and to provide chronological ties for clearer comparisons with other cores. Further discussions of the interpretation of the rapid changes in the Greenland ice are given by Hecht et al. (this volume).

A further feature of the ice zone, where the rapid variations occur, is that from laboratory studies of the ice core and borehole tilt measurements the ice flow properties appear to also vary in bands (Shoji and Langway

1985, 1988). This may give rise to a banded distortion of the ice layers with a corresponding distortion to the ice time-depth scale. It should be noted that this type of banding has not been found in the Vostok core. Staffelbach et al. (1988) have pointed out that the specially varying flow properties of ice could lead to enhancement of the distortion of the time depth scale, through a process known as "boudinage" formation, which involves stretching and necking of the soft and hard layers.

An analysis given at this workshop (by Weertman) indicated that for ice with a flow law index of $n=3$, necking of the layers can be expected in regions where the longitudinal (stretching) stress dominates over the horizontal shear stress. Where the shear stress dominates, the boudinage formation is suppressed because the effective power flow index is reduced to a value of $n \simeq 1$. In regions where horizontal shear stress dominates, differential shear in layers of spatially varying flow properties might also distort the time-depth scale. Drill sites at ice divides may be a poorer choice if boudinage formation (i.e., inhomogeneous flow of ice on a mesoscale) really occurs in ice layers.

For Dye 3, the large rapid variations occur from the depth of 1853 m (i.e., from 184 m above the bed) downwards, similar to those of Camp Century through the ice age period. A dirty layer horizon found in the Dye 3 core, 87 m from the bottom, indicates that there should be caution in the interpretation of any age scale through the basal layers of the ice.

Seasonal features seem to exist at the depths where the rapid changes are encountered. It is recommended that annual layer dating be attempted on some of these features, with special attention to the section where the rapid changes take place. The influence of possible boudinage on the age profile should be considered and the maximum error on the relative dating should be estimated.

The complication to the flow and age scales for regions near the bed implies that in order to clarify the past histories, new cores are needed in regions where the older ice exists at much greater distance above the bed.

THE DETERMINATION OF ANNUAL LAYERS FROM MULTIPLE PARAMETER TECHNIQUES

For regions of sufficiently high accumulation rates, annual layers can usually be clearly detected in many features of the ice core, e.g., $\delta^{18}O$, dust, chemical composition, conductivity, and particulates. As the age and depth increase, the annual layers become thinner and less discernible.

Multiple parameter measurements are commonly used to detect seasonal signals where sufficient annual accumulation exists. It is however, possible to infer seasonality even when the accumulation is extremely low. Very few data exist; however, from detailed multiparameter analysis in areas where the seasonal signal is conserved, it should be possible to infer what one can expect for low accumulation areas. The subject needs to be investigated further.

In some locations $\delta^{18}O$ shows annual variations, but microparticles and chemical composition show approximately semiannual variations. Irregular, approximately semiannual variations also occur in precipitation at some locations. A major complication is that with increasing depth and age there may be changes in regime from annual to semiannual variation (or vice versa), especially across major transitions like that of the glacial-interglacial changes.

The variations of dust in the ice may prove to be the most valuable and stable indicator of annual changes in old ice. By making use of new equipment, with special fine probes to carry out high resolution profiling along the core, it may be possible to extend the detection of annual layers back much further than has been possible so far. An example of fine band layering within the deep glacial age of the Dye 3 core is shown in Fig. 2.

Even if annual cycles are not clearly detectable in an ice core, it would still be useful to obtain indirect indicators of changes in seasonality. At different times through the astronomical cycles, the relative amplitude of the annual cycles change with the phase of the perihelion. Any indicator of seasonality change would therefore be useful for the clarification of the chronology, as well as for the interpretation of the physical processes involved.

DATING BY RADIOACTIVE ISOTOPES

Radioactive dating is most valuable because it provides an independent absolute check on dates derived from other techniques. A range of different isotopes are available, with different half-lives, which prove suitable for ice of different ages, as given by Stauffer (this volume). This section addresses the problems with radioactive dating and the use of a suite of isotopes, and ratios of isotopes, with some new techniques to provide greater accuracy to the ice core chronology.

For radioactive dating the present and initial concentration of an isotope have to be known. The present concentration can be measured by decay counting, accelerator mass spectroscopy (AMS), or resonance ionization mass spectrosocopy (RIMS). The initial concentration has to be estimated based on the experience with fresh snow and ice. The estimate of the initial concentration is especially difficult for radioisotopes which are attached to aerosols, since their concentration shows large variations with time and geographical location.

For example, ^{10}Be shows variations in the range of a factor of eight. The variations of ^{36}Cl are in the same order. It was hoped that the ratio $^{36}Cl/$

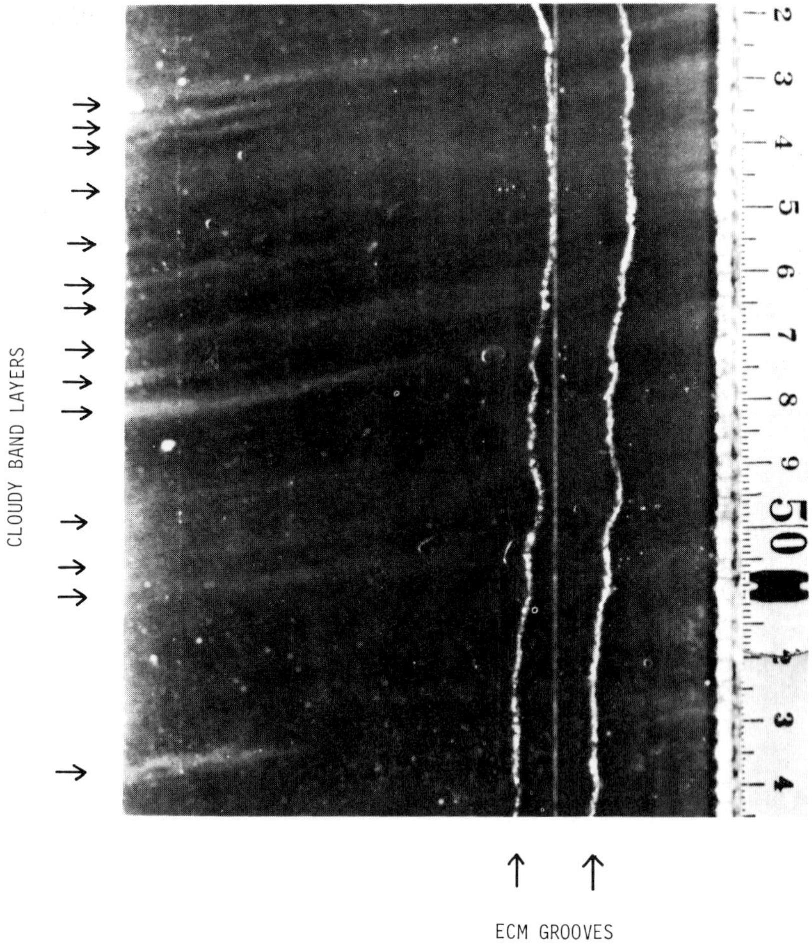

Fig. 2—The photo clearly shows the cloudy band layering slightly inclined from the horizontal plane of the ice sheet. This vertical section sample was obtained from a depth of 1814 m at Dye 3, Greenland in 1981. Multiple parameter measurements revealed that the cloudy band location coincides with a lower (near-zero) level in surface conductivity (ECM) and a higher dust content. Scratch lines showing on the surface of the vertical section are trace grooves made by ECM measurements. Photograph by Hitoshi Shoji.

^{10}Be, which decreases with an apparent half-life of approximately 370 000 yrs, would show much less variation. The fluctuations in present annual snow samples are, however, in the range of a factor of 2 to 5 and they do not become significantly smaller if instead of annual samples, larger samples representing several years, are analyzed. It is possible that the peculiar geochemical behavior of Cl is responsible for these fluctuations and it is expected that the $^{26}Al/^{10}Be$ ratio (apparent half-life 1.38 Ma) shows less fluctuation and may therefore be better suited for dating.

An excellent isotope for dating old ice is ^{81}Kr with a half-life of 213 000 yrs. Its concentration in the atmosphere and in the bubbles of ice is so small that the measurement of its concentration is difficult. With the RIMS technique the ^{81}Kr concentration ($^{81}Kr/Kr$) in total Kr, extracted from 20 kg of ice, can be measured with an accuracy of 10%, corresponding to an age uncertainty of about 30 000 yrs.

With ^{14}C, dating of ice in the range of 5000–20 000 yrs is most useful and might be possible from 1000–50 000 yrs, by using large volume ice samples from locations such as surface ablation zones. ^{14}C is therefore especially suited to date ice samples from the late part of the last glaciation and the transition from the glaciation to the Holocene. At present, for a single dating over a smaller range, about 10 kg of ice is needed. The accuracy of such a single dating is about 600 yrs. Ages obtained by ^{14}C dating, compared with ages obtained by annual layer counting, are on the average about 500 years too young. The *in situ* production of ^{14}C in the snow layer could be responsible for the ages being too young.

Dating by radioisotopes cannot compare in accuracy with annual layer counting, when annual layers are present. A combined dating, however, would guarantee that there are no large mistakes in annual layer counting, e.g., caused by missing out a large series of missing layers or the apparently annual layers being different from one year.

In many sites the clear detection of annual layers is difficult or even impossible. In such cases the ^{14}C dating method is especially useful. Furthermore, it allows direct comparison with other records which have time scales based on ^{14}C measurements.

Dating by radioisotopes is especially useful for dating old ice reaching the surface in marginal areas of ice sheets. For dating ice older than 1000 yrs ^{14}C, ^{81}Kr, and $^{26}Al/^{10}Be$ seem to be most promising.

PROSPECTS FOR POSSIBLE NEW DATING TECHNIQUES

With new technological developments, over a range of different analytical techniques, it is becoming possible to measure lower concentrations of radioactive isotopes which allows trends for greater accuracy and the use of smaller samples. This may make possible the application of a number of new techniques to ice dating. For usual ice cores, a sample size on the order of 10 kg may be practical to a limited extent. For surface samples from ablation zones, it may be practical to obtain sample sizes in the range of tons. For detailed high resolution measurements along cores, it would be most useful to have sample sizes as small as 1 g of ice.

Some of the new techniques which may be of use to ice dating have been described by Stauffer (this volume) and Finkel (presentation at this workshop). These include:

—electron spin resonance (ERS) for silty layers in ice,
—neutron activation (NA) of reference horizons, such as tephra,
—fission track dating of volcanic particles,
—proton induced X-Ray Emission (PIXE),
—continuous chemical analysis,
—amino acid racemization,
—thermoluminescene (TL) of clays or tephra from solid impurities in the ice,
—accelerator mass spectrometry (AMS), and
—resonance ionization mass spectrometry (RIMS).

Table 1 indicates the broad ranges of ages suitable for the different techniques and the sample sizes of ice required for the measurements. Further development and application of these new techniques should be regarded as a high priority for the future development of more accurate chronologies for ice cores and the dating of past environmental changes.

FURTHER RESEARCH FOR ICE CHRONOLOGY

Future progress in determining the physical processes involved in the past environmental changes recorded in glaciers and ice sheets will be greatly aided by the development of improved chronologies for the ice cores. The major future research activities highlighted from the group discussions include the following:

1. More ice cores are needed from strategic locations chosen to optimize the determination of the chronology of the record.
2. The new dating techniques of AMS and RIMS need to be further developed for routine application to the dating of ice, especially for the older ice.
3. Other promising new dating techniques should be developed and tested for ice dating over the full range of the ice age scales required. This should include ice sheet modeling integrated with data from other sources.
4. Greater use needs to be made of the high potential for the dating of old ice in the ablation zones of ice sheet margins where large samples can be obtained, and it should be possible to connect the surface dating to the chronology of the deep interior ice.
5. Further research is needed on the chronology of the apparently rapid

Table 1 Prospective new chronological techniques

amount of ice required ↓ / approximate time resolution →	1 yr	10^3 yr	10^5 yr
1 g	* Continuous chemical analysis (e.g., HNO_3, *dust*, H_2O_2) (A & R) * Neutron activation to characterize reference horizons (R)	* Fission track dating of tephra particles (A)	° Electron spin Resonance (ESR) for silty layers (A) * Uranium-series dating of banded tephra in ice (A)
1 kg		* ^{81}Kr ($t_{1/2}$=210,000 yr) (A) ° ^{41}Ca ($t_{1/2}$=100,000 yr) (A) ° Thermoluminescene (TL) of clays or tephra (A)	* K-Ar dating of volcanic tephra (A) * $^{10}Be/^{26}Al$ ($t_{1/2}=1.38\times10^6$ yr) (A)
1 ton		° Amino acid racemization, ~5 mg organic material required (A) * ^{14}C dating of biologic tissues (A)	

* likely to be implemented in the next few years
° speculative techniques which would require much further work
(A) Absolute chronology
(R) Relative chronology

changes of environmental indicators in the ice core records, including effects of possible inhomogeneous ice flow.

6. A major goal should be to extend the record of annual layers in the ice to greater ages, hopefully back to the initiation of the Last Interglacial period.
7. Clear global reference horizons need to be established for intercomparisons of ice cores and other records by use of features such as $\delta^{18}O$ of the entrapped air, methane, carbon dioxide, clearly finger-printed dust, and possibly ^{10}Be.
8. A major challenge of high priority is to establish an accurate ice core chronology (to ±1 or 2 ka) reaching back beyond the Last Interglacial period.

If at least one deep ice core can be accurately dated with a continuous record over the last complete ice age cycle then it would provide a unique control for the timing of environmental changes through history. By use of clear reference horizons in common with other records it should also then be possible to control the precise timing of the past environmental changes over this period around the globe.

Acknowledgements. The rapporteur wishes to acknowledge with gratitude the assistance and contributions from the other members of the Group in the construction of this paper. Particular thanks are also due to the Dahlem Konferenzen team for their assistance with the production of the paper and for the organization of the workshop.

REFERENCES

Aitken, M.J. 1985. Thermoluminescence dating. New York: Academic Press.

Andrews, J.T.; Shilts, W.W.; and Miller, G.H. 1983. Multiple deglaciations of the Hudson Bay Lowland, Canada, since deposition of the Missinaibi (last interglacial?). *Quatern. Res.* **19**: 18–37.

Berger, A. 1978. Long term variations of daily insolation and Quaternary climatic changes. *J. Atmos. Sci.* **35**: 2362–2367.

Berger, A. 1984. Accuracy and frequency stability of the Earth's orbital elements during the Quarternary. In: Milankovitch and Climate, eds. A. Berger et al., NATO ASI series C, vol. 126, Part 1, pp.3–39. Dordrecht: Reidel.

Budd, W.F., and Young, N.W. 1983. Techniques for the analysis of temperature depth profiles in ice sheets, and, Application of modelling techniques to measured profiles of temperatures and isotopes. In: The Climatic Record in Polar Ice Sheets, ed. G. de Q. Robin, pp.145–179. Cambridge: Cambridge Univ. Press.

Fireman, E.L. 1986. Uranium-series dating of ice at the Allan Hills site. *J. Geophys. Res.* **91 (B4)**: 539–334; **91 (B8)**: 8393.

Lorius, C. 1968. A physical and chemical study of the coastal ice sampled from a core drilling in Antarctica. IUGG General Assembly of Berne. *IAHS-AISH Publ.* **79**: 141–148.

Lorius, C.; Jouzel, J.; Ritz, C.; Merlivat, L.; Barkov, N.I.; Korotkevich, Ye.S.; and Kotlyakov, V.M. 1985. A 150 000 year climatic record from Antarctic ice. *Nature* **316 (6029)**: 591–596.

Millar, D.H.M. 1981 Radio-echo layering in polar ice sheets and past volcanic activity. *Nature* **292**: 441–443.

Robin, G. de Q. 1983. The Climatic Record in Polar Ice Sheets. Cambridge: Cambridge Univ. Press.

Shoji, H., and Langway, C.C., Jr. 1985. Mechanical properties of fresh ice core from Dye 3, Greenland. *Geophys. Monog.* **33**: 39–48. Washington, D.C.: Amer. Geophys. Union.

Shoji, H., and Langway, C.C., Jr. 1988. Flow law parameters of the Dye 3, Greenland deep ice core. *Ann. Glaciol.* **10**: 146–150.

Staffelbach, T.; Stauffer, B.; and Oeschger H. 1988. A detailed analysis of the rapid changes in ice-core parameters during the last ice age. *Ann. Glaciol.* **10**: 167–170.

Stauffer, B.; Hofer H.; Oeschger H.; Schwander, J.; and Siegenthaler, U. 1984. Atmospheric CO_2 concentration during the last glaciation. *Ann. Glaciol.* **5**: 160–164.

The Environmental Record in Glaciers and Ice Sheets
eds. H. Oeschger and C.C. Langway, Jr., pp. 193–205
John Wiley & Sons Limited

Temporal Variations of Trace Gases in Ice Cores

M.A.K. Khalil and R.A. Rasmussen

Institute of Atmospheric Sciences
Oregon Graduate Center
Beaverton, OR 97006, U.S.A.

Abstract. Analyses of air bubbles embedded in polar ice reveal the composition of the preindustrial and ancient atmospheres. So far extensive measurements of carbon dioxide, methane, nitrous oxide, and some chlorocarbons have been made on ice cores from both polar regions. The results provide a remarkable record of the magnitude and timing of human influences on the global cycles of these gases. Except for the chlorocarbons, for which there is no evidence of any substantial preindustrial concentrations, the other gases (CO_2, CH_4, and N_2O) started increasing only during the last 200 years as a result of the growing population and increasing needs for energy and food. The increase of N_2O probably started only a few decades ago. The record shows that CO_2 concentrations were about 280 ppmv 200 yrs ago while methane and nitrous oxide concentrations were about 700 ppbv and 285 ppbv, respectively. Now there is 25% more CO_2 in the atmosphere, 8% more N_2O, and there is more than twice as much methane in the atmosphere. At present longer records are available for CO_2 and CH_4 which show large natural variations during glacial and interglacial periods.

INTRODUCTION

During the last decade it has been clearly demonstrated that a large number of man-made trace gases are increasing rapidly in the Earth's atmosphere. Moreover, CO_2 and trace gases such as methane, nitrous oxide and carbon monoxide, which have large natural sources, are also increasing because of increasing anthropogenic contributions to their global cycles. The continued buildup of these gases in the atmosphere eventually will change the Earth's climate and may deplete stratospheric ozone thus affecting human health and global habitability.

Analyses of polar ice cores are perhaps the only means to determine the concentrations of trace gases in the preindustrial and ancient atmospheres. Shallow ice cores provide data on preindustrial levels of CO_2 and trace gases, spanning the last two or three centuries, and show the influence of the recent explosive growth of human population and industrial and agricultural activities. Deep cores have provided data on CO_2 levels extending back to some 160 000 years, showing the connection between CO_2 and long-term natural climatic variations, such as periods of glaciation and warming. Here we review the salient results of the research on trace gases in bubbles of polar ice that represent a direct record of atmospheric concentrations centuries ago. The present data consist of a long record of CO_2 and shorter records of methane (CH_4), nitrous oxide (N_2O), and some chlorocarbons (CCl_3F, CCl_2F_2, CH_3CCl_3).

CARBON DIOXIDE

The increase of carbon dioxide and its potential for changing the Earth's climate has been a focus of research for a century. Efforts to understand the global cycle of CO_2 have intensified over the last 30 years as atmospheric measurements have been taken. Carbon dioxide is essential for the growth of plants and contributes to the natural atmospheric greenhouse effect. Over the last 30 years, CO_2 has increased at about 1 ppmv/yr (0.3%/yr) so that the present concentration is about 350 ppmv (Keeling et al. 1976; GMCC 1986). Figure 1 shows the concentrations of CO_2 over the last 200–300 yr (from Friedli et al. 1986; Siegenthaler and Oeschger 1987).

The analyses of various ice cores show that the preindustrial concentration of CO_2 was about 280 ppmv (see, for example, Delmas et al. 1980; Neftel et al. 1985; Pearman et al. 1986). The increasing use of fossil fuels, which are the most significant direct anthropogenic source of CO_2, can explain the increase being observed at present. However, other processes that can contribute to an increase of CO_2 include the effects of deforestation and changes in land use. Recently Siegenthaler and Oeschger (1987) have used extensive measurements of CO_2 in ice cores and their isotopic abundances to estimate that during the last 200 years there has been a substantial contribution to the global CO_2 cycle from biogenic (nonfossil fuel) sources probably related to deforestation or other changes of land use. The main features of the record are low concentrations of about 280 ppmv in the 18th century and a gradual increase to about 293 ppmv at the beginning of the 20th century. Neither the increase in the 19th century nor the total change since the 18th century can be explained by the fossil fuel source alone. While only a small fraction of the fossil fuel emissions took place before the 20th century, they estimate that about half the nonfossil emissions occurred before 1900, but over the last 25 yrs the nonfossil source may have been much smaller, amounting to between 0% and 25% ($\leq$ 1 Gt/yr) of the average annual production from combustion of fossil fuels (about 4 Gt/yr). According to these estimates fossil fuels dominate the present anthropogenic

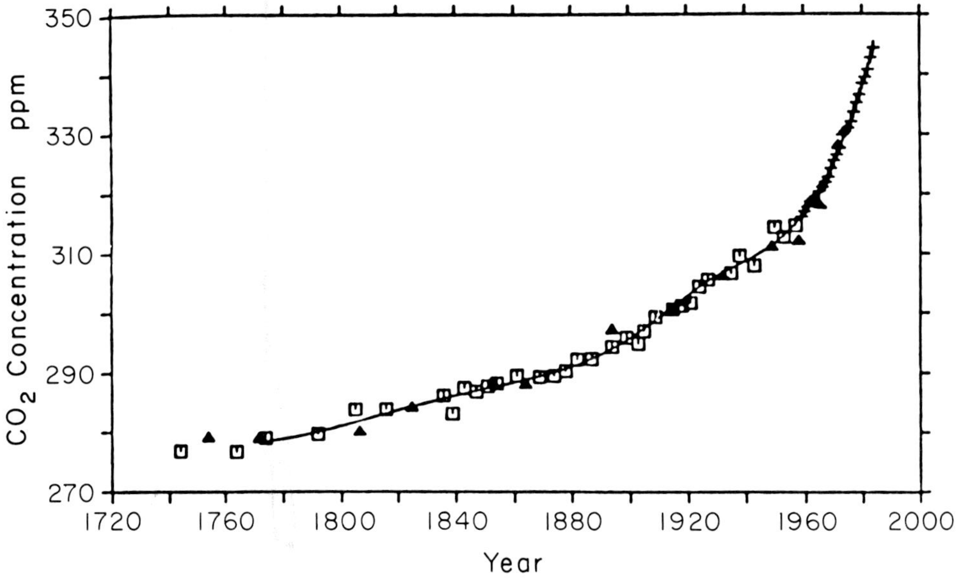

Fig. 1—The atmospheric concentration of CO_2 obtained from analyses of air bubbles in a polar ice core from Siple station in Antarctica. The later part of the record is from atmospheric measurements taken by Keeling. (The figure is reproduced from the papers of Friedl et al. 1986 and Siegenthaler and Oeschger 1987.)

emissions and are the primary reason for the continuing increase of CO_2 (see also, Broecker et al. 1979). Other estimates put the present nonfossil source at about 2 Gt/yr, but uncertainties are very large (Clark 1982).

A deep (2083 m) ice core was drilled by the Russian Antarctic expeditions at Vostok in East Antarctica. Barnola et al. (1987) have constructed a 160 000 year record of CO_2 from the analyses of the Vostok ice core as shown in Fig. 2. Over these much longer time scales the concentration of CO_2 has undergone large variations with low concentrations of about 190–200 ppmv during glacial periods and around 260–280 ppmv during interglacial epochs. Moreover, there appears to be a 20 000 yr cycle which may correspond to the precession of the Earth's orbit.

Explanations for the fluctuations of CO_2 with glacial cycles are generally based on changes of biological productivity in the world's oceans (see Sundquist and Broecker 1985).

METHANE

During the last two decades methane has been increasing at about 1% per year or about 16 ppbv/yr (Rasmussen and Khalil 1981; Blake et al. 1982;

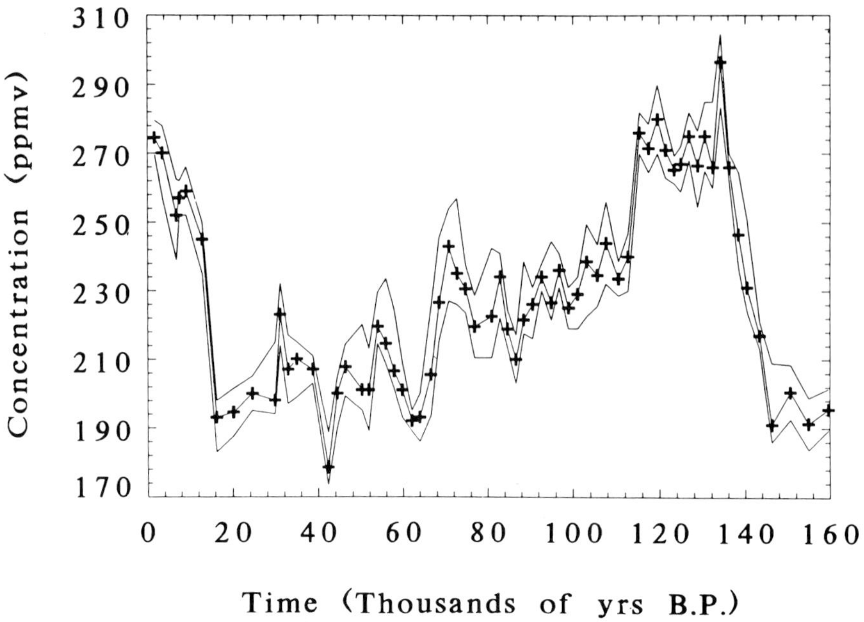

Fig. 2—The long-term variation of atmospheric CO_2 obtained from analysis of the Vostok core. (The figure is adapted from the paper of Barnola et al. 1987.)

Khalil and Rasmussen 1985). The increase began some 100–200 years ago and has slowly escalated to the present rate. Before that time concentrations of methane appear to have remained constant for several thousand years. Now there is more than twice as much methane in the atmosphere compared to the levels 100 years ago and earlier. The average concentration of methane in the troposphere is about 1670 ppbv (1987: Rasmussen and Khalil, unpublished data); it was only about 700 ppbv 200 years ago (see Khalil and Rasmussen 1987). On even longer time scales, spanning the last 10000 to 100000 years, Stauffer et al. (1988) have shown that there was only about 350 ppbv of methane left in the atmosphere during the last glaciation.

These conclusions are based on three types of data currently available. During the last decade intensive systematic measurements of methane have been taken at various locations spanning latitudes from the Arctic to the Antarctic. These measurements clearly show the increasing concentrations. Before the late 1970s data were available only from sporadic measurements reported by various scientists. These measurements tend to be scattered reflecting the uncertainties or disagreements on the absolute concentration of methane. Still these measurements extend the atmospheric record back to the early 1960s and show an increasing trend comparable to the present rate of increase. Work on determining concentrations of methane and other trace gases from spectroscopic records also confirms the rate of increase to be about 1%/yr in recent times (Rinsland et al. 1985). Ice core measurements are at present the exclusive source of information on the global concentrations and trends of methane before the middle of this century. The ice core record is quite intensive and complete up to about 1000 yrs B.P. For earlier times, up to about 3000 yrs B.P., there are larger gaps of up to several hundred years. For even earlier times, recent measurements span the period between 3000 years and 100 000 years and cover the time of the last glaciation (see Khalil and Rasmussen 1987 and references; Robbins et al. 1973; Khalil and Rasmussen 1982; Craig and Chou 1982; Stauffer et al. 1985; Pearman et al. 1986. Stauffer et al. 1988). All the available measurements, spanning the last 100 000 years are shown in Fig. 3.

A more detailed record is shown in Fig. 4, taken from our recent paper (Khalil and Rasmussen 1987). It shows the concentrations of methane in

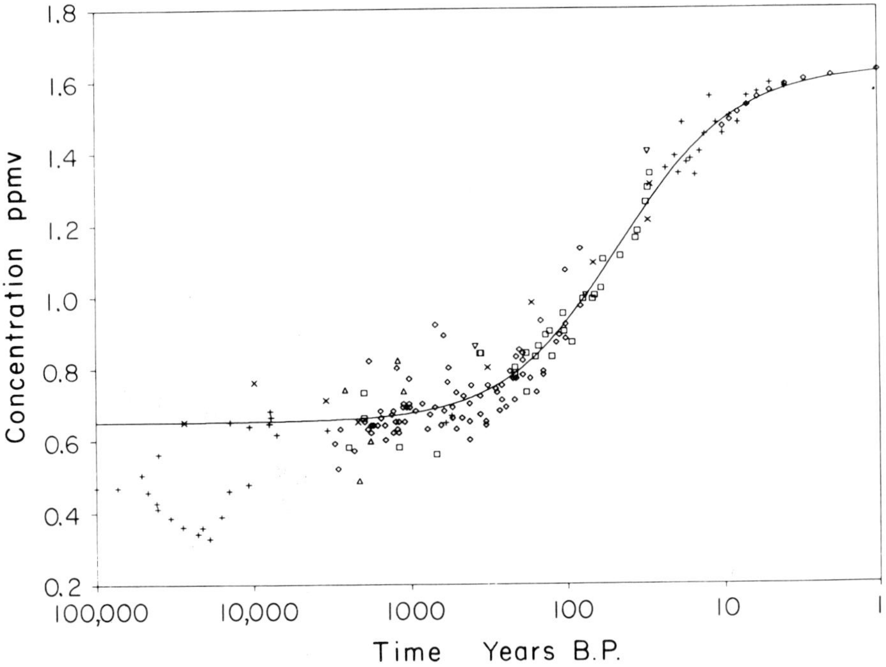

Fig. 3—Atmospheric methane over the last 100 000 years. The solid line is a plot of the equation: $C(ppbv)=[6.7\ T^{1.2} + 1675]/[0.0103\ T^{1.2} + 1]$, where T is the time B.P. Recent data show that concentrations of methane may have dipped down to 350 ppbv during the last glaciation.

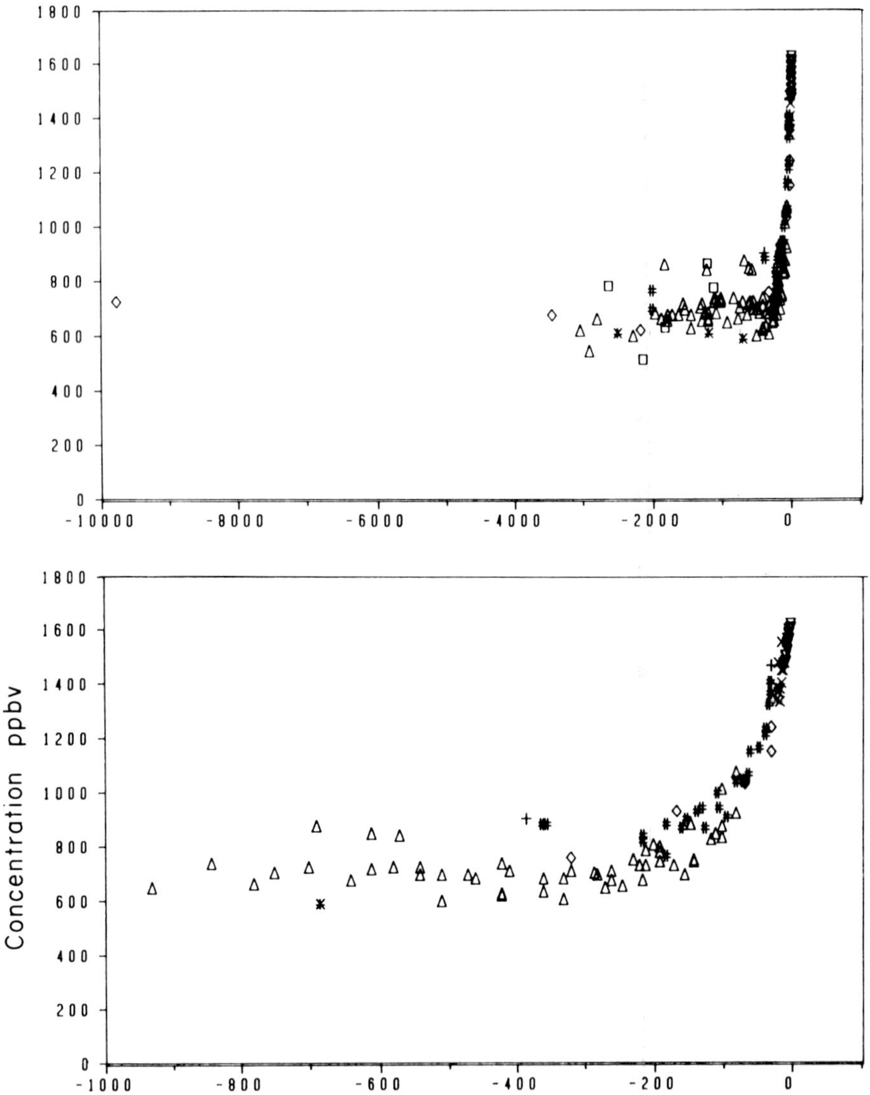

Fig. 4—Atmospheric methane over the last 10 000 years. Successive panels span the last 1000 years and the last 100 years (from Khalil and Rasmussen 1987).

ice cores and in the atmosphere over the last 10 000 yrs, 1000 yrs, and 100 yrs. The ice core measurements are from various groups and show excellent agreement wherever there are overlaps. The measurements of different groups tend to fill gaps left behind by earlier studies, so in additon to overlaps each data set contains unique information about the long-term trends of methane.

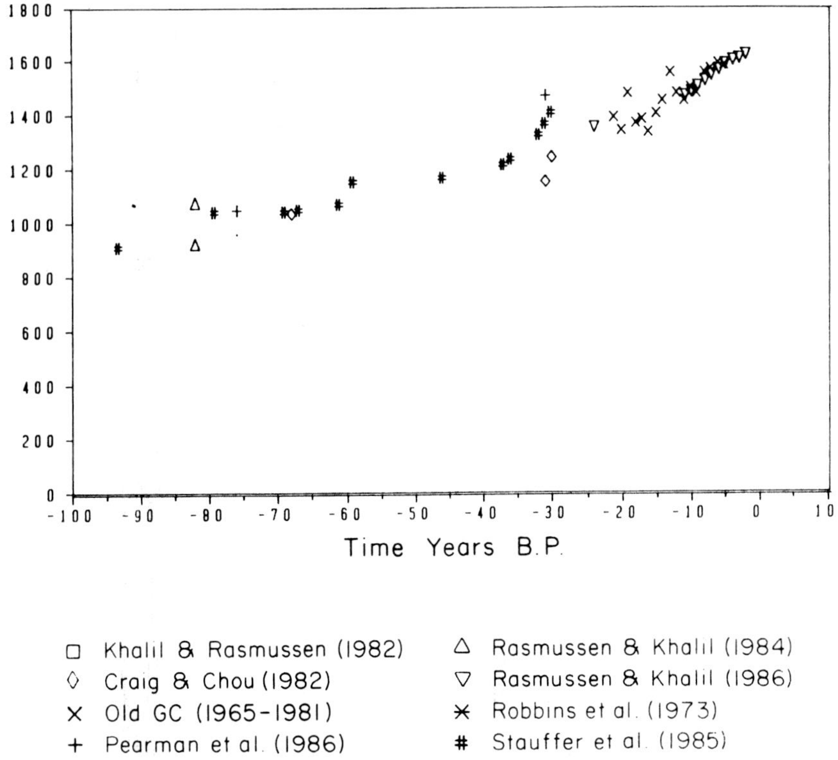

Ice core measurements also show a small but significant difference of concentration between the two hemispheres. Larger concentrations are found in the Greenland ice cores compared to the Antarctic cores (Rasmussen and Khalil 1984). Such an asymmetry lends credibility to the accuracy of the measurements since it is expected from the asymmetrical distribution of land between the hemispheres from which most methane originates.

There has been considerable discussion on the probable causes of increasing methane. It is believed that methane is released to the atmosphere primarily from biogenic processes. However, some of these processes are directly affected by human activities. For instance, even the earliest budgets of methane suggested that cattle and rice paddies were important global sources of methane. As a consequence of the growth of human population the number of cattle has increased as has the land used for growing rice. There are also industrial and urban sources, and methane from landfills that have increased over the years. The present global emissions are estimated to be about 500 ± 100 Tg/yr. In all, it appears that now about half of the annual emissions of methane are from human activities. The other half are of natural origin, the most significant being those from the wetlands.

The observed increase of methane can be explained as resulting from increased sources if the anthropogenic sources are scaled to the growth of human population over the last 200 years (Khalil and Rasmussen 1985). A second explanation is based on more subtle mechanisms that cause increased emissions from primarily natural sources. For instance, more methane could be released into the atmosphere by increased turnover of decaying biomass accompanying the increase of CO_2 and deforestation (Guthrie 1986), or warming trends may release more methane from natural ecosystems such as melting permafrost in the tundra. However, there is yet another, entirely different possible explanation. Methane is removed from the atmosphere primarily by reacting with OH radicals. If OH concentrations decline, the concentration of methane would increase. In the (nonurban) atmosphere, OH radicals are produced by sunlight, ozone, and water vapor and are removed from the atmosphere primarily by reacting with CO and methane. Increasing anthropogenic emissions of CO from combustion processes and increased methane could reduce OH concentrations. In fact, carbon monoxide is also a by-product of the oxidation of methane (and other hydrocarbons). On the other hand production of ozone by more CO, CH_4 and NO_x can tend to increase OH. Therefore the cycles of CH_4, CO, OH, and O_3 are inextricably tied together but each component is subject to external infuences. The observed increase of methane could be due entirely to a reduction in the oxidizing capacity of the atmosphere caused by a net depletion of OH. The change in the concentration of methane in the atmosphere over the last century is so large that this possibility is unlikely but it can be tested by analyses of other trace gases in polar ice cores. Gases such as methylchloride and the light hydrocarbons may be used as tracers of past OH concentrations. At present what appears to be more likely is that much of the increase of methane over the last 200 years has occurred because of increasing emissions linked to agricultural processes that have increased with the human population. A smaller but significant part of the increase may have been caused by a reduction in OH concentrations. A calculation based on emissions from anthropogenic sources increasing proportional to the human population and a 20% decrease of OH over the last century yields an explanation for the observed time series shown in Fig. 4. Fig. 5 shows the rates of increase both measured and calculated with the model mentioned above (from Khalil and Rasmussen 1987).

NITROUS OXIDE

The average atmospheric concentration of nitrous oxide is increasing at a relatively slow rate of about 0.3% per yr or 0.8 ppbv/yr (Weiss 1981; Khalil and Rasmussen 1983). Ice core measurements show that the preindustrial

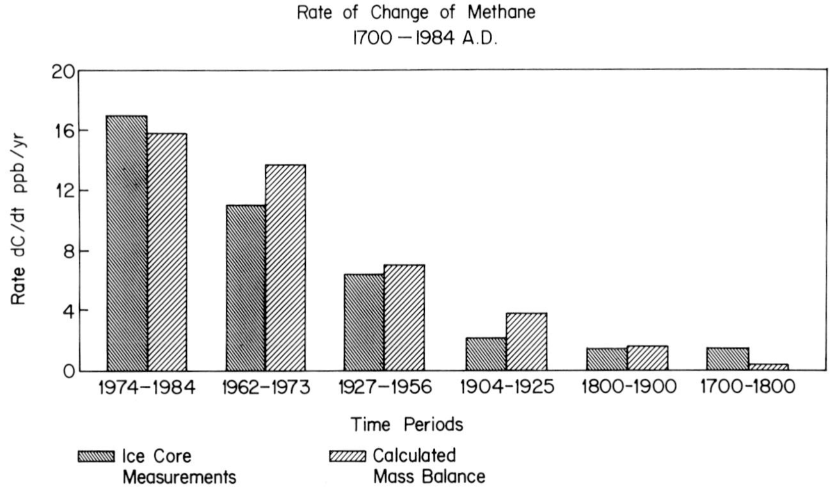

Fig. 5—The rates of increase of methane over the last 200 years. The measured rates correspond to the composite data from various sources as shown in Figs. 3 and 4. The calculated rates are obtained from a mass balance model in which we leave natural emissions constant and scale anthropogenic sources with the growth of human population. A 20%/century reduction of OH is also included.

concentrations were about 285 ppbv compared to present levels of about 307 ppbv (1984). The net increase of N_2O over the last century has been only about 8% (Khalil and Rasmussen 1988).

The present global cycle of nitrous oxide is not as well understood as the cycles of CH_4 and CO_2; however, it appears that oceans and soils are the main natural sources, and combustion processes, including the burning of fossil fuels, are the main anthropogenic sources although the growing use of nitrogren fertilizers in recent years may become a significant anthropogenic source in the future. The present global emissions are about 20–30 Tg/yr. Nitrous oxide is removed primarily by photochemical processes in the stratosphere giving it an average atmospheric lifetime of about a hundred years.

The observed increase can be explained by the increasing use of fossil fuels and especially by the increasing number of coal fired power plants. It appears that the rapid increase may have begun as recently as 30 years ago. Fig. 6 shows the observed increase and its explanation, using a global mass balance model, with an increasing source from use of fossil fuels. The model and the assumptions regarding the changing sources are discussed in more detail by Khalil and Rasmussen (1988).

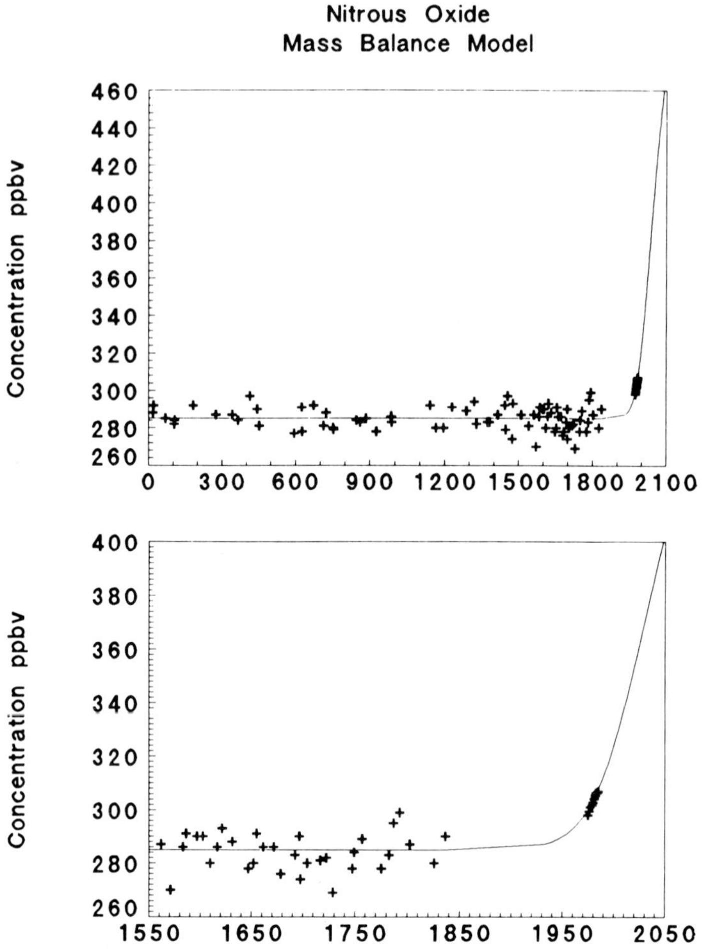

Fig. 6—Atmospheric nitrous oxide over the last 3000 years (from Khalil and Rasmussen 1987).

CHLOROCARBONS: CH_3CCl_3, CCl_3F, CCl_2F_2

Chlorofluorocarbons, particularly CCl_3F (F-11) and CCl_2F_2 (F-12), are considered harmful to the stratospheric ozone layer. It has been conjectured that there may be natural sources of these fluorocarbons from volcanoes and of methylchloroform from the oceans. Measurements in polar ice cores show that the concentrations of F-11, F-12, and CH_3CCl_3 are below detectable limits of around 5 pptv. At other times very high concentrations are measured which we believe are caused by contamination of ice cores while in storage in refrigerators that use these fluorocarbons as refrigerants.

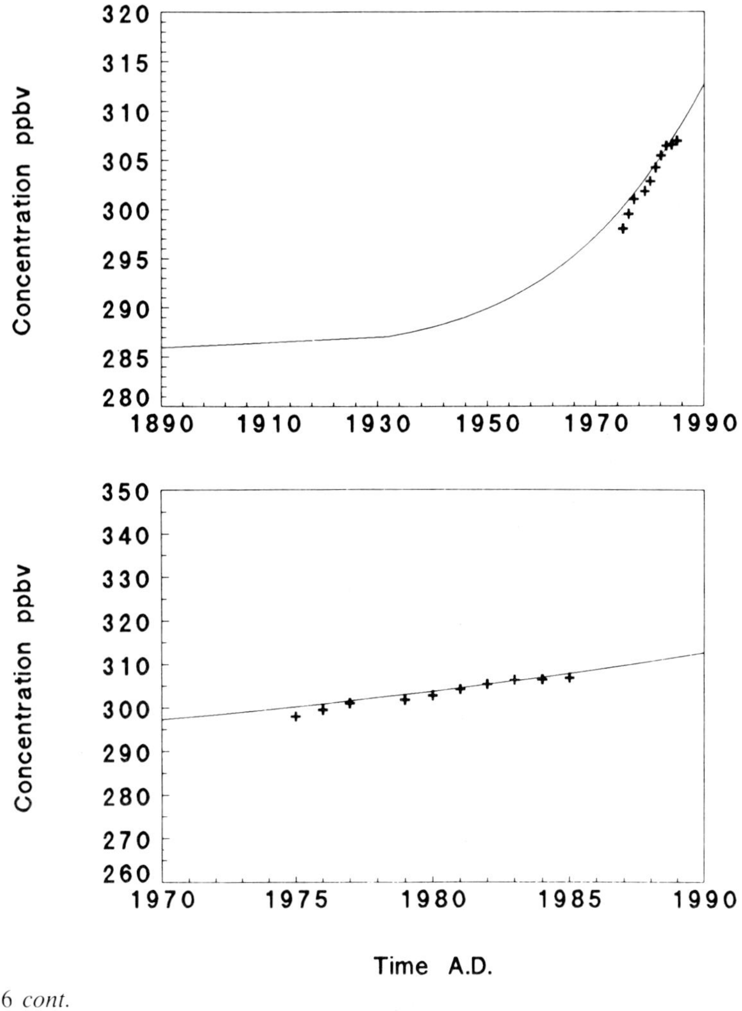

Fig. 6 *cont.*

Moreover, there is no clearly demonstrated natural source of these compounds that would support their existence in the preindustrial atmosphere while the present concentrations and trends can be fully explained by the industrial emissions. The possibility that small amounts of these chlorocarbons were in the preindustrial atmosphere cannot be completely ruled out based on ice core analyses and present budgets.

PRESENT METHODS AND FUTURE PROSPECTS

The simplest method for measuring trace gas concentrations in bubbles of polar ice is to melt a sample in a sealed container and measure the concentration in the head space. N_2 and O_2 are measured simultaneously to obtain a mixing ratio of the trace gas in the bubbles. A simple mass balance between the amount in the headspace and the melt water gives the mixing ratio of the gas in the bubbles. Our measurements of methane, nitrous oxide, and chlorocarbons were made with this method. However, only a few gases of current interest can be measured this way. Measurements of other gases require dry extraction so that only the air from the bubbles is removed for analysis so as to avoid the influence of the gases dissolved in the ice or otherwise incorporated in the ice lattice.

In our opinion, promising lines of research include the following:

1. There is evidence that the methods for extracting the ice cores and the air from the bubbles produce artifacts in the measured concentrations. An evaluation of the wet and various dry extraction procedures is necessary to be certain about the accuracy of the measured concentrations in representing atmospheric levels.
2. Measurements of nitrous oxide and methane extending back to the last 160 000 years can now be done relatively easily on the Vostok ice core. Such a record would provide valuable information on the natural cycles of these gases and the links between terrestrial biological activity and the climate.
3. Measurements of CO, nonmethane hydrocarbons, and CH_3Cl are needed to provide information on the possible changes in average OH concentrations. While desirable, the measurements of CO and some important hydrocarbons may not be possible with the existing techniques.
4. A careful analysis of halocarbons is needed to establish the lack or presence of any small natural sources.
5. An analysis of OCS may reveal natural fluctuations of the nonvolcanic component of the stratospheric sulfate layer.
6. Although considerable progress has been made in recent years, experiments are needed to determine the time for the closing of pores as firn becomes ice. This provides information on the difference of ages betweeen the ice and the embedded air bubbles. The firn-ice transition time is of particular importance for more recent ice (see Schwander et al. 1988).

The question of the age of the air compared to the age of ice also includes other variables that are not fully evaluated. For instance, when bubbles close off they are perhaps 70 m below the surface. Whether the air being sealed off represents the atmosphere of the time or a mixture of air from earlier times depends on the rapidity of mixing of air in the firn layer.

We have reviewed the considerable progress that has been made on the measurements and interpretation of long-term variations of CO_2, CH_4, N_2O, and a few chlorocarbons in the atmosphere based on the analyses of polar ice cores. These studies have provided firmer estimates of natural emissions and the role of human activities in changing the trace gas composition of the atmosphere. New extraction methods and sensitive measurement techniques that are now becoming available make it possible to analyze many more trace gases in the ice cores.

Acknowledgements. This work was supported in part by grants from the National Science Foundation (ATM 8414020, DPP 8207470), the Department of Energy (DE-FG06-85ER6031), and the National Aeronautics and Space Administration (NAGW-280). Additional support was provided by the resources of Biospherics Research Corporation and the Andarz Company.

REFERENCES

Barnola, J.M.; Raynaud, D.; Korotkevich, Y.S.; and Lorius, C. 1987. Vostok ice core provides 160 000-year record of atmospheric CO_2. *Nature* **329**: 408–414.

Blake, D.R.; Meyer, R.E.; Tyler, S.; Makide, YU.; Montague, D.C.; and Rowland, F.S. 1982. Global increase of atmospheric methane concentration between 1978 and 1980. *Geophys. Res. Lett.* **82**: 477–480.

Broecker, W.S.; Takahashi, T.; Simpson, H.J.; and Peng. T.-H. 1979. Fate of fossil fuel carbon dioxide and the global carbon budget. *Science* **206**: 409–418.

Clark, W.C., ed. 1982. Carbon Dioxide Review. New York: Oxford Univ. Press.

Craig, H., and Chou, C.C. 1982. Methane: record in polar ice cores. *Geophys. Res. Lett.* **9**: 1221–1224.

Delmas, R.J.; Ascencio, J.M.; and Legrand, M. 1980. Polar ice evidence that atmospheric CO_2 20 000 B.P. was 50% of present. *Nature* **284**: 155–157.

Friedli, H.; Lotscher, H.; Oeschger, H.; Siegenthaler, U.; and Stauffer, B. 1986. Ice core record of $^{13}C/^{12}C$ ratio of atmospheric CO_2 in the past two centuries. *Nature* **324**: 237–238.

GMCC. 1986. Geophysical Monitoring for Climatic Change, vol. 14, ed. R.C. Schnell. Boulder, CO: NOAA/GMCC/ARL.

Guthrie, P.D. 1986. Biological methanogenesis and the CO_2 greenhouse effect. *J. Geophys. Res.* **91**: 10847–10851.

Keeling, C.D.; Bacastow, R.B.; Bainbridge, A.E.; Ekdahl, C.A.; Guenther, P.R.; Waterman, L.S.; and Chin, J.F.S. 1976. Atmospheric carbon dioxide variations at Mauna Loa Observatory. *Tellus* **28**: 538–551.

Khalil, M.A.K., and Rasmussen, R.A. 1982. Secular trends of atmospheric methane. *Chemosphere* **11**: 877–883.

Khalil, M.A.K., and Rasmussen, R.A. 1983. Increase and seasonal cycles of nitrous oxide in the Earth's atmosphere. *Tellus* **35B**: 161–169.

Khalil, M.A.K., and Rasmussen, R.A. 1985. Causes of increasing methane: depletion of hydroxyl radicals and the rise of emissions. *Atmos. Envir.* **19**: 397–407.

Khalil, M.A.K., and Rasmussen, R.A. 1987. Atmospheric methane: trends over the last 10 000 years. *Atmos. Envir.* **21**: 2445–2452.

Khalil, M.A.K., and Rasmussen, R.A. 1988. Nitrous oxide: trends and global mass balance over the last 3,000 years. *Ann. Glaciol.* **10**: 73–79.

Neftel, A.; Moor, E.; Oeschger, H.; and Stauffer, B. 1985. Evidence from polar ice cores for the increase in atmospheric CO_2 in the past two centuries. *Nature* **315**: 45–47.

Pearman, G.I.; Ethridge, D.; deSilva, F.; and Fraser, P.J. 1986. Evidence of changing concentrations of CO_2, N_2O and CH_4 from air bubbles in antarctic ice. *Nature* **320**: 248–250.

Rasmussen, R.A., and Khalil, M.A.K. 1981. Increase in the concentration of atmospheric methane. *Atmos. Envir.* **15**: 883–886.

Rasmussen, R.A., and Khalil, M.A.K. 1984. Atmospheric methane in the recent and ancient atmospheres. *J. Geophys. Res.* **89**: 11599–11605.

Rinsland, C.P.; Levine, J.S.; and Miles, T. 1985. Concentration of methane in the troposphere deduced from 1951 solar spectra. *Nature* **318**: 245–249.

Robbins, R.C.; Cavanagh, L.A.; Silas, L.J.; and Robinson, E. 1973. Analysis of ancient atmospheres. *J. Geophys. Res.* **78**: 5341–5344.

Schwander, J.B.; Stauffer, B.; and Sigg, A. 1988. Air mxing in firn and the age of the air at pore close-off. *Ann. Glaciol.* **10**: 141–145.

Siegenthaler, U., and Oeschger, H. 1987. Biospheric CO_2 emissions during the past 200 years reconstructed by deconvolution of ice core data. *Tellus* **39B**: 140–154.

Stauffer, B.; Fischer, G.; Neftel, A.; and Oeschger, H.; 1985. Increase of atmospheric methane recorded in antarctic ice. *Science* **229**: 1386–1388.

Stauffer, B.; Lochbronner, E.; Oeschger, H.; and Schwander, J. 1988. Methane concentration in the glacial atmosphere was only half of the preindustrial Holocene. *Nature*, in press.

Sundquist, E.T., and Broecker, W.S., eds. 1985. The Carbon Cycle and Atmospheric CO_2: Natural Variations Archean to Present. Washington, D.C.: Amer. Geophys. Union.

Weiss, R.F. 1981. The temporal and spatial distribution of atmospheric nitrous oxide. *J. Geophys. Res.* **86**: 7185–7195.

The Environmental Record in Glaciers and Ice Sheets
eds. H. Oeschger and C.C. Langway, Jr., pp. 207–223
John Wiley & Sons Limited

Trace Metals and Organic Compounds in Ice Cores

D.A. Peel

British Antarctic Survey, Natural Environment Research Council, Cambridge, CB3 0ET, U.K.

Abstract. Polar ice sheets offer a unique opportunity to reconstruct historical records of large-scale air pollution. Progress has, however, been hampered because the concentrations of primary pollutants found in polar snow are the lowest detected on Earth. Recent improvements in technique have enabled an important advance in understanding the causes of variations in Pb content of polar ice through the past 155,000 years, hence establishing a firm reference against which to evaluate modern trends. Since the start of the industrial period a very marked human impact in Greenland has contrasted with a weak but significant impact in Antarctica. Few reliable data are available for the other heavy metals. For Cu and Zn it appears that any human impact on a global scale is extremely limited and although Cd appears substantially enriched, natural enrichment processes may prove responsible.

INTRODUCTION

The Environmental Record in Ice Cores

The permanently ice-covered regions of the world have preserved a unique and datable record of the past composition of the Earth's atmosphere. It is the purpose of this paper to examine the evidence in snow and ice for anthropogenic pollution in the primary atmospheric aerosol. The particular cases of sulfate and nitrate, which are mainly gas derived, are dealt with elsewhere (Clausen and Langway, this volume) and will not be considered further here. The primary aerosol particles have residence times ranging from a few days up to several weeks (exceptionally) in the lower troposphere, and up to 1–2 years in the stratosphere. By studying their distribution and temporal variations in remote areas, it is possible to gain insight into the processes and time scales of long-range transport of pollution. The majority of pollutants, especially those of a metallic or gaseous nature, are also involved in natural biogeochemical cycles. By comparing the composition of modern snow with that of ancient ice deposited before the industrial period, the human impact in modern times can be assessed directly.

Global Pollutants

Most extensively, ice-covered areas on Earth are far removed from major sources of industrial pollution and impact studies in these areas are concerned with large-scale, if not global-scale, air pollution. Two principal, mainly insoluble components of the atmospheric aerosol are involved: a carbonaceous fraction that consists mainly of sooty carbon and organics, and an inorganic fraction comprising a complex range of elements in various chemical forms. The carbonaceous fraction arises predominantly from the incomplete combustion of fossil fuels but also includes a wide range of trace organic pollutants that are resistant to chemical degradation during atmospheric transport. The inorganic component is dominated by fly ash and by the emissions from a variety of manufacturing processes, smelting, and mining operations. Most interest has focused on the toxic metals and metalloids, e.g., Pb, Cd, Hg, and As. These elements appear to be strongly enriched in the atmosphere and there is vigorous debate on the extent to which pollution may be responsible.

Geographic Considerations

Most studies have been carried out in the polar regions where persistently low temperatures have helped to preserve the most reliable stratigraphic records. Greenland and the Arctic ice fields, which lie relatively close to the industrialized areas of Europe, USSR and North America, receive pollution from air masses originating in mid-latitudes, especially in the period December to April when there is a more efficient S-N transport (Barrie 1986a). The Antarctic continent on the other hand is protected from the mid-latitude air masses by a persistent zonal circulation around the southern oceans. Any pollution reaching the continent will represent a hemispheric-scale influence.

Few data have been reported for pollution studies on glacier ice in mid latitudes; reflecting a more regional scale human impact, they could provide a useful complement to the polar studies (Wagenbach, this volume).

IDENTIFYING AND QUANTIFYING AN ANTHROPOGENIC IMPACT IN ICE CORES

Transfer Functions

In order to interpret data from snow and ice, in terms of variations in the composition of the atmosphere, it is essential to understand the complex processes by which aerosol is removed from the atmosphere. Junge (1977) suggests that rather simple air/snow relationships are likely to exist at remote, clean-air locations, although few reliable studies have sampled aerosol and snowfall simultaneously, in order to test the relationship directly. The most recent studies from both Greenland and Antarctica, albeit carried out over very restricted periods, suggest that different components of the aerosol are deposited with varying efficiency. For example, Ng and Patterson (1981) concluded that there was a preferential deposition of small Pb-rich anthropogenic aerosols. Davidson et al. (1985) found the inverse situation, with an order of magnitude reduction in the scavenging ratio for Pb compared with the crustal elements. If such a fractionation proves to be significant over the averaged conditions recorded in ice cores, then it could lead to a misinterpretation of the relative source strengths of the various components in the atmosphere. In the absence of reliable and representative data it is usually assumed that the concentrations in air and snowfall for all species follow the same linear relationship and that this has not changed significantly with time. This problem is considered in detail by Davidson (this volume).

Natural vs. Anthropogenic Sources

Several empirical approaches have been developed to identify the relative contributions of natural and anthropogenic emissions to the trace metal content of both precipitation and the atmospheric aerosol.

Enrichment factors. In remote areas it is expected that the Earth's crust and the sea will be the principal natural sources of the atmospheric aerosol. To obtain a first impression of the relative amount of a particular element that can be attributed to each source, it is common practice to calculate an enrichment factor (EF) for the element (X) relative to an unambiguous reference element (REF) from the source material:

$$[\mathrm{EF}] = \frac{[\mathrm{X}]_{\text{snow, aerosol}}}{[\mathrm{REF}]_{\text{snow, aerosol}}} \Bigg/ \frac{[\mathrm{X}]_{\text{marine, crust, volcanoes}}}{[\mathrm{REF}]_{\text{marine, crust, volcanoes}}}$$

Elements with EF significantly greater than 1 are called "enriched" and probably have another major source, which could be anthropogenic or an unrecognized natural source. The approach and its limitations have been discussed in detail by Rahn (1976). Variations on this approach are used extensively to interpret the present-day distribution of trace elements in polar ice.

Historical trends. The most direct and, in principle, simplest method to demonstrate human impact is to determine whether there is a historical trend in the composition of the ice cover that can be related to trends in anthropogenic emissions. This is the goal for most studies on the polar ice sheets although progress has been limited by the very low concentrations encountered.

Unequivocal anthropogenic tracers. Unambiguous evidence for human impact is obtained from studies of pollutant species which have no natural origin. Many manufactured organic compounds have no known natural occurrence. Species that are resistant to degradation in sunlight (and these are likely to be the persistent pollutants of major environmental concern such as the organochlorine compounds DDT and the PCBs) should provide valuable tracers which can be related both to the type of activity for which they were manufactured and to the geographical origin.

Isotope ratios. The stable isotope composition of the Pb (especially the $^{206}Pb/^{207}Pb$ ratio) in the atmospheric aerosol can be used as a tracer of atmospheric air masses (Maring et al. 1987). So far, the approach has not been applied to ice core impurities but offers potential in Greenland where concentrations of Pb in modern snow appear sufficient for successful analyses.

Elemental tracer system. A seven element (As, Se, Sb, Zn, In, noncrustal Mn, and noncrustal V) regional tracer system has been developed (Lowenthal and Rahn 1985) to determine the sources of pollution aerosol penetrating the Arctic. Characteristic signatures from areas of North America, Europe, and Asia can be followed successfully up to several thousand km downwind from the source. Such a procedure has great potential. The very limited data sets available for snow and ice preclude its application at present, although with the future development of multi-element techniques, such as ICP/MS, it could be applied to studies on the Greenland ice sheet and in other ice covered areas of the Arctic.

CRITERIA FOR DATA SELECTION

Because of the large distances normally involved in transport of pollutants to remote snowfields, concentrations in snow and ice are extremely low (< 1–200 pg/g). Consequently, many of the reported data have suffered from contamination problems and the reports often lack details of the sampling and analytical procedures, now known to be essential if the data are to be evaluated seriously.

Boutron (1986) has reviewed the procedural stages from sample collection through to laboratory subsampling and analysis that must be adopted if reliable measurements of lead and other heavy metals are to be achieved.

With appropriate changes to the composition of some of the materials that actually contact the samples, a similar analytical regime will in the future be required for measurement of other pollutant species.

It is clear that whatever precautions are taken to collect snow or ice cores cleanly, they will be more or less contaminated on the outside, and must be decontaminated before analysis. The efficiency of this process must be proved by measuring a profile of concentration from the surface towards the center of the sample (Fig. 1). Using such procedures, for Pb analysis at least, it has proved possible to work successfully with snowblocks, dry firn cores, and deep thermally drilled ice cores where drilling fluid has been used (Boutron et al. 1987).

TIME TRENDS OF LEAD IN GREENLAND AND ANTARCTICA

Pb is the only nongaseous pollutant species for which there are reliable, albeit incomplete, data extending from preindustrial periods to the present.

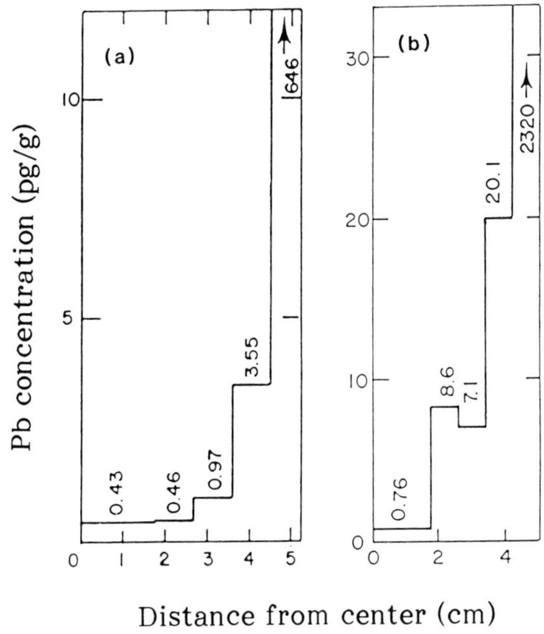

Fig. 1—Cross-core profiles of Pb concentration for two sections of the 905m Dome C Antarctic ice core which was thermally drilled in a dry hole: (*a*) from 452 m depth (12160 yr B.P.) showing clear plateau in interior of core (from Boutron and Patterson 1986), (*b*) from 173 m depth (3846 yr B.P.) showing penetration of contamination into the interior of the core (from Boutron et al. 1988).

During the past 20 years much greater effort has focused on improving the quality of this dataset than for any other species. Extremely refined procedures that have proved essential for successful analysis in remote areas will undoubtedly serve as a model for future investigations on other species.

Greenland

The principal time series available is that reported by Murozumi and others (1969) from an outstanding study which gave clear evidence for an increase in Pb concentration in snow and ice of around 200–fold during the past 2000 years, with an increase of 5–fold during the past 100 years (Fig 2). Despite intensive controversy (MARC 1985; Wolff and Peel 1985b; Boutron 1986) and several attempts by other groups to confirm this result, the data are still widely cited as convincing evidence for long-range transport of

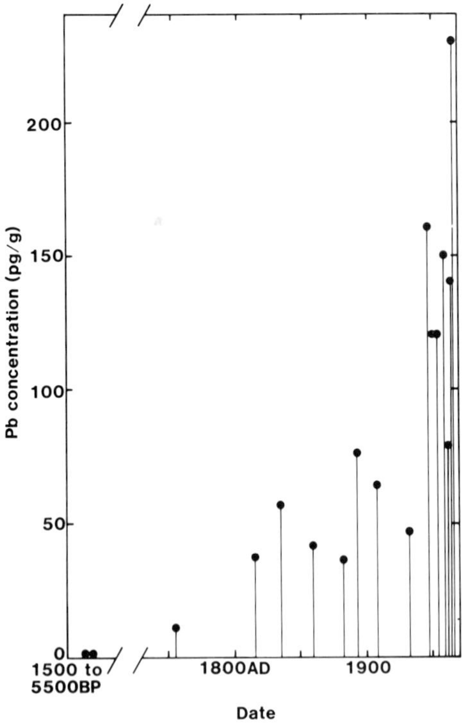

Fig. 2—Pb concentration in Greenland ice from prehistoric times to the present (Wolff and Peel 1985b).

pollution. Several independent measurements of concentrations in present-day snows have confirmed levels in the range of 150–400 pg Pb/g in several parts of Greenland and the Arctic, although a recent study (Wolff and Peel 1988) suggests that concentrations may have started to decline. Comparative values for preindustrial ice, measured on the Camp Century core by Ng and Patterson (1981), were the first data to be supported by cross-core concentration profiles. Although this showed that some contamination had penetrated towards the interior of the core, it established a convincing upper limit of 1.4 pg Pb/g. This level is consistent with the known natural contributions of Pb to crustal dusts and volcanic debris trapped in the ice (Table 1).

A weakness of the Greenland data set is that it is a compilation of data from widely separated sites, covering nonoverlapping time periods. Nevertheless, the overall increase in Pb concentrations in snow, of about 200 times since the preindustrial period, seems firmly established. Additionally, in common with several other impurities in snow, the

Table 1 Sources of lead in ancient and modern polar snow and ice

			Calculated Pb Contribution pg/g				
Location	Sample Date	Measured Pb pg/g	crustal[a] dusts	Volcanoes[b]	Oceans[c]	Residual (anthro-pogenic?)	EF crust
Antarctica[1]	1983	2.3	0.35	0.08	0.0008	1.9	6.6
Antarctica[2]	1980	6.3	0.12	0.09	0.0003	6.2	68
Greenland[3]	1965	200	0.4	0.4	0.00002	~ 200	500
Greenland[4]	1986	28	3.0	0.37	0.0001	25	9.3
Antarctica[5]	3,800–13,000 yr B.P.	0.58	0.63	0.29	0.0001	–	~ 1
Greenland[6]	5,500 yr B.P.	< 1.4	0.4	0.4	0.00002	–	< 3.5

[1] Boutron and Patterson, 1987
[2] Wolff and Peel, 1985a
[3] Murozumi et al., 1969
[4] Wolff and Peel, 1988
[5] Boutron and Patterson, 1986
[6] Ng and Patterson, 1981
[a] Calculated as $[\text{crustal Pb}] = 1.7\times10^{-4}[\text{Al}]$
[b] Calculated as $[\text{volcanic Pb}] = 0.32\times10^{-5}[\text{excess } SO_4]$
[c] Calculated as $[\text{oceanic Pb}] = 0.46\times10^{-8}[\text{Na}]$

concentration of Pb exhibits a clear winter maximum (Murozumi et al. 1969; Wolff and Peel 1988). This occurs when atmospheric conditions favor the injection of polluted air masses from Eurasia, and to a lesser extent from North America.

Antarctica

The situation in Antarctica was, until recently, much less clear. More remote from heavily industrialized continents, concentrations of Pb are very much lower than in Greenland; consequently, contamination control and analytical sensitivity present more severe problems.

Steady improvement in blank assessment, together with more routine measurement of cross-sample contamination profiles during the past five years, has led to a satisfactory consensus on Pb levels in present-day snow. Values around 6 pg Pb/g have been reported from sites in both West and East Antarctica and both near the coast and in the interior (Wolff and Peel 1985b). They do not seem to be significantly regulated by snow accumulation rate (over an order of magnitude) or by seasonal processes.

The most reliable time series through the past 200 years has been reported (Boutron and Patterson 1983) from a site about 180 km inland of Dumont d'Urville in East Antarctica. The data suggest that there may have been up to two- to three-fold overall increases during this period although the signal is very noisy. Boutron and Patterson (1987) have reappraised the data and suggest that all but the three lowest concentration points of this series (0.9, 1.1, and 1.2 pg Pb/g dated between 1917 and 1947) may have suffered slight contamination due to partial melting of the samples in transit.

The whole problem has been put into a firm perspective by a series of extremely careful studies on the Vostok and Dome C cores (Boutron et al. 1988), which encompass the whole of the last glacial cycle including the end of the previous ice age. This period was characterized by major changes in airborne dust and sea salt loadings in the global atmosphere and is, therefore, an ideal period to detect the relative contributions of alternative sources of natural Pb. Although our knowledge of the history of fluxes of airborne Pb through the past 155,000 years is based on only 20 data points, it is now apparent that most of the major fluctuations within this preindustrial period can be accounted for simply from the natural average abundance of Pb in wind blown dusts (Fig 3). Only in periods with low dust levels, such as the present Holocene, do volcanic emissions make a significant contribution of up to half the natural Pb content. Contributions from sea spray are uniformly insignificant (Table 1).

The concentration of Pb in preindustrial, Holocene ice averages around 0.6 pg Pb/g (Boutron and Patterson 1987), with values down to 0.3 pg Pb/g (which can now be determined with ±50% precision). In comparison with

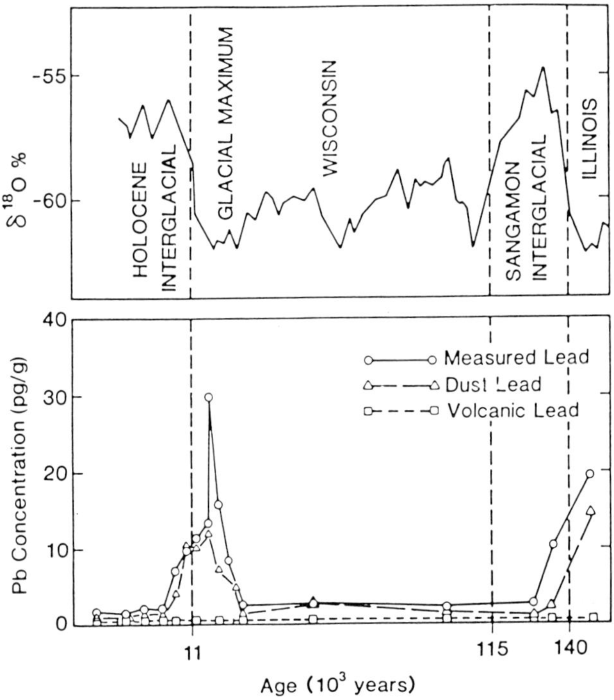

Fig. 3—Composite data from the Antarctic Dome C and Vostok ice cores showing variations of total Pb, calculated coastal and volcanic Pb across the Wisconsin/Holocene Illinois/Sangamon boundaries. Contributions from marine aerosol are negligible (from Patterson and Boutron 1988).

the data from the industrial period, this suggests that Pb concentrations may have increased 2–3 times during the period up to 1940, with a further increase of 4–5 times subsequently. Whether this latter increase is representative of large-scale pollution of the Antarctic tropospheric cell is debatable. Boutron and Patterson (1987) now consider that only one data point (2.3 pg Pb/g) for present Antarctic snow, from a site 433 km inland from Dumont d'Urville, is unaffected by local emissions of Pb from Antarctic stations or from field activities. If verified, this implies that local emissions of Pb are of comparable magnitude to fluxes entering from outside the continent and indicates that post–1940 increases that can be attributed to "global" pollution may be limited to a factor of 2.

The much weaker influx of Pb to Antarctica is due to several factors:

1. Pb emissions, especially from combustion of leaded fuels, occur predominantly in the Northern Hemisphere.

2. The atmospheric residence time of Pb aerosol in the troposphere is short in comparison with the time required for interhemispheric mixing.
3. The south polar convergence zone is a much stronger and more coherent feature than that in the Northern Hemisphere, hence an effective barrier to meridional transport.

Clearly, new and detailed time series through the industrial period must now be obtained from single sites, far remote from any form of human activity, on both the Greenland and Antarctic ice sheets.

OTHER HEAVY METALS

Modern Levels in Snow

Concentrations of other heavy metals in polar snow fall around or below those observed for Pb. Consequently, the effects of contamination are at least as severe as those now recognized for Pb and in some cases (e.g., for Hg) are compounded by additional difficulties of analysis. Many data have been reported from both the Greenland and Antarctic ice sheets. They have been reviewed comprehensively by MARC (1985), Wolff and Peel (1985b), and Boutron (1986), all of whom concluded that many of the data, even for modern snow, were probably wrong. The data were put into a more global context of atmospheric deposition by Galloway and others (1982).

Table 2 summarizes the most reliable data available from the polar ice sheets. In Antarctica, for only two sites in the Antarctic Peninsula, and probably for single snowfall events, have data been supported by satisfactory cross-sample concentration profiles (Wolff and Peel 1985a). The reported concentrations of Cd, Cu, and Zn for surface snow are at least 10 times smaller than any other values reported for Antarctic snow and ice, including values reported for a block of blue ice from East Antarctica, more than 12,000 years old (Boutron et al. 1984).

Table 2 Comparison of heavy metal concentrations in north and south polar snows (units: pg/g)

	Cd	Cu	Pb	Hg	Ni	Ag	Zn
Greenland	0.7[1]–5	6[1]–97	28[1]–200	10	195	10	27[1]–200
Antarctica	0.3	1.8	2.3–5	2	?	0.5	3.3

[1] Data from Wolff and Peel, 1988
Remaining data compiled by Wolff and Peel, 1985b

Recent data from Greenland (36 km from Dye 3), reported by Wolff and Peel (1988) and spanning a single accumulation year (1983/84), also show much lower levels of Cd, Cu, and Zn than almost all previously reported values for both modern snow and ancient ice. There is, however, partial agreement with a series of measurements made by Mart (1983) in the East Arctic Ocean. Wolff and Peel use relatively mild sample dissolution conditions (pH 3); nevertheless, they present several strands of evidence to support the belief that their technique does measure the total anthropogenic metal content. Indeed, the Measurement of Metals Workshop (Barrie 1986b) concluded that many anthropogenic metals are predominantly in a fine-particle form that is acid-soluble.

It is evident that nearly all the data for heavy metals apart from Pb can only be regarded as upper limits and cannot firmly provide a basis for assessing a human impact. The tortuous improvements in technique that have been found essential for successful analysis of Pb must now be adopted routinely for measurements of the other heavy metals in both Greenland and Antarctica.

Enrichment Factors (EF)

In the absence of time series data for these elements, preliminary limits on the extent of a human impact can be set on the basis of their crustal EFs. Large EFs have been reported generally for a wide range of trace metals in aerosol and precipitation from both polar and oceanic environments. With improvements in contamination control the reported values have progressively declined in recent years.

Estimated enrichment factors for recent snow from Greenland and Antarctica, using the most reliable data, are given for Pb, Cd, Cu, and Zn in Table 3. The average EF for Pb, taking into account the long-term increase in concentration with time and the fact that Pb is not enriched in ancient ice, most probably arises from anthropogenic sources. Of the

Table 3 Crustal enrichment factors for heavy metals in modern polar snow

Location	Cd	Cu	Pb	Zn
Antarctica[1]				
S. Palmer Land	182	4.5	68	6.3
Greenland[2]				
Dye 3	83	1.5	20	9

[1] Wolff and Peel, 1985a
[2] Wolff and Peel, 1988

remaining elements, only for Cd is there a substantial EF. This could reflect an anthropogenic input, although natural enrichment mechanisms such as volcanism cannot be ruled out for this volatile element. Cd also occurs at the lowest concentration so that it is more susceptible to contamination than other species.

Considerably smaller EFs for Cu and Zn may give upper limits for anthropogenic increases. However, particularly in Greenland, Cu is barely enriched above crustal abundance. So far, for both these elements, it is conceivable that global cycling has not been significantly perturbed by pollution.

SOOT AND ORGANIC COMPOUNDS

Soot

Particulate elemental carbon (EC) has been extensively studied in the Arctic atmosphere since it became apparent that it is a primary constituent of Arctic haze. It can be produced only by combustion processes, hence it may serve as a convenient tracer for anthropogenic emissions.

Studies on the aerosol lack historical perspective and need to be complemented with time series measurements which now can only be obtained from analysis of the snow cover. Such work has, so far, been neglected. Clarke and Noone (1985) have, however, established the seasonal range of variability in EC (4 to 127 ng g^{-1}) at several sites in the Arctic. These concentrations are sufficient to affect the snow albedo significantly.

Antarctica does not suffer from an "Arctic haze' problem, lacking efficient tropospheric transport pathways from neighboring continents throughout the year. Consequently, without very greatly improved analytical methods it is unlikely that EC of any large-scale significance could be detected in Antarctic snow. However, measurements made relatively close to manned stations could provide a useful tracer for pollution from these sites, as there are no natural sources of graphite carbon on the continent. Preliminary studies have already started at South Pole station (Clarke and Warren, personal communication).

Organic Compounds

Studies on organic pollutants in snow and ice are also very sparse, and no time trends can be evaluated from either hemisphere. This is unfortunate; some of these compounds could yield more direct indices of anthropogenic pollution than is possible with heavy metals. Many manufactured organic substances are unknown in nature; provided compounds are identified that are resistant to solar radiation, they could yield unambiguous evidence for a human impact. Moreover, the range of compounds is so great that it is possible to identify species that are mainly connected with one type of human activity, or have a restricted geographical pattern of usage.

Among the persistent polychlorinated hydrocarbons there is a wide range of pesticides and fungicides, including DDT, BHC and dieldrin, that is used primarily in agriculture and malarial control. The polychlorinated biphenyls (PCB), on the other hand, are mainly released by incineration processes. While DDT is still used extensively throughout the Southern Hemisphere, legislation has severely restricted its use in the Northern Hemisphere.

The few data available from Antarctica have been reviewed briefly by MARC (1985) and relate mainly to measurements of DDT and its metabolites, together with the PCBs in near-surface snow. Table 4 gives an impression of the quantities detected. Only the more recent data were supported by mass spectrometer to give a rigorous qualitative analysis. The most recent measurements, which are comparable with values currently being detected in Arctic snow (McNeely and Gummer 1984), suggest that the earlier data, certainly for DDT, probably suffered from contamination. This is not surprising since the concentrations are about 400 times lower than present-day concentrations of lead.

It seems that both DDT and PCBs probably have been detected in Antarctic snow, but this cannot be regarded as proven until the data are supported by cross-sample concentration profiles and rigorous blank assessment. If the finding is confirmed by such procedures, then historical trends should be determined. Because these must originate from a zero baseline, they will provide a particularly strong basis for evaluating pollution trends in relation to the global inventory of production and emission rates.

CONCLUSIONS

Human Impact on the Primary Aerosol

Lead is the only species for which there are sufficient, reliable time series data to make some assessment of the human impact, judged against a broad

Table 4 Typical concentrations of organochlorine compounds detected in north and south polar snows (units: fg/g)

	ΣBHC	ΣDDT	PCBs
Arctic	< 2–26[1]	< 2–4[1]	—
Antarctica	2.3[2]	0.015[2]	0.16[2]

[1] Data from McNeely and Gummer (1984)
[2] Data from Tanabe et al. (1983)

perspective of purely natural fluctuations through the past glacial cycle. Nevertheless, this data set has been severely reduced by present understanding of contamination problems and the methods needed to expose them. Less than 20 reliable data points are available for the whole of Antarctica during the post–1900 period, of which only 3 points pre-date 1980. Lead concentrations in Antarctic snow appear to have increased by 2–3 times from the early Holocene up to 1940, and a further 2–5 times between 1940 and the present at most sites sampled so far in Antarctica. Part of this recent increase may be a result of emissions from Antarctic stations rather than a general pollution of the Antarctic tropospheric cell. Substantially higher concentrations of Pb than observed at present were deposited during the previous glacial maximum where there was a greatly increased dust loading in the atmosphere; these values can be well accounted for assuming a normal average crustal abundance of Pb in the dust. There is no evidence for any unknown natural enrichment processes during the past 155,000 years.

Concentrations of Pb are at present some 30 times higher in Arctic snows and there is a satisfactory consensus around a value of 200 pg/g^{-1} for present-day snow in Greenland, Alaska, and the eastern Arctic Ocean. The only reliable time series was determined 20 years ago for Greenland snow, although a value for early Holocene ice of around 1.4 pg/g has been confirmed more recently and is consistent with the levels found in ancient Antarctic ice and is mainly attributable to unenriched crustal dusts. There seems little doubt that Pb concentrations have increased around 200–fold in Greenland snow since prehistoric times.

A contrasting flux of pollution onto the Antarctic and Greenland ice sheets is to be expected because 90% of global Pb emissions occur in the Northern Hemisphere and interhemispheric exchange is restricted. However, if future research shows that a significant part of the even modest increase in Pb from Antarctica is of local origin, this will then suggest that the south polar convergence zone is an even more important barrier.

Measurements on a range of other toxic metals and metalloids (e.g., Hg, As, Se, Ag), which occur at concentrations similar to or lower than that of Pb, need to be repeated using the criteria that have proved essential for analysis of Pb at remote locations. If high enrichment factors continue to be observed, then it will be essential to trace the development of this enrichment through the past several hundred years to confirm that it is linked to anthropogenic activity and not to natural processes. A range of organic compounds such as DDT and the PCBs, which have no natural source, has been identified in modern snows from both Antarctica and the Arctic, but at concentrations well below those of Pb. While it seems that their presence has been detected, giving unequivocal evidence for a human impact, future measurements must be validated by cross-sample analyses and more complete details of blank assessment procedures. Having proved a present-day signal in this way, it will then be vital to achieve a detailed time series. At present, similar concentrations of PCB and DDT, at levels not far above detection limits, have been determined in both Antarctica and Greenland. This result is hard to rationalize in the face of evidence from Pb for the small influence of Northern Hemisphere pollution on the composition of the Antarctic ice sheet.

Main Problems that Must be Resolved

"Anomalously enriched" elements. The most reliable measurements of Pb, Cu, and Zn in Greenland and Antarctic snow no longer indicate that these elements are grossly enriched, to the extent that unknown (unproven) natural processes play a significant role. Other elements such as Cd, Hg, Se, and As, that appear to be anomalously enriched in both the aerosol and snowfall studies, must now be viewed with the same suspicion as earlier data for Pb. If residual enrichment is still observed, then time series data from ice cores may be the only way to reconstruct time trends through the past few centuries to determine whether the enrichment is due to pollution or to natural processes.

Snow/air scavenging ratios. Interpretation of ice core data must currently assume that scavenging ratios have been steady through the measurement period. Reliable direct studies of aerosol and snowfall have been conducted during rather restricted periods, and at present it is not possible to assess quantitatively how the scavenging ratio may vary with changing weather patterns (e.g., snow accumulation rate and seasonal patterns). This uncertainty may present problems in areas of very low snow accumulation rate (<20 g cm^{-1} a^{-1}) in the interior of Antarctica, where 60–70% of impurities in snow may be deposited by dry deposition.

Importance of local pollution in Antarctica Experience with Pb now shows that the extent of global impact is so limited, in Antarctica, that the regional scale influence of local pollution from manned stations and field activities must be assessed accurately so that the net effect of long-range pollution can be quantified. While a comprehensive inventory of waste disposal is an essential starting point, the process of dispersion of pollution into the surrounding snowfield must also be monitored and modeled at representative coastal and inland stations. Soot or stable organic combustion products such as benzopyrene may be suitable tracers.

Concept of global background aerosol There are large differences in levels of pollution between the two remotest ice-covered areas of the world and at least order of magnitude differences between these regions and more temperate ice-covered areas. Hence, no single zone can be considered to represent a global pollution level or a global natural background. Nevertheless, if the relative changes over time in an area such as Antarctica can elucidate the natural controls on a particular species, then such relationships may then be applied to ice cores from other areas to reveal a true human impact. Clearly, in order to understand fully the global cycling of trace substances in the atmosphere, sampling must be conducted on a wide geographic scale. While data from temperate regions may reflect more local or regional scale behavior, they will complement the larger scale histories derived from polar ice cores.

REFERENCES

Barrie, L.A. 1986a. Arctic air pollution: an overview of current knowledge. *Atmos Envir.* **20(4):** 643–663.

Barrie, L.A. ed. 1986b. Measurement of Metals in Precipitation. Ontario, Canada: Atmospheric Environment Service.

Boutron, C. 1986. Atmospheric toxic metals and metalloids in the snow and ice layers deposited in Greenland and Antarctica from prehistoric times to present. In: Toxic Metals in the Atmosphere, eds. J.O. Nriagu and C.I. Davidson, pp. 467–505. Chichester: Wiley.

Boutron, C.; Leclerc, M.; and Risler, N. 1984. Atmospheric trace elements in Antarctic prehistoric ice collected at a coastal ablation area. *Atmos. Envir.* **18**: 1947–1953.

Boutron, C., and Patterson, C.C. 1983. The occurrence of lead in Antarctic recent snow, firn deposited over the last two centuries and prehistoric ice. *Geochim. Cosmo. Acta* **47**: 1355–1368.

Boutron, C., and Patterson, C.C. 1986. Lead concentration changes in Antarctic ice during the Wisconsin/Holocene transition. *Nature* **323:** 222–225.

Boutron, C., and Patterson, C.C. 1987. Relative levels of natural and anthropogenic lead in recent Antarctic snow. *J. Geophys. Res.* **92**: 8454–8464.

Boutron, C.; Patterson, C.C.; Lorius, C.E.; Petrov. V.N.; and Barkov, N.I. 1988. Atmospheric lead in Antarctic ice during the last climatic cycle. *Ann. Glaciol.* **10:** 5–9.

Boutron, C.; Patterson, C.C.; Petrov, V.N.; and Barkov, N.I. 1987. Preliminary data on changes of lead concentrations in Antarctic ice from 155000 to 26000 years B.P. *Atmos. Envir.* **21**: 1197–1202.

Clarke, A.D., and Noone, K.J. 1985. Soot in the Arctic snowpack: a cause for perturbations in radiative transfer. *Atmos. Envir.* **19**: 2045–2053.

Davidson, C.I.; Santhanam, S.; Fortmann, R.C.; and Olson, M.P. 1985. Atmospheric transport and deposition of trace elements onto the Greenland ice sheet. *Atmos. Envir.* **19**: 2065–2081.

Galloway, J.N.; Thornton, J.D.; Norton, S.A.; Volchok, H.L.; and McLean, R.A.N. 1982. Trace metals in atmospheric deposition: a review and assessment. *Atmos. Envir.* **16**: 1677–1700.

Junge, C.E. 1977. Processes responsible for the trace content in precipitation. In: Isotopes and Impurities in Snow and Ice, Proc. IUGG Symp. Grenoble, 1975. *IAHS-AISH Publ.* **118**: 63–77.

Lowenthal, D.H., and Rahn, K.A. 1985. Regional sources of pollution aerosol at Barrow, Alaska, during winter 1979–80 as deduced from elemental tracers. *Atmos. Envir.* **19**: 2011–2024.

MARC. 1985. Historical Monitoring. London: Monitoring and Assessment Research Centre, Report No. 31.

Maring, H.; Settle, D.M.; Buat-Menard, P.; Dulac, F.; and Patterson, C.C. 1987. Stable lead isotope tracers of air mass trajectories in the Mediterranean region. *Nature* **300**: 154–156.

Mart, L. 1983. Seasonal variations of Cd, Pb, Cu and Ni levels in snow from the eastern Arctic Ocean. *Tellus* **35B**: 131–141.

McNeely, R., and Gummer, W.D. 1984. A reconnaissance survey of the environmental chemistry in east-central Ellesmere Island, NWT. *Arctic* **37**: 210–233.

Murozumi, M.; Chow, T.J.; and Patterson, C.C. 1969. Chemical concentrations of pollutant lead aerosols, terrestrial dusts and sea salts in Greenland and Antarctic snow strata. *Geochim. Cosmo. Acta* **45**: 1247–1294.

Ng, A., and Patterson, C.C. 1981. Natural concentrations of lead in ancient Arctic and Antarctic ice. *Geochim. Cosmo. Acta* **45**: 2109–2121.

Patterson, C.C., and Boutron, C. 1988. Lead records in Antarctic ice: changes in global atmospheric concentrations during the past 150,000 years. *Antarc. J. US*, Review Issue, in press.

Rahn, K.A. 1976. The chemical composition of the atmospheric aerosol. Technical Report, Graduate School of Oceanography, Univ. of Rhode Island, U.S.A.

Tanabe, S.; Hidaka, H.; and Tatsukawa, R. 1983. PCBS and chlorinated hydrocarbon pesticides in Antarctic atmosphere and hydrosphere. *Chemosphere* **12**: 272–288.

Wolff, E.W., and Peel, D.A. 1985a. Closer to a true value for heavy metal concentrations in recent Antarctic snow by improved contamination control. *Ann. Glaciol.* **7**: 61–69.

Wolff, E.W., and Peel, D.A. 1985b. The record of global pollution in polar snow and ice. *Nature* **313**: 535–540.

Wolff, E.W., and Peel, D.A. 1988. Concentrations of cadmium, copper, lead and zinc in snow from near Dye 3 in South Greenland. *Ann. Glaciol.* **10**: 193–197.

The Environmental Record in Glaciers and Ice Sheets
eds. H. Oeschger and C.C. Langway, Jr., pp. 225–247
John Wiley & Sons Limited

The Ionic Deposits in Polar Ice Cores

H.B. Clausen* and C.C. Langway, Jr.**

**Glaciological Department, Geophysical Institute*
2200 Copenhagen, Denmark
*** Ice Core Laboratory, Department of Geology*
State University of New York at Buffalo
Amherst, NY 14226, U.S.A.

Abstract. The atmospherically transported or generated water soluble ions, such as H^+, Na^+, Cl^-, NO_3^-, and SO_4^{2-}, are deposited on polar ice sheets in time-unit sequences. These ionic constituents of the total chemistry content are derived from a variety of natural sources of terrestrial, marine, and atmospheric origin, with an increasing component attributed to man's activities. Ionic studies have been made on several ice cores from both polar ice sheets and permit an interhemispheric comparison to be made of the ionic depositional trends as related to climatic change, volcanic activity, and other palaeoenvironmental data.

Measurements of the strong mineral acid concentrations of H_2SO_4 and HNO_3 in carefully dated ice cores from the Dye 3 region in S.E. Greenland, show that today's yearly deposit of sulfuric and nitric acid is 4 and 2 times the yearly amount deposited prior to A.D. 1900 and A.D. 1950 respectively.

For several reasons this increase can be attributed to anthropogenic activities. At present, the yearly sulfate deposit appears to have leveled off but the yearly nitrate deposit continues to increase.

Limited data from Antarctic ice cores do not show similar increasing trends in strong acid concentrations.

INTRODUCTION

Chemically, polar ice is predominantly solid H_2O, but trace quantities of water soluble and insoluble substances are also present in the strata. These trace substances are atmospherically transported and incorporated during snowfall as crystal nuclei. They are also incorporated by wash out and during nonpreciptating periods as dry fallout. Since glaciers are fundamentally products of atmospheric processes, one may approach the study of the chemical composition of an identified season, annual, or longer time interval of deposited snow as being representative of the average bulk concentration of the same constituents in the atmosphere for the same time period. With this in mind, the study of the major ionic species found in various polar ice cores provides unique chronological records of palaeodata in terms of temporal and spatial changes in the gross chemical composition of the atmosphere. Of particular interest are data sets for short- and long-term trends and alternating periods of atmospheric stability and instability as related to volcanic disturbances and climate change.

State of the art glaciochemistry studies are most often collaborative, multidisciplinary scientific efforts where multiple parameter physical and chemical analyses are made on laterally adjacent samples. As a minimum, meaningful glaciochemistry studies, particularly time series analyses, require parallel studies of the stable isotopes and the solid electrical conductivity for cross-correlation purposes.

This paper combines and interprets the limited existing data available for ionic chemistry studies made on polar ice cores from Greenland and Antarctica. It focuses on the depositional patterns and trends of the dominant ions which have occurred for the past 10,000 years or during the post-Wisconsin glacial age (Holocene).

ICE CORES

The chemistry of ice sheets was a sadly neglected subject until about thirty years ago. At that time an ice coring drill capable of recovering continuous ice cores to intermediate depths (about 400 m) was developed and this permitted the first systematic examination of the inner nature and compostion of an ice sheet. Later versions of ice coring drills were used to auger successfully down to bedrock in both Greenland and Antarctica. Since 1966 only three ice cores have been obtained that completely penetrate the inland thickness of the Greenland and Antarctica ice sheets (Camp Century, 1480 m; Byrd Station, 2164 m; Dye 3, 2037 m). Two other cores have reached depths of about 1000 m and 2083 m in Antarctica (Dome C and Vostok). Until now less than ten ice cores have been obtained from intermediate depths and about two dozen from shallow depths (about 100 m). Not all recorded ice cores were measured for ionic constituents.

At least two physical processes affect the behavior of a deposited polar snow layer: diffusion and densification. However, in areas where the annual rate of net accumulation is more than 250 kg/m^2, the seasonal variations in the oxygen isotope rations ($\delta^{18}O$) are not obliterated by the diffusion process and also survive the densification process which transforms snow into firn and glacier ice. Under these conditions an annual snow layer thickness, λ, is represented by the distance between two sequential $\delta^{18}O$ minima, which are usually more distinct than the maxima. In Greenland we define this layer to represent the calendar year January to January; in Antarctica from mid-July to mid-July. The annual rate of deposit of a given chemical species is determined by multiplying the concentration level with the layer thickness, λ_H, and the density of ice. Because thinning of λ occurs with depth due to deformation, annual layer thicknesses are referred to the original annual layer thickness, λ_H, at the time of deposition. The chemical concentration levels of annual layers are not altered, at least during the Holocene period to which we limit our discussions.

SOURCES

The ionic impurities found in the polar snow layers have a widespread and, in some cases, multiple source origin. As would be expected, both natural and anthropogenic constituents are found. Deposition on the ice sheets represents the final step in the not very well established atmospheric transport, residence, and photochemical reaction processes which precede incorporation of ions into snow layers. The predominant ions which have been measured are H^+, SO_4^{2-}, NO_3^-, and Cl^-, with some data available for other cations such as Na^+, NH_4^+, and Ca^{2+}. Most of the sufficinetly complete ionic studies are associated with deep, intermediate, and shallow ice core drilling projects, but even here data are restricted.

H^+, SO_4^{2-}, and NO_4^- have a closely associated relationship in glacier chemistry. H^+ is the overwhelmingly dominant cation found in Holocene glacier ice. Its concentration level $[H^+]$ is determined mainly by the amount of H_2SO_4, HNO_3, and in some cases, HCl present. It is primarily the concentration of H_2SO_4 and HCl, measured in spaced intervals of ice cores, that reveals the magnitude and periodicity of past volcanic activity. High HNO_3 peaks in the cores are usually associated with higher summer temperatures or melt phenomena. All of these acid signals extend well above the unperturbed background level.

Sulfate found in ice layers has been attributed to a variety of natural and man-made marine and terrestrial sources. These include ocean spray, volcanoes, organic decay, and combustion products of fossil fuels. It is customary to refer to the $[SO_4^{2-}]$ for polar snows as $[SO_4^{2-}]_{ex}$, where ex is the excess sulfate concentration corrected for the marine contribution (0.14 $\times$ $[Cl^-]$ by weight). All sulfate discussed in this paper is given as excess sulfate.

Nitrate sources have equally complex sources. For the Northern Hemisphere they include continental, atmospheric, and stratospheric contributions and consist of biomass burning, soils, oxidation of NH_3, photochemical processes, and lightning. The remoteness of Antarctica from land masses seems to preclude NO_3^- contributions from any significant sources other than the atmosphere. No evidence is available that indicates volcanic activity is an important source of the high HNO_3 acid signals found in ice cores.

Chloride concentrations found in polar deposits are primarily of marine origin, although on rare occasions some major volcanic eruptions contribute considerable Cl^- above baseline levels in form of hydrochloric acid.

Sodium is predominantly derived from marine sources. Calcium is regarded to be of continental origin. Ammonium concentrations above background are considered to originate from industrial and agricultural activities.

Anionic and cationic measurements are most often made today by ion chromatography. Earlier cation measurements were made mainly by atomic absorption spectrometry and specific electrode techniques. Concentration levels of ion constituents found in glaciers are strongly influenced by the availability of source materials, the distance from the source, and meteorological parameters such as wind circulation patterns, accumulation, and temperature.

HYDROGEN, SULFATE, AND NITRATE IONS

At this time H^+ is the only ion measurable by a rapid technique; all others require time-consuming and detailed sample preparation procedures and measurements. Hammer (1980) invented and perfected a solid electrical conductivity method (ECM) which provides an assessment of the $[H^+]$ level when the measured electrical current signals are calibrated with pH measurements. In principle, ECM determines the acidity of solid ice whereas pH determines the acidity of melted ice. The difference is usually insignificant for the Holocene ice layers in Greenland and for the Holocene and Wisconsin ice layers in Antarctica. The high concentration levels of alkaline carbonaceous impurities in the Greenland Wisconsin age ice affects the acidity signals and consequently the method does not produce unambiguous results. The original ECM pH calibration curve by Hammer (1980) holds for several ice cores including the Byrd and the Crête intermediate core. The ECM provides an important continuous record of $[H^+]$, with a resolution of a few mm along the length of the core. This invaluable method identifies the position and magnitude of strong acid levels which are subsequently analyzed by ion chromatography to identify the anions related to the acid signal. In Holocene ice the $[H^+]$ inferred by ECM is generally close to the sum of $[NO_3^-]$ and $[SO_4^{2-}]_{ex}$. There are deviations from this simple model in which the sums do not balance. In this case, major cation concentration analyses for Na^+, NH^+, and Ca^{2+} are necessary. Deviations may occur, for instance, where the acidity of the ice layer is increased due to an influx of HCl from volcanic eruptions. Large amounts of volcanic HCl are in some cases associated with a smaller amount of HF.

Table 1 presents the anion concentration levels found in polar ice cores for specified time periods and Fig. 1 shows site locations of the samples. The data in Table 1 are grouped by site locations and sub-locations for

Table 1 Average anion concentrations and annual deposition rates in ice cores

	1	2	3	4	5	6	7	8	9	10	11	12
Site location	Elevation	Time interval	Years in record	Annual rate of accum.	Cl^- conc.	Cl^- Annual deposit	NO_3^- conc.	NO_3^- Annual deposit	SO_4^{2-} conc.	Excess SO_4^{2-} conc.	Excess SO_4^{-2} Annual deposit	Reference
	m.a.s.l.	A.D.		kg/m^2	ng/g	$kgCl^-/km^2$	ng/g	$kgHNO_3/km^2$	ng/g	ng/g	kgH_2SO_4/km^2	
GREENLAND												
Dye 3	2486	Holocene	220(a)	496	21	10	52	26	24	21	11	Herron 1982b; Finkel et al. 1986
Dye 3 18C	2617	1780–1909	30	459	26	12	55	26	29	25	12	Langway & Goto-Azuma 1988
Dye 3 18C	2617	1910–1949	40	459	45	21	61	28	55	49	23	Langway & Goto-Azuma 1988
Dye 3 4B	2491	1920–1930	11	487	30	14	52	26	52	48	24	Finkel et al. 1986
Dye 3 4B	2491	1944–1954	11	487	21	10	69	34	72	70	35	Finkel et al. 1986
Dye 3 18C	2617	1950–1959	10	459	34	16	77	36	71	66	31	Langway & Goto-Azuma 1988
Dye 3 18C	2617	1960–1969	10	459	27	12	90	42	81	77	36	Langway & Goto-Azuma 1988
Dye 3 20D	2620	1968–1984	17	459	26	12	115	53	87	83	39	Mayewski et al. 1986
Dye 3 18C	2617	1970–1984	15	459	32	15	100	47	90	86	40	Langway & Goto-Azuma 1988
Dye 3	2486	1973–1978	6	496	21	11	86	43	90	87	44	Neftel et al. 1985
Dye 3	2486	1975–1981	7	496	21	11	114	57	85	82	42	Herron 1982b
Dye 3	2486	1979–1980	2	496	21	11	137	69	87	84	43	Herron 1982b
Milcent	2410	1256–1262	7	487	33	16	61	30	61	57	28	Langway et al. 1988

cont.

Table 1 (cont.)

Site location	1 Elevation m.a.s.l.	2 Time interval A.D.	3 Years in record	4 Annual rate of accum. kg/m^2	5 Cl^- conc. ng/g	6 Cl^- Annual deposit $kgCl^-/km^2$	7 NO_3^- conc. ng/g	8 NO_3^- Annual deposit $kgHNO_3/km^2$	9 SO_4^{2-} conc. ng/g	10 Excess SO_4^{2-} conc. ng/g	11 Excess SO_4^{-2} Annual deposit kgH_2SO_4/km^2	12 Reference
Milcent	2410	1782–1785	4	487	28	14	–	–	31	27	13	Herron 1982a
Crête	3172	1000–1700	60	275	–	–	60	17	–	–		Risbo et al. 1981
Crête	3172	1258–1259	2	275	13	3.6	80	22	52	50	14	Langway et al. 1988
Site A	3092	1815–1816	2	285	13	3.7	64	19	37	35	10	Clausen & Hammer 1988
Site A	3092	1782–1784	3	285	14	4.0	45	13	24	22	6	Clausen & Hammer 1988
Site A	3092	1891–1902	12	285	11	3.0	69	20	37	36	10	Steffensen 1988
Site D	3018	1891–1902	12	340	7	2.4	72	25	37	36	13	Steffensen 1988
North Cent.	2930	1442–1450	9	129	11	1.4	83	11	80	79	10	Herron 1982b
North Cent.	2930	1900–1920	21	129	–	–	112	15	–	–		Clausen et al. 1986
North Cent.	2930	1973–1977	5	129	11	1.4	164	21	130	129	17	Herron 1982b
Camp Cen.	1880	Holocene	100[b]	349	40	14	71	25	40	34	12	Herron & Langway 1985
Camp Cen.	1880	1256–1265	10	349	58	20	46	16	29	20	7.2	Langway et al. 1988
Camp Cen. I	1880	1788–1800	13	349	66	23	65	23	42	33	12	Herron, 1982b
Camp Cen. II	1910	1906–1918	13	349	–	–	52	18	–	–		Rasmussen et al. 1984
Camp Cen.	1880	1976–1977	2	349	64	22	120	43	120	111	40	Herron 1982b

ANTARCTICA												
South Pole	2835	1256–1262	7	73	28	2.1	92	6.9	65	61	4.6	Langway et al. 1988
South Pole	2835	1811–1820	10	73	31	2.3	67	5.0	64	60	4.5	Langway et al. 1988
South Pole	2835	1876–1885	10	73	30	2.2	70	5.2	53	49	3.7	Langway et al. 1988
South Pole	2835	1923–1936	14	73	29	2.1	74	5.5	45	41	3.1	Herron, 1982a
South Pole	2835	1959–1969	11	85	–	–	88	7.6	–	–		Legrand & Delmas 1986
East Antarctica												
D 80	2525	1959–1969	11	237	26	6.2	55	13	37	34	8.2	Legrand & Delmas 1986
D 57	2050	1780–1820	40	170	–	–	37	6.4	22	22	3.8	Zanolini et al. 1985
D 55	2050	1959–1969	11	70	–	–	33	2.3	–	–		Legrand & Delmas 1986
Dome C	3240	1760–1980	220	34	–	–	15	0.5	48	48	1.7	Legrand & Delmas 1986, 1987
Dome C	3240	1969–1979	11	34	–	–	10	0.3	–	–		Legrand & Delmas 1986
Vostok	3490	Holocene	(c)	23	–	–	54	1.3	–	–		Legrand et al. 1984
West Antarctica												
Byrd	1530	Holocene	(d)	101	62	6.3	41	4.2	41	33	3.4	Herron 1982b
Byrd	1530	1254–1261	8	101	68	6.9	40	4.1	39	29	3.0	Langway et al. 1988
Siple	1000	Post 1900	(e)	400	83	33	28	11	35	24	9.8	Herron 1982b
GAP Gomez	1130	1942–1980	39	343	83	29	25	8.7	36	25	8.7	Mulvaney & Peel 1988

cont.

Table 1 (cont.)

Site location	1 Elevation	2 Time interval	3 Years in record	4 Annual rate of accum.	5 Cl^- conc.	6 Cl^- Annual deposit	7 NO_3^- conc.	8 NO_3^- Annual deposit	9 SO_4^{2-} conc.	10 Excess SO_4^{2-} conc.	11 Excess SO_4^{-2} Annual deposit	12 Reference
	m.a.s.l.	A.D.		kg/m^2	ng/g	kgCl^-/km^2	ng/g	kgHNO_3/km^2	ng/g	ng/g	kgH_2SO_4/km^2	
GAP St25	1835	1970–1976	7	441	61	27	–	–	–	–		Mumford & Peel 1982
GAP Dolleman	398	1973–1985	13	404	417	168	37	15	245	189	78	Mulvaney & Peel 1988
GAP Dolleman	398	1978–1980	3	404	726	293	27	11	165	67	28	Mulvaney & Peel 1988
RISP J9	55	1245–1290	45	92	128	12	52	4.9	82	64	6.0	Langway et al. 1988
RISP C16	70	1590–1598	9	100	144	14	44	4.5	84	65	6.6	Herron 1982a
RISP C16	70	1971–1977	7	100	144	14	39	4.0	61	42	4.2	Herron 1982a
RISP Q13	50	1590–1598	9	140	348	49	46	6.5	98	51	7.3	Herron 1982a
RISP Q13	50	1973–1977	5	140	348	49	41	5.8	103	56	8.1	Herron 1982a

(a) 220 years in the period 8400 B.C.–A.D. 1900
(b) Some 100 years in the Holocene period
(c) Unspecified amounts of years from the depth interval 50–950 m, corresponding to an age interval 1500–60000 B.P.
(d) Unspecified amounts of years before A.D. 1350
(e) Unspecified amounts of years

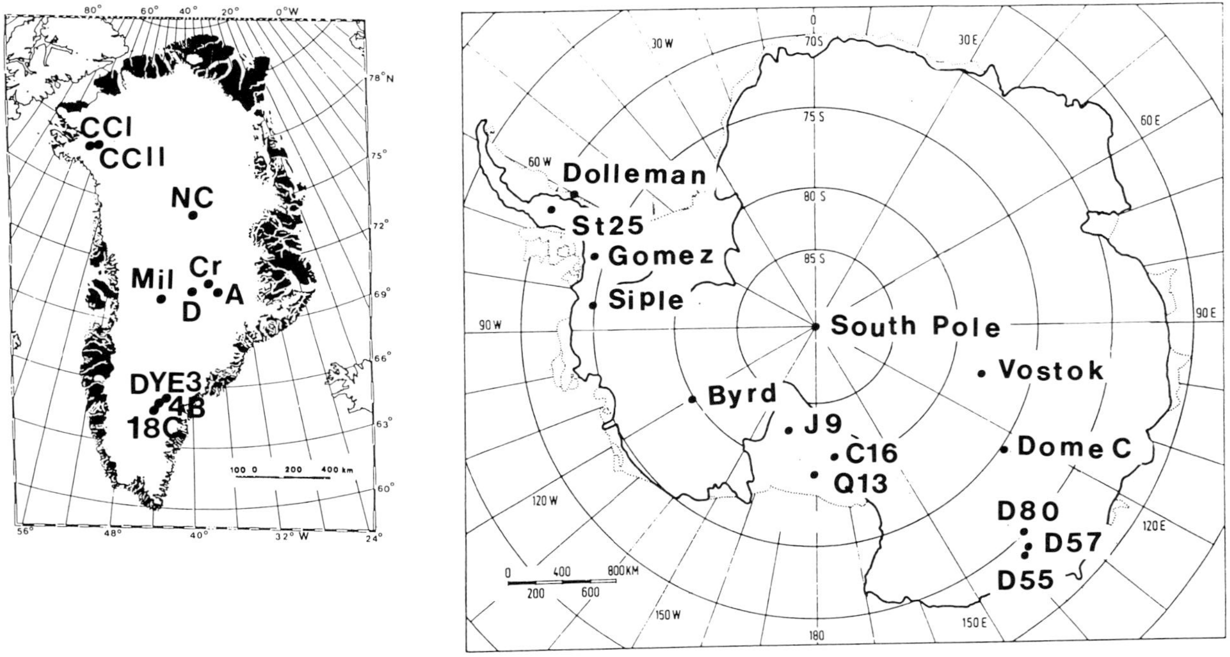

Fig. 1—Map of drill sites in Greenland and Antarctica.

Table 2 Summary of anion concentrations and depositions from Table 1

1	2	3	4	5	6	7	8
Site location	Time interval	Cl^- conc.	Cl^- Annual deposit	NO_3^- conc.	NO_3^- Annual deposit	Excess SO_4^{2-} conc.	Excess SO_4^{2-} Annual deposit
	A.D.	ng/g	kgCl^-/km^2	ng/g	kgHNO_3/km^2	ng/g	kgH_2SO_4/km^2
GREENLAND							
Dye 3	Holocene	21	10	52	26	21	11
Dye 3	1780–1959	34	16	61	29	46	22
Dye 3	1960–1984	26	12	105	49	83	40
Milcent	1256–1785	31	15	61	30	46	22
Crête	1000–1902	10	3.0	63	18	35	11
North Cent.	1442–1920	11	1.4	103	13	79	10
North Cent.	1973–1977	11	1.4	164	21	129	17
Camp Cen.	Holocene	40	14	71	25	34	12
Camp Cen.	1260–1918	63	22	55	19	27	9.5
Camp Cen.	1976–1977	64	22	120	43	111	40

ANTARCTICA							
South Pole	1256–1936	30	2.2	74	5.4	51	3.7
South Pole	1959–1969	–	–	88	7.5	–	–
East Antarctica							
D Sites	1815–1969	26	6.2	40	6.2	25	4.8
Dome C	1760–1980	–	–	15	0.5	–	–
Dome C	1969–1979	–	–	10	0.3	–	–
Vostok	Holocene	–	–	54	1.3	–	–
West Antarctica							
Byrd	Holocene	65	6.6	40	4.1	31	3.1
Siple/Gomez	1900–1980	83	31	26	10	24	9.2
GAPDolleman	1973–1985	417	168	37	15	189	78
RISP J9/C16	1245–1977	132	13	49	4.7	62	5.9
RISP Q13	1590–1977	348	49	43	6.2	54	7.7

Greenland and Antarctica for various time intervals as listed in column 2. It should be noted that the number of ice core measurements for each time interval entry varies considerably. For example, the Dye 3 Holocene entry shows average annual ion concentration values, columns 5, 7 and 9, representing the mean of several hundred measurements, whereas the Site A Greenland entry (1815–1816) is the product of 20 measurements. Other locations represent even fewer measurements. The annual ion deposits in column 8 and 11 are expressed in the strong acid units since this is the form in which they are present in the ice. Further detailed and sample information are contained in the referenced papers. The time intervals shown for individual sites are ranked from oldest to youngest. Table 2 is a condensed and summarized form of the data listed in Table 1. Both tables show various trends and depositional patterns. An inspection of Table 1 shows the following general patterns:

1. The nitrate concentrations are higher than sulfate concentrations in A.D. pre-1900 deposits in Greenland and high altitude Antarctic sites.
2. There is a different time pattern in the nitrate and sulfate increase found in Greenland since A.D. 1900. Sulfate increases first with the nitrate rise starting about 50 years later.
3. The yearly sulfate deposit in Greenland now is between 2 to 4 times the background level and appears to have ceased to increase.
4. The yearly nitrate deposit in Greenland now is about 2 times the background level and appears to be still increasing.
5. No sulfate or nitrate concentration increases, similar to those found in Greenland, are observed in Antarctica.

In Greenland the mean annual deposit of excess H_2SO_4 for the entire Holocene period at Dye 3 is 11 kg/km^2 (Table 2, column 8). This value excludes all high peaks caused by major volcanic eruptions for the numerous time intervals measured and is taken as the background value for the location. As is shown, the excess H_2SO_4 gradually increases about fourfold from about A.D. 1910 to A.D. 1984. The increase above background is attributed to anthropogenic input from fossil fuel consumption. Among others, Barrie et al. (1985) has compiled data indicating a yearly increase in European anthropogenic SO_2 emissions since the beginning of this century. The emissions correlate well with the excess H_2SO_4 data for Dye 3 and their studies from the Canadian Arctic. Similar trends are reported by other authors. Finkel et al. (1986) not only showed this same increasing trend, but also reported that it was accompanied by a seasonal time shift in the depositional patterns of NO_3^- and SO_4^{2-} in the Dye 3 layers during the 1950s compared to the 1850s, relative to the $\delta^{18}O$ and Cl^- depositional patterns. Since the Cl^- peaks were still contained in the mid-winter $\delta^{18}O$ minima, the conclusion was that the increase in excess SO_4^{2-} was not due to a climatic influence and therefore was most likely caused by an anthropogenic effect. The anthropogenic inputs of nitrate and sulfate peak during early spring as does the arctic haze (Barrie et al. 1985). Preindustrial nitrate concentrations peak in summer. Tables 1 and 2 show that from about the 1960s until now the annual excess H_2SO_4 deposit has been about 30 kg/km^2 (subtracting the natural background of 11 kg/km^2).

The Camp Century annual excess H_2SO_4 depositional pattern, gathered from much less data, is similar to Dye 3 for the Holocene and modern deposits. Although Camp Century is nearly 1300 km north of Dye 3 it is at approximately the same elevation. Camp Century is the only Greenland location besides Dye 3 that shows a significant increase in SO_4^{2-} during the last decades.

The remaining Greenland locations have a much smaller data base. The background annual H_2SO_4 deposit for North Central A.D. 1450 is 10 kg/km^2; the same as Dye 3 and Camp Century. The modern deposit represented by the five year period 1973–1977 is 17 kg/km^2/yr. The 7 kg/km^2/yr above background may well be within the error limits and background envelope for this location. In any event, the deposit is much less than found at Dye 3 and Camp Century. Because of very limited measurements, the Milcent and Crête (Site A and D) data for H_2SO_4 provides only background values, which favorably compare with the other locations in Greenland.

The mean annual deposit of HNO_3 for the Holocene background at Dye 3 is 26 kg/km^2. These data show a general increasing trend which started in about 1950 and has nearly doubled until today. The increasing trend differs from H_2SO_4 in that it begins later. The same concentration levels and increasing trend for HNO_3 is observed at Camp Century. The HNO_3 depositional pattern for North Central is the only other Greenland location where measurements are available for background and modern deposits. Here the increasing trend corresponds to one-half the values for natural background. The observed increase of nitrate corresponds well with the present-day observations of ozone increase in the troposphere in the Northern Hemisphere (Volz and Kley 1988). It appears that today's ozone budget is strongly influenced by photochemical processes caused by the increased level of anthropogenic emission of nitrous gases.

The Antarctic locations shown in Tables 1 and 2 have even less ionic data available than Greenland. These data are also grouped by age and geographical locations. On the average, ion deposits in Antarctica are much lower than in Greenland. As in Greenland, ionic concentration levels are influenced by distance from the source and elevation. The background of annual deposited excess H_2SO_4 for the Holocene at Byrd is 3.1 kg/km^2; no modern values exist for this location. Other west Antarctic locations at Siple, Gomez, and especially Dolleman have much higher values for excess

H_2SO_4. This is attributed to these locations being close to a very active marine sulfurous gas source (Mulvaney and Peel 1988). The annual deposits of HNO_3 vary considerably at each location but the average concentration levels are much lower than anywhere found in Greenland (except for Dolleman).

The overall geographical distribution of the anion concentration levels also shows that greater amounts are deposited in Greenland than in Antarctica. Although the post–1900 excess H_2SO_4 annual deposit for inland Greenland is about 30 kg/km^2, except for the region in the central sector of the ice sheet (North Central), the available data indicate that modern deposits for inland Antarctica are at least an order of magnitude lower. From another comparative viewpoint, the total amount of volcanic H_2SO_4 deposited in the Dye 3 area, Greenland by the major eruptions of (*a*) Tambora, Indonesia in A.D. 1815 (Clausen and Hammer 1988); (*b*) Laki, Iceland in A.D. 1783 (Clausen and Hammer 1988); (*c*) an unknown equatorial eruption in A.D. 1259 (Langway et al. 1988), and (*d*) Thera, Greece in 1645 B.C. (Hammer et al. 1987) is 60, 180, 120, and 205 kg/km^2 respectively. This total amount of volcanic material was deposited over a one year period in the case of Laki and over a two year period for the other three eruptions. The amount of H_2SO_4 deposited in the Dye 3 region attributable to the Tambora eruption is 30 kg/km^2 (one-half of the two year deposit) and is about equal to the present-day annual background deposition rate of anthropogenic sulfate. In a table compiled by Neftel et al. (1985), the value given for the total annual anthropogenic sulfur emission for the entire Northern Hemisphere ranges from 69 to 98 TgS/yr. This amount is about 7 to 10 $\times$ 10^9 times more than the 29 kg/km^2 yearly anthropogenic sulfuric acid deposited at Dye 3. From the same table, the amount of anthropogenic nitrogen emission for the Northern Hemisphere ranges from 13 to 57 TgN/yr which is equal to 3 to 10 $\times$ 10^9 times more than the 23 kg/km^2 yearly anthropogenic nitric acid deposited at Dye 3. In spite of the high uncertainty in the estimates for S- and N-sources, the higher values of S than N indicate a faster removal of the S component than N from the atmosphere before these constituents reach Dye 3.

In addition to the ionic studies made at Dye 3, Camp Century and North Central, studies were made of the total beta activity from atmospheric thermo-nuclear bomb tests and of the strong acids from major volcanic eruptions (Clausen and Hammer 1988). Table 3 lists the influx rate for the various events normalized to the deposit rates measured for the same events at North Central.

The HNL notation in Table 3 refers to the USSR bomb tests at high northern latitudes in the early 1960s; the LNL refers to the U.S.A. bomb tests at low northern latitudes around A.D. 1954. The general atmospheric transport for various natural and artificial species from source to sink varies

Table 3 Relative deposition rates of volcanic acid, radioactive debris, and anthropogenic sulfate and nitrate

	Bomb produced total					
	Volcanic acid		β activity		Anthropogenic	
Site	Laki	Tambora	HNL	LNL	Sulfate	Nitrate
Dye 3	1.8	1.2	1.2	2.1	4.1	2.9
Camp Century	2.9	1.3	1.4	2.0	4.0	2.3
North Central	1	1	1	1	1	1

with the injection height of the atmospheric plume: for the huge Tambora eruption and nuclear bomb debris it is mainly via the stratosphere; for the more moderate Laki eruption via the mid- and high troposphere; and for other anthropogenic matter generally via the mid-troposphere. The data in Table 3 suggest that North Central may be an atmospherically "sheltered" area in reference to species transported via the mid-troposphere where most pollutants are carried. Another indication that the atmospheric transport process might now be different at North Central than at Dye 3 and Camp Century is revealed in the ECM records. The upper 100 m (200 years) of snow accumulation for the Crete core (Hammer et al. 1980) was determined for acidity by pH and liquid conductivity measurements; the lower 300 m (1200 years) of the core by the ECM. The last century of these data shows no significant increase in $[H^+]$. These data also show that Crête (3200 m elevation) has a different ionic depositional pattern than does Dye 3 (2400 m elevation). This may be caused by the elevation difference which forms a threshold and results in an atmospherically "sheltered" region for the higher inland ice sheet.

Table 4 shows a comparison of the average annual HNO_3 deposition rates and $[NO_3^-]$ from locations in Greenland and Antarctica having similar snow accumulation rates and surface temperature conditions. The Antarctica locations with snow accumulation rates below 90 kg/km^2 and with very low mean annual surface temperatures which might influence HNO_3 deposition (South Pole, D 55, Dome C. Vostok) were excluded. These data are presented as 1 σ deviations from the statistical mean. If we assume that the average atmospheric source of nitrate is equal to a global background of about 9 kg HNO_3/km^2, as is found in Antarctica, the global atmospheric production of HNO_3 would amount to 1 TgN/yr. This is of the order of magnitude as the amount reported for by natural lightning sources and by stratospheric photochemical processes. Lightning as a nitrate source is believed to be more active at lower latitudes zones, and therefore a much

Table 4 Comparison of HNO_3 deposition rates in Greenland and Antarctica

	Average Annual Deposit HNO_3 kg/km^2	Average Concentration [NO_3^-] ng/g
Greenland	25 ± 10	71 ± 18
Antarctica	9 ± 3	43 ± 11

larger production rate would be needed to deposit this amount in polar glaciers due to the high HNO_3 depletion, which must occur during its long transport to the poles. Considering all the uncertainties involved, we conclude that lightning and stratospheric photochemical processes are both sources for atmospheric nitrate.

Some investigators have suggested that large NO_3^- concentrations measured in ice cores are due to volcanic activity. Table 5 lists volcanic eruptions detected in various ice cores as high [H^+] signals by the ECM. Subsequent anion analyses made on laterally adjacent samples stipulated that the signals were caused only by the strong acids H_2SO_4, HCl and HF.

Furthermore, the continuous ECM record of the Dye 3 deep core shows a number of very high acidity signals which are clearly not related to volcanism. An example is given in Fig. 2. This figure shows an acidity peak measured as HNO_3 in the A.D. 1484 layer of the deep core compared to the volcanic sulfate peak from the 1645 B.C. Thera eruption in Greece. The A.D. 1484 chemical data are from Shoji and Langway (1988) and the 1645 B.C. data from Hammer et al. (1987). The "1484 peak" was, however, connected to a 5 cm thick ice layer that originally formed at the surface during a very warm period. The thickness of an ice layer is a function of the ambient surface temperature and is dependent on elevation and latitude of the location. The higher, central polar ice sheet is colder with much less surface melting and results in less ECM signal-scatter, permitting a simpler identification of the volcanic signals. The background noise in the annual [NO_3^-] oscillations in the Dye 3 ice core is much higher than at inland stations such as Crête. The annual [NO_3^-] cycle at Crête is so regular, systematic, and predictable that it can be determined by measuring only four samples per annum (Risbo et al. 1981). In short, Table 5 and Fig. 2 illustrate the importance of cross-correlating various chemical and physical property measurements to verify findings and to avoid the misinterpretation of stratigraphic indicators, in this case spurious volcanic signals.

The North Central (NC) Greenland area has the closest resemblance to general Antarctica environmental conditions. The annual snow accumulation at NC is 129 kg/km^2; the annual NO_3^- deposit is about 13 kg HNO_3/km^2 with less pronounced seasonal [NO_3^-] variations than at Crête. The [NO_3^-] in an A.D. 1916 ice layer at NC was one order of magnitude higher than the

Table 5 Ionic measurements of strong acid signals found in ice cores

Name	Location	Year of Eruption	Strong Acid	Reference
Hekla	Iceland	A.D. 1947	H_2SO_4	Langway & Goto-Azuma 1988
Katmai	Alaska	A.D. 1912	H_2SO_4	Langway & Goto-Azuma 1988
Krakatoa	Indonesia	A.D. 1883	H_2SO_4	Langway et al. 1988
Tambora	Indonesia	A.D. 1815	H_2SO_4	Clausen & Hammer 1988; Langway et al. 1988
Laki	Iceland	A.D. 1783	H_2SO_4	Clausen & Hammer 1988
Unknown		A.D. 1600	n.i.*	Risbo et al. 1981
Unknown	Equatorial	A.D. 1258	H_2SO_4	Langway et al. 1988
Katla	Iceland	A.D. 1178	n.i.*	Risbo et al. 1981
Eldja	Iceland	A.D. 934	HCL, HF	Herron 1982b
Unknown		A.D. 516	HCL, HF	Herron 1982b
Unknown		44 B.C.	H_2SO_4	Herron 1982b
Thera	Greece	1645 B.C.	H_2SO_4	Hammer et al. 1987
Unknown		18 k yrs B.P.	HCl, HF	Hammer et al. 1985

*n.i. – not identified, only nitrate was measured.

natural background of 100 ng/g (Clausen et al. 1983). The ECM and pH measurements confirmed that nitrate was due to HNO_3. The amount of HNO_3 contained in this single ice layer is comparable to the amount of H_2SO_4 deposited at this site by the A.D. 1912 Katmai eruption. The 1916 "event" was not due to contamination or melting. This extreme nitrate peak was observed in one ice core out of three spaced less than 100 m apart.

CHLORIDE, SODIUM, AND OTHER CATIONS

Tables 1 and 2 list the $[Cl^-]$ and Table 6 lists the various cations measured in polar ice. The cation data are especially sparse and distributed over the ice sheets at very limited locations.

Contrary to $[SO_4^{2-}]$ and $[NO_3^-]$, $[Cl^-]$ shows no significant increase in ice cores during the last decades. For the Dye 3 region the background mean concentration for the last 10 000 years amounts to some 25 ng/g (Table 1, column 5).

Chloride and sodium ions have a close relationship in ice cores. In seawater, $[Cl^-]/[Na^+]$ by weight is 1.82 and $[SO_4^{2-}]/[Cl^-] = 0.14$. The latter

Table 6 Average cation concentrations and annual deposition rates for polar ice cores

1 Site location	2 Time interval A.D.	3 Years in record	4 Annual rate of accum. kg/m^2	5 Na^+ conc. ng/g	6 Na^+ annual deposit kg Na^+/km^2	7 Other cations	8 Other cations conc. ng/g	9 Other cations annual deposit kg $Cat./km^2$	10 Reference
Greenland									
Dye 3	1680–1845	6	496	12	6.0	NH_4^{+}[(a)] Ca^{2+}[(b)]	13 6.5	6.5 3.2	Busenberg & Langway 1979
Milcent	1170–1885	42	487	7[(b)]	3.6				Herron et al. 1977
Milcent	1971–1973	3	487	12[(b)]	5.8				Herron et al. 1977
Antarctica									
South Pole	1923–1936	14	73	14	1.0				Herron 1982a
South Pole	1958–1970	13	85			NH_4^+	3.0	0.3	Legrand et al. 1984
D 80	1959–1969	11	237	11	2.5	NH_4^+ Mg^{2+}	1.3 1.6	0.3 0.4	Legrand & Delmas 1986
Dome C	1969–1979	11	34			NH_4^+	1.3	0.04	Legrand et al. 1984
Vostok	Holocene	(c)	23			NH_4^+	2.0	0.05	Legrand et al. 1984
RISP C16	1971–1977	7	100	67	6.7				Herron 1982a
RISP Q13	1971–1977	7	140	202	28				Herron 1982a
GAP St25	1970–1976	7	441	42	19	Mg^{2+}	2.8	1.2	Mumford & Peel 1982
GAP Dolleman	1978–1980	3	404	382	154	Mg^{2+} Ca^{2+} K^+	46 15 15	18.5 5.9 6.1	Mulvaney & Peel 1988

[(a)] Measured by selective electrode, [(b)] Measured by Atomic Absorbtion Spectrometry, [(c)] Same comment as in Table 1, the values lie in the range (0.5–0.4) ng/g

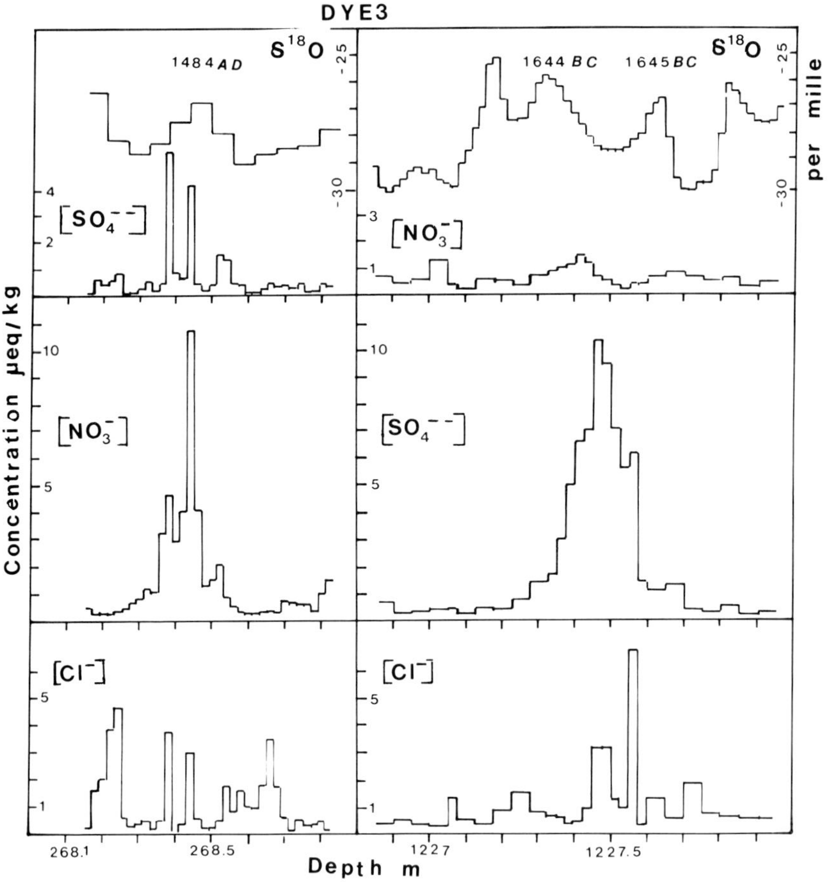

Fig. 2—The measured anion concentrations (SO_4^{2-}, NO_3^-, and Cl^-) in two Dye 3 ice core sequences which showed high acidity level in the annual layers of A.D. 1484 and 1644–1645 B.C. The upper curves show the $\delta^{18}0$ variations (scale to the right in per mille). The other curves show the anion concentrations (scale to the left in μ equivalents per kg of ice). Depth scale at the bottom in m below the 1979 summer surface. The nitrate peak is due to the high summer temperatures at the surface in A.D. 1484 and the sulfate peak is due to a volcanic eruption in 1645 B.C. (see text).

ratio is often used to correct the measured sulfate concentration for the marine contribution of [SO_4^{2-}], assuming all $C1^-$ to be of marine origin. $C1^-$ has, besides its marine origin, volcanism as a natural source. As Na^+ is considered to be entirely of marine origin, it would seem more appropriate to correct the measured [SO_4^{2-}] for its marine sulfate contribution by applying the seawater ratio of [SO_4^{2-}]/[Na^+] = 0.25 to the measured sulfate value.

Table 7 Marine sulfate contribution

Site	$[Cl^-]/[Na^+]$	% deviation from seawater $[Cl^-]/[Na^+]$ ratio 1.82	Marine sulfate contr. calc. on basis of 0.14 × $[Cl^-]$ ng/g	0.25 × $[Na^+]$ ng/g	Reference
Dye 3	2.17	+19	4	3	Busenberg & Langway 1979
South Pole	2.07	+14	4	4	Herron & Langway 1985
D80	2.36	+30	4	3	Legrand & Delmas 1986
RISP C16	2.15	+18	20	17	Herron & Langway 1985
RISP Q13	1.72	− 5	49	51	Herron & Langway 1985
GAP St.25	1.45	−20	9	11	Mumford & Peel 1982
GAP Dolleman	1.90	+ 4	102	96	Mulvaney & Peel 1988

Table 7 shows the correction for marine sulfate based on the assumption that either Cl^- or Na^+ are entirely of marine origin. The data are from ice samples where both Cl^- and Na^+ measurements have been performed (Table 1 and 6). Though the relative deviations of the measured $[Cl^-]/[Na^+]$ from the seawater ratio lay in the range −20 to +30 per cent, the data show that it really does not matter which method is applied when the measured $[SO_4^{2-}]$ is corrected for marine sulfate contribution. Even for locations with high marine contribution, the largest absolute difference between the two methods of 6 ng SO_4^{2-}/g will only introduce an error of 0.1 μeq/kg in an ion balance calculation. This error is negligible compared to the general errors in this kind of calculation. Therefore, because this correction is small, especially in ice samples from high elevated polar sites, we apply the correction based on the Cl^- measurements to the SO_4^{2-} values. The various cation data other than Na^+ are not plentiful enough to speculate about comparative relationships of higher concentration at coastal and lower altitude sites and a decrease inland towards high altitude sites.

SUMMARY AND CONCLUSIONS

An ice sheet can be envisioned as a sedimentary depositional environment. With understandable variance, ice sheets are nourished and affected by atmospheric processes and events which are faithfully recorded in the sequences that they occur. The first hurdle in the attempt to investigate the valuable palaeoenvironmental records from the body of an ice sheet was successfully overcome about thirty years ago with the advent of ice core drilling tools and techniques. Since then new ice core studies for scientific purposes have advanced simultaneously with the recovery of each new shallow and intermediate core. The development of a drill capable of augering through the entire 3000 to 4000 m thickness of a polar ice sheet was a glaciological turning point. The scientific potential of a complete investigation of the full vertical profile of the Greenland and Antarctica ice sheets could now be realized. The published results from studies of the three deep ice cores obtained to bedrock in Greenland and Antarctica in 1966, 1968, and 1981 and two other major deep ice cores from Antarctica have provided exciting previews of what is possible from a well planned and organized scientific study of deep polar ice cores.

This paper focused on the results of several studies on the major ionic species found in the polar ice sheets as measured in ice cores. The cores were recovered from various geographical locations and from different depths in Greenland and Antarctica. Core samples have been measured for H^+, SO_4^{2-}, NO_3^-, Cl^-, Na^+, Ca^{2+}, NH_4^+, Mg^{2+}, and K^+. There are limited data for all species but most information exists for the major ionic components which include H^+, SO_4^{2-}, NO_3^-, and Cl^-. All data are discontinuous and sporadic but some ionic curves extend back 10 000 years and further. Ionic sources of the polar deposits consist of all the terrestrial, marine, and extraterrestrial components of natural and artificial origin that exist in the atmosphere. Atmospheric transport and deposit mechanisms are not yet fully understood. Postdepositional processes do not noticeably alter ionic concentrations originally established at the surface. Reliable age of a sample is established by cross-correlating stable isotope and "acidity" measurements along the core with ionic and radioactive bomb debris (e.g., total β-activity) analyses.

As would be expected, the concentration levels and annual deposit rates of ionic species determined from ice core studies vary in time and space. These data are incomplete but the records show systematic trends and general patterns on a polar and hemispheric basis. Major and minor (10^5 years to yearly) climate changes are reflected as differences in concentration levels of the various ionic species. The hydrogen ion is the most dominant species found in polar deposits. Its concentration level is determined mainly by the amount of strong acids present. NO_3^- peaks are usually associated with higher summer temperatures and melt layers and show a strong seasonal preference. Sulfate in ice deposits has a multiple source origin. Chloride and sodium are primarily of marine origin. Calcium is mainly derived from continental sources and excess ammonium originates from industrial and agricultural activities.

The deep ice cores from Dye 3, Camp Century, and Byrd have sufficient ion concentration data starting back before the beginning of the Holocene age (10 000 yrs B.P.) and, with the exception of Byrd, extends to the near present snow surface. The Holocene annual deposit rates, which generally includes all available measurements to about A.D. 1900, are taken as the natural background data base. Excess sulfate concentration levels after about A.D. 1900 gradually increase to a fourfold increase in A.D. 1984. The excess sulfate increase correlates with fossil fuel combustion. The increase in nitrate starts about A.D. 1950 and has nearly doubled since then. Chloride deposition shows no average increasing trend except that associated with distance from marine sources. In general, the Antarctic average annual ion deposits are much lower than in Greenland. This is in part attributed to its isolation from land masses and differences in wind circulation. Specific major volcanic eruptions can be identified first by the ECM technique and then by ion measurements, back to thousands of years in both ice sheets. Greenland appears to have a "sheltered" central area where a lesser concentration of natural and anthropogenic ionic deposits accumulate and in different patterns. This presumably occurs because of the higher elevation in the central area which acts as a threshold or barrier to select atmospheric transport processes. Lightning does not appear to be the major producer of atmospheric HNO_3 for polar deposits. Volcanic HNO_3 emanations do not make major contributions to the strong acid signals detected by the ECM technique, but warm periods and melt layers do have elevated nitrate concentration levels.

The most striking conclusion evident from this review is the surprisingly limited amount of ionic chemistry data available for the polar ice sheets. This is hard to understand, considering the importance of ice sheets as unique chronological archives of global, hemispheric, regional, and local palaeodata. The importance of these records in revealing significant new evidence on climate change, volcanic activity, precipitation and gas chemistry, atmospheric transport and circulation, anthropogenic input, and other palaeoenvironmental information is well recognized.

Acknowledgements The authors would like to thank their respective national funding agencies, the Danish Committee on Scientific Research in Greenland and the U.S. National Science Foundation, for their support in many of the studies mentioned in this paper. H.B. Clausen received a travel grant from SUNY at Buffalo to spend a month at the University to make his contribution to this paper.

REFERENCES

Barrie, L.A.; Fisher, D.; and Koerner, R.M. 1985. Twentieth century trends in Arctic air pollution revealed by conductivity and acidity observations in snow and ice in the Canadian high Arctic. *Atmos. Envir.* **19**: 2055–2063.

Busenberg, E., and Langway, C.C. Jr. 1979. Levels of ammonium, sulfate, chloride, calcium and sodium in snow and ice from Southern Greenland. *J. Geophys. Res.* **84 (C4)**: 1705–1709.

Clausen, H.B., and Hammer, C.U. 1988. The Laki and Tambora eruptions as revealed in Greenland ice cores from 11 locations. *Ann. Glaciol.* **10**: 16–22.

Clausen, H.B.; Rasmussen, K.; and Risbo, T. 1983. Varitions of nitrate concentration in ice cores from a Greenland low accumulation area – the 1916 "event". CACGP Symposium on tropospheric chemistry, p. 50. Oxford: Pergamon.

Finkel, R.C.; Langway, C.C. Jr.; and Clausen, H.B. 1986. Changes in precipitation chemistry at Dye 3, Greenland. *J. Geophys. Res.* **91 (D9)**: 9849–9855.

Hammer, C.U. 1980. Acidity of polar ice cores in relation to absolute dating, past volcanism, and radio-echoes. *J. Glaciol.* **25 (93)**: 359–372.

Hammer, C.U.; Clausen, H.B.; and Dansgaard, W. 1980. Greenland ice sheet evidence of post-glacial volcanism and its climatic impact. *Nature* **288 (5788)**: 230–235.

Hammer, C.U.; Clausen, H.B.; Friedrich, W.L.; and Tauber, H. 1987. The Minoan eruption of Santorini in Greece dated to 1645 BC? *Nature* **328 (6130)**: 517–519.

Hammer, C.U.; Clausen, H.B.; and Langway, C.C. Jr. 1985. The Byrd ice core: continuous measurements and solid electrical condictivity measurements. *Ann. Glaciol.* **7**: 214.

Herron, M.M. 1982a. Glaciochemical dating techniques. In: Nuclear and Chemical Dating Techniques: Interpreting the Environmental Record, *ACS Sympos. Ser.* **176**: 303–318.

Herron, M.M. 1982b. Impurity sources of F^-, Cl^-, NO_3^- and SO_4^{2-} in Greenland and Antarctic precipitation. *J. Geophys. Res.* **87 (C4)**: 3052–3060.

Herron, M.M., and Langway, C.C. Jr. 1985. Chloride, nitrate and sulfate in the Dye 3 and Camp Century, Greenland ice cores. In: Greenland Ice Core: Geohysics, Geochemistry and the Environment, *Geophys. Monog.* **33**: 77–84. Washington, D.C.: Amer. Geophysical Union.

Herron, M.M.; Langway, C.C. Jr.; Weiss, H.V.; and Cragin, J.H. 1977. Atmospheric trace metals and sulfate in the Greenland ice sheet. *Geochim. Cosmo. Acta* **41**: 915–920.

Langway, C.C. Jr.; Clausen, H.B.; and Hammer, C.U. 1988. An interhemispheric volcanic time-marker in ice cores from Greenland and Antarctica. *Ann. Glaciol.* **10**: 102–108.

Langway, C.C. Jr. and Goto-Azuma, K. 1988. Temporal variations in the deep ice core chemistry record from Dye 3, Greenland. *Ann. Glaciol.* **10**: 209.

Legrand, M.; De Angelis, M.; and Delmas, R.J. 1984. Ion chromatographic determination of common ions at ultratrace levels in Antarctic snow and ice. *Analyt. Chim. Acta* **156**: 181–192.

Legrand, M.R., and Delmas, R.J. 1986. Relative contribution of tropospheric and stratospheric sources to nitrate in Antarctic snow. *Tellus* **38B**: 236–249.

Legrand, M.R., and Delmas, R.J. 1987. A 220–year continuous record of volcanic H_2SO_4 in the Antarctic ice sheet. *Nature* **327 (6124)**: 671–676.

Mayewski, P.A.; Lyons, W.B.; Spencer, M.J.; Twickler, M.: Dansgaard, W.; Koci, B.; Davidson, C.I.; and Honrath, R.E. 1986. Sulfate and nitrate concentrations from a South Greenland ice core. *Science* **232**: 975–977.

Mulvaney, R., and Peel, D.A. 1988. Anions and cations in ice cores from Dolleman Island and the Palmer Land plateau, Antarctic Peninsula. *Ann. Glaciol.* **10**: 121–125.

Mumford, J.W., and Peel, D.A. 1982. Microparticles, marine salts and stable isotopes in a shallow firn core from the Antarctic Peninsula. *Br. Antarc. Surv. Bull.* **56**: 37–47.

Neftel, A.; Beer, J.; Oeschger, H.; Zürcher, F.; and Finkel, R.C. 1985. Sulphate and nitrate concentrations in snow from South Greenland 1895–1978. *Nature* **314 (6012)**: 611–613.

Rasmussen, K.L.; Clausen, H.B.; and Risbo, T. 1984. Nitrate in the Greenland ice sheet in the years following the 1908 Tunguska event. *Icarus* **58**: 101–108.

Risbo, T.; Clausen, H.B.; and Rasmussen, K.L. 1981. Supernovae and nitrate in the Greenland ice sheet. *Nature* **294 (5842)**: 637–639.

Shoji, H., and Langway, C.C. Jr. 1988. Flow-law parameters of the DYE 3, Greenland, deep ice core. *Ann. Glaciol.* **10**: 146–150.

Steffensen, J.P. 1988. Analysis of the seasonal variations of dust, Cl^-, NO_3^-, and SO_4^{2-} in two Central Greenland firn cores. *Ann. Glaciol.* **10**: 171–177.

Volz, A., and Kley, D. 1988. Evaluation of the Montsouris series of ozone measurements made in the nineteenth century. *Nature* **332**: 240–242.

Zanolini, F.; Delmas, R.J.; and Legrand, M. 1985. Sulphuric and nitric acid concentrations and spikes along a 200 m deep ice core at D 57 (Terre Adèlie, Antarctica). *Ann. Glaciol.* **7**: 70–75.

The Environmental Record in Glaciers and Ice Sheets
eds. H. Oeschger and C.C. Langway, Jr., pp. 249–266
John Wiley & Sons Limited

The Impact of Observed Changes in Atmospheric Composition on Global Atmospheric Chemistry and Climate

P. J. Crutzen and C. Brühl

Max-Planck-Institut für Chemie
6500 Mainz, F.R. Germany

Abstract. Human activities cause significant changes in the background concentrations of radiatively and chemically active trace gases that influence atmospheric ozone and temperature distributions as well as global climate. Ozone concentrations are expected to decline strongly in the upper stratosphere and to increase in the troposphere of the Northern Hemisphere. In the lower stratosphere the situation is less clear. The sudden appearance of the stratospheric springtime "ozone hole" in Antarctica also indicates that our current capability to predict ozone changes may be severely limited. Extrapolations of the effects of human activities are also uncertain due to insufficient knowledge about trace gas budgets. Improved analyses about preindustrial atmospheric trace gas contents are an important contribution leading to a better understanding of the natural components of the trace gas budgets.

INTRODUCTION

The atmosphere consists of a mixture of gases, dominated in abundance by N_2 (78.08% by volume in dry air), O_2 (20.948%), and Ar (0.934%), leaving only some 0.035% or 350 ppm to the sum of all other gases in dry air. None of the major gases show any measurable variability in the atmosphere, indicating that they have very long lifetimes. For all practical purposes they can, therefore, be treated as nonvariable "background gases." Argon is of course a totally inert gas and the same almost applies to nitrogen, except for the fact that a very small fraction of it is fixed by microorganisms on land and in the oceans, in fires and in lightning discharges leading, e.g., to NH_3, NO, and NO_2, which play major roles in the chemistry of the atmosphere. Molecular oxygen is also rather inert. However, in the stratosphere and troposphere a fraction ranging between 10^{-7} to 10^{-5} of it is converted to ozone. The formation of the fixed nitrogen and ozone compounds is of the utmost importance for biosphere, atmospheric chemistry, and climate.

The most reactive gases in the atmosphere are the variable and minor constituents. First of all, there is water vapor. Its volume mixing ratio varies from a few ppm in the stratosphere to a few percent above tropical oceans. Water in its various phases plays major roles both in atmospheric chemistry and in climate. Carbon dioxide, with a current volume mixing ratio of about 345 ppm, is the next most abundant gas. It is of major importance in biospheric processes and, as a "greenhouse" gas, also in the Earth's climate. Due to fossil fuel combustion, deforestation, and other human influenced activities, the abundance of CO_2 in the atmosphere has been and still is steadily increasing at a current rate of about 0.4% per year (Keeling et al. 1984). Except for some influence on the acid-base chemistry of hydrometeors, CO_2 has a negligible effect on the chemistry of the atmosphere. For the atmospheric chemist the most interesting gases are those in the ppm (10^{-6}), ppb(10^{-9}), ppt(10^{-12}), or smaller range. An overview of global "background" atmospheric chemistry and how it has been and will be affected by anthropogenic activities, based on "reasonable" scenarios for these, is given below.

SHORT OVERVIEW OF ATMOSPHERIC CHEMISTRY

The Stratosphere

In the stratosphere ozone is formed by the photodissociation of molecular oxygen, followed by the recombination of O atoms with O_2 molecules.

$$O_2 + h\nu \rightarrow 2\,O, \quad \lambda < 240\text{ nm} \tag{1}$$

$$O + O_2 + M \rightarrow O_3 + M \quad (2x) \tag{2}$$

$$\text{net: } 3\,O_2 \rightarrow > 2\,O_3$$

The great majority of ozone (~99.9%) is also destroyed in the stratosphere by catalytic sets of reactions which may be summarized as follows:

$$O_3 + h\nu \rightarrow O + O_2, \quad \lambda < 1140\text{ nm} \tag{3}$$

$$X + O_3 \rightarrow XO + O_2 \tag{4}$$

$$O + XO \rightarrow X + O_2 \tag{5}$$

$$\text{net: } 2\,O_3 \rightarrow 3\,O_2.$$

In this reaction cycle the catalyst X may be NO or Cl. Within the natural stratosphere, catalytic reactions of NO and NO_2 are the most important ones for the ozone balance. NO mainly comes from the oxidation of nitrous oxide (N_2O), a product of microbiological processes in soils, which is also formed during the combustion of N-containing fossil fuels (Hao et al. 1987) and plant matter (Crutzen et al. 1985). The present atmospheric N_2O content of about 303 ppb is growing at a rate of about 0.2% per year (Schnell 1986).

The most important human impact on stratospheric ozone is due to the rapid growth of Cl and ClO (ClO_x) catalysts which are produced by the photodissociation of $CFCl_3$, CF_2Cl_2, CCl_4, $C_2F_3Cl_3$, CH_3CCl_3, and other industrial products. Together, these gases have until now added about 2 ppb of chlorine to the natural background level of 0.6 ppb, originating from CH_3Cl oxidation in the stratosphere. Consequently, important changes in the stratospheric ozone distribution are taking place. As an additional problem, several of the above chlorofluorocarbon gases also have strong absorption bands in the atmospheric infrared window region from 7–12 μm. Their growth in the atmosphere by about 5% annually can, therefore, also contribute in major ways to the global "greenhouse" warming.

Here we also wish to point to the following important interactions between the NO_x and ClO_x catalysts, which protect stratospheric ozone against otherwise more rapid destruction by the ClO_x catalysts:

$$ClO + NO_2 + M \rightarrow ClONO_2 + M \quad (6)$$

$$ClO + NO \rightarrow Cl + NO_2 \quad (7)$$

$$Cl + CH_4 \rightarrow HCl + CH_3. \quad (8)$$

Reaction 6 converts both NO_x and ClO_x catalysts into $ClONO_2$ which does not react with ozone, while reactions 7 and 8 convert active ClO_x into unreactive HCl with NO and NO_2 acting as catalysts. In addition, below about 25 km, NO_x catalyzes the destruction of OH radicals in the stratosphere, e.g., via

$$OH + NO_2 (+M) \rightarrow HNO_3 (+M) \quad (9)$$

$$OH + HNO_3 \rightarrow H_2O + NO_3 \quad (10)$$

$$NO_3 + h\nu \rightarrow NO_2 + O \quad (11)$$

- -

$$\text{net: } 2\,OH \rightarrow H_2O + O,$$

again leading to ozone protection as it lowers the production of reactive ClO_x from HCl

$$HCl + OH \rightarrow Cl + H_2O. \quad (12)$$

Through reaction 8, CH_4 is important for the photochemistry of the stratosphere, as it is the main reaction which converts active chlorine into HCl. Furthermore, the oxidation of about 1.6 ppm of CH_4 leads to about 3.2 ppm of H_2O, causing a doubling of water vapor mixing ratios going from low to high altitudes in the stratosphere.

The Troposphere

Although only about 10% of all ozone is located in the troposphere, it is of fundamental importance for the chemical composition of the atmosphere as it is photolyzed by solar ultraviolet radiation at wavelengths below about 310 nm, producing electronically excited oxygen atoms

$$O_3^{+} h\nu \rightarrow O(^1D) + O_2,\ \lambda \leqq 310\ \text{nm}, \quad (13)$$

which can react with water vapor to form hydroxyl radicals

$$O(^1D) + H_2O \rightarrow 2OH. \quad (14)$$

Attack by hydroxyl radicals is the first step in the oxidation of most compounds that are emitted into the atmosphere by natural and human activities.

In the "background" atmosphere OH reacts mainly with CO ($\approx$ 70%) and CH_4 ($\approx$ 30%). Of particular importance for tropospheric chemistry are interactions involving CH_4, CO, NO_x, O_3, and OH. For example, in NO-rich environments the oxidation of CH_4 may proceed via

$$CH_4 + OH\,(+O_2) \rightarrow CH_3O_2 + H_2O \quad (15)$$

$$CH_3O_2 + NO \rightarrow CH_3O + NO_2 \quad (16)$$

$$CH_3O + O_2 \rightarrow CH_2O + HO_2 \quad (17)$$

$$HO_2 + NO \rightarrow OH + NO_2 \quad (18)$$

$$2\times (NO_2 + h\nu \rightarrow NO + O,\ \lambda < 400\ \text{nm}) \quad (19)$$

$$2\times (O + O_2 + M \rightarrow O_3 + M) \quad (2)$$

- -

$$\text{net: } CH_4 + 4\,O_2 \rightarrow CH_2O + H_2O + 2\,O_3,$$

or in NO-poor environments via

$$CH_4 + OH\,(+O_2) \rightarrow CH_3O_2 + H_2O \quad (15)$$

$$CH_3O_2 + HO_2 \rightarrow CH_3O_2H + O_2 \quad (20)$$

$$CH_3O_2H + h\nu \rightarrow CH_3O + OH \quad (21)$$

$$CH_3O + O_2 \rightarrow CH_2O + HO_2 \quad (17)$$

- -

$$\text{net: } CH_4 + O_2 \rightarrow CH_2O + H_2O.$$

As CH_3O_2H also reacts with OH, the oxidation of methane in NO-poor environments leads to the destruction of odd hydrogen (OH, HO_2) radicals. Because one of the reaction paths of the photolysis of CH_2O causes the production of two HO_2 radicals,

$$CH_2O + h\nu \rightarrow H + CHO \quad (22)$$

$$H + O_2 + M \rightarrow HO_2 + M \quad (23)$$

$$CHO + O_2 \rightarrow CO + HO_2 \quad (24)$$

with about 30% probability, the oxidation of methane critically influences the ozone and odd hydrogen radical chemistry of the atmosphere. Here we will refrain from listing all possible chemical routes for CH_4 oxidations (for a more detailed discussion see, e.g., Crutzen 1987), but will only give the important average net results, which are as follows.

During the oxidation of CH_4 to CO, there will be:

1. in NO-poor environments, a net destruction of 2–3.5 odd hydrogen radicals and 0.7 ozone molecules,
2. in NO-rich environments, a net production of 0.5 odd hydrogen radicals and 2.7 ozone molecules.

In addition, considering that CO oxidation to CO_2 in NO-rich environments occurs through

$$CO + OH \rightarrow H + CO_2 \quad (25)$$

$$H + O_2 + M \rightarrow HO_2 + M \quad (23)$$

$$HO_2 + NO \rightarrow OH + NO_2 \quad (18)$$

$$NO_2 + h\nu \rightarrow NO + O,\ \lambda < 400\text{ nm} \quad (19)$$

$$O + O_2 + M \rightarrow O_3 + M \quad (2)$$

- -

$$\text{net: } CO + 2O_2 \rightarrow CO_2 + O_3,$$

and in NO-poor environments either via

$$CO + OH \rightarrow H + CO_2 \quad (25)$$

$$H + O_2 + M \rightarrow HO_2 + M \quad (23)$$

$$HO_2 + O_3 \rightarrow OH + 2O_2 \quad (26)$$

- -

$$\text{net: } CO + O_3 \rightarrow CO_2 + O_2$$

or via

$$2\times (CO + OH \rightarrow H + CO_2) \quad (25)$$

$$2\times (H + O_2 + M \rightarrow HO_2 + M) \quad (23)$$

$$HO_2 + HO_2 \rightarrow H_2O_2 + O_2 \quad (27)$$

- -

$$\text{net: } 2\,CO + O_2 + 2\,OH \rightarrow 2\,CO_2 + H_2O_2,$$

the oxidation of CO to CO_2 leads to the production of one ozone molecule in NO-rich or to the potential loss of one ozone molecule in NO-poor environments. H_2O_2, a potent oxidant in cloud droplets, is another important by-product of these reactions.

The borderline between NO-rich and NO-poor is roughly defined by an NO to O_3 ratio of about 2×10^{-4} at the Earth's surface, corresponding to a volume mixing ratio of about 5 ppt. Because the lifetime of NO_x is only of the order of a few days and major emissions take place from fossil fuel burning in the midlatitude regions of the Northern Hemisphere, NO-rich environments are mainly located there (see Table 1).

Table 1 Estimated global emissions of NO to the atmosphere in units of 10^{12} g N per year (Crutzen et al. 1985; Ehhalt and Drummond 1982; Galbally and Roy 1978; Logan 1983; WMO 1986)

Anthropogenic Emissions		Natural Emissions	
Industrial	13–29	Soil exhalations	5–15
Aircraft	0.3	Lightning	2–10
Biomass burning	3–7	Marine	0·15
Agriculture (fertilizers)	0.5–1	Flow from stratosphere	0·5

PAST AND FUTURE CHANGES IN ATMOSPHERIC COMPOSITION

Significant changes are taking place in the "background" concentrations and emissions of various trace gases that are important for global atmospheric photochemistry and climate. Here we will discuss those that are due to CO_2, CH_4, N_2O, the chlorofluorocarbons, CO, and NO_x on the basis of information of their past emissions or atmospheric concentrations.

To estimate the past, ice core data are available for CO_2, CH_4 and N_2O, which show that increases in the volume mixing ratios (μ) for CO_2 (Andreae et al. 1988; Barnola et al. 1987; Neftel et al. 1985), CH_4 (Bolle et al. 1986; Craig and Chou 1982; Rasmussen and Khalil 1984; Stauffer et al. 1985), and N_2O (Pearman et al. 1986; Weiss and Craig 1976; Zardini et al. 1987) can be described reasonably well by the formulae

$$\mu(CO_2) = 268 + 70\,R(t)\ \text{ppm} \tag{28}$$

$$\mu(CH_4) = 0.53 + 1.07\,R(t)\ \text{ppm} \tag{29}$$

$$\mu(N_2O) = 273 + 33\,R(t)\ \text{ppm}, \tag{30}$$

where R(t) denotes the ratio between past and present world populations (UN 1984; Clark 1982). The reason for the increase in CO_2 is mainly fossil fuel burning and emissions due to agricultural soil organic matter oxidation and deforestation (e.g., Oeschger and Siegenthaler 1988). The currently observed growth is about 0.35% per year.

For the future, CO_2 concentrations will be estimated by considering the "base case" scenario of Mintzer (1987), which projects a growth of about 0.6% per year over the next century.

Methane emissions to the atmosphere are relatively more influenced by human activities, so that the current growth rate is of the order of about 1% per year. Various human activities contribute strongly to the atmospheric methane source, such as animal husbandry, coal mining, landfills, tropical biomass burning, and probably most of all rice cultivation (Bingemer and Crutzen 1987; Cicerone et al. 1983; Holzapfel-Pschorn and Seiler 1986). Table 2 gives a CH_4 budget which especially reflects current uncertainties about the wetland sources. For the future we extrapolate CH_4 concentrations by scaling to the projected world population as in equation 29.

Regarding the N_2O budget, large emissions arise from the burning of nitrogenous fossil fuels and biomass (Crutzen et al. 1985; Pierotti and Rasmussen 1976; Weiss and Craig 1976). Interestingly, the N_2O to CO_2 molecular emission ratios for coal, oil, and biomass combustion are all about equal to 2×10^{-4}. Consequently, the estimated N_2O source from these combustion sources may add up to about 3 Tg($1\,Tg=10^{12}$g) N_2O–N per year (Bolle et al. 1986; Crutzen et al. 1985; Hao et al. 1987). Our estimates on

Table 2 Estimated sources and sinks of methane in units of 10^{12} g per year (Bingemer and Crutzen 1987)

Sinks		Sources	
Tropospheric reactions with OH	290–350	Domestic animals (ruminants)	70–80
Stratospheric reactions with OH	25–35	Natural gas leaks	$\leqq$ 35
Uptake by aerobic soils	10–30	Coal mining	30–40
Annual growth (~ 1% per year)	50	Landfills	30–70
Total required source (sum of the above)	375–465	Natural fauna	$\leqq$30
		Rice fields/natural wetlands (by difference)	135–320

the global N_2O source are the following. The current loss of N_2O in the stratosphere is in the range of 6–10 Tg/year (Crutzen and Schmailzl 1983). With a global increase of about 0.2%/year, an additional 3 Tg N is added annually to the atmosphere, requiring together a global source of 9–13 Tg/year. Because the preindustrial atmosphere contained some 280 ppb of N_2O (Pearman et al. 1986) the natural source of N_2O can be estimated to have been 5.5–9.2 Tg N/yr, although due to deforestation activities this can well be an underestimate for present conditions, considering that tropical forests are an important source for N_2O (Keller et al. 1986). Taking into account the production of 3 Tg N/yr from fossil fuel and biomass combustion, at most 1 Tg N/yr must be supplied by other sources, such as release from soil organic matter lost due to agriculture and denitrification and nitrification of nitrogen fertilizer. With a total use of N-fertilizer of some 60 Tg annually (FAO 1984), the average global yield of N_2O–N would, therefore, have been at most 1.5%. Such low N_2O yields agree with available field data (Bolle et al. 1986). However, available data are yet too ambiguous to allow this conclusion with certainty.

For the future, we will adopt the N_2O emissions as derived by Kavanaugh (1987), which are based on the assumption that the growth in N_2O emissions is dominated by the combustion of coal and oil (Table 3).

No ice core data are yet available on carbon monoxide. Some information from the Northern Hemisphere indicates the possibility of a trend by 0.5–4% per year from 1950 to 1977 (Rinsland and Levine 1985). An upward trend by 2% per year has been observed for the winter months, but no long-term trends were observed during summer (Dianov-Klokov and Yurganov 1981). No significant trends are reported in the Southern Hemisphere. Because of paucity of data we will try to describe the atmospheric CO concentrations by estimated emission rates and photochemical modeling. The emissions

Table 3 Scenarios for future emissions of CO, N_2O, and NO_x in units of 10^{12} g CO and 10^{12} g N per year

Year	CO	N_2O–N	NO_x–N
1985	350	3.5	20
2000	280	9.2	30
2025–2075	320	13.7	46

from natural and industrial sources and from biomass burning in molecules $m^{-2}\ s^{-1}$ prior to 1985 are estimated by

$$I_{CO}(t) = I_{CO,nat} + I_{CO,ind} + I_{CO,bio} \tag{31}$$

with

$$\begin{aligned} I_{CO,nat} &= 1.3 \times 10^{15} \\ I_{CO,bio} &= 7 \times 10^{14} \\ I_{CO,ind} &= 5 \times 10^{14}\ R_{CO2}(t). \end{aligned}$$

Natural CO inputs to the atmosphere were estimated to be equal to about 10^{15} g CO, mostly due to the oxidation of hydrocarbons that are emitted by vegetation. The biomass burning contribution $I_{CO,bio}$, which comes mostly from developing countries, is based on data given in Crutzen et al. (1985), assuming an average volume emission ratio between CO and CO_2 of 12% and a global CO_2 production rate of 2×10^{15} g C/yr due to biomass burning. Since most hydrocarbon emissions from forests and most biomass burning occur in the tropics, and deforestation activities decrease one but increase the other source, we will assume that the sum of the natural and biomass sources of CO have been constant during the past and will remain constant for the future. The industrial source for CO in 1980 corresponding to 3.5×10^{14} g has been taken from Kavanaugh (1987). For the past, the CO source will be extrapolated from equation 31 with R_{CO2} (t) the CO_2 emission rate by fossil fuel burning relative to the year 1980 as given in Clark (1982). For the future, CO emissions are taken as shown in Table 3.

For calculations which simulate processes in the two hemispheres, natural emissions are distributed according to the relative land areas so that fluxes in the Northern Hemisphere are multiplied by 1.4 and those in the Southern Hemisphere by 0.6. Because biomass burning and hydrocarbon emissions take place mostly in the tropics, global average emission fluxes are multiplied by 1.1 and 0.9 in the Northern and Southern Hemisphere, respectively.

Industrial emissions are assumed to be confined to the Northern Hemisphere.

NO_x emissions at the Earth's surface are treated analogous to those of CO, so that in units of molecules $m^{-2}s^{-21}$

$$I_{NO,nat} = 2 \times 10^{13}$$

$$I_{NO,ind} = 5.4 \times 10^{13} \text{ in A.D. } 1980$$

$$I_{NO,bio} = 1.4 \times 10^{13} R_d(t). \qquad (32)$$

The natural emissions correspond to a total global emission from soils of about 7.5 Tg N/yr (WMO 1986; Galbally and Roy 1978). Our lower estimate takes into account that reabsorption of NO_2, formed from NO oxidation by ozone, may occur very rapidly by vegetation, so that a substantial fraction of the NO emissions will not be available for reactions in the atmosphere. Lightning discharges are another natural source of NO for which we assume a global production rate of 3.5 Tg N/year, distributed uniformly between the Earth's surface and 7 km altitude (Chameides et al. 1977; WMO 1986).

For the emissions originating from industry and transportation for 1980, a rate of 20 Tg N/yr (e.g., Ehhalt and Drummond 1982; Logan 1983) is adopted. Future industrial emissions of NO_x follow those projected by Kavanaugh (1987), as shown in Table 3. The NO_x production by biomass burning, about 5 Tg N/yr, agrees with the estimates by Crutzen et al. (1985) and Andreae et al. (1988). For the past we weighted the current emission by populations in the developing world. For the future we keep this source constant as probably much of the tropics is already now affected by agricultural biomass burning practices.

Since NO_x emissions occur mostly in the Northern Hemisphere and the mean atmospheric lifetime of NO_x is only a few days, we assume that hemispheric average emissions describe the distributions of NO_x and its influence on ozone. For the calculations, the natural emissions of NO_x from soils are, therefore, multiplied by factors of 1.4 and 0.6 for the Northern and Southern Hemisphere, respectively, since they are mainly dependent on the relative fractions of land areas. The lightning source of NO_x is distributed according to the lightning frequency distribution in the two hemispheres (Turman and Edgar 1982). All industrial emissions of NO_x are assumed to take place in the Northern Hemisphere. NO_x emissions from biomass burning follow the approximate ratios of tropical land areas in the two hemispheres, so that emissions are multiplied by 1.1 and 0.9 in the Northern and Southern Hemisphere, respectively.

The natural background of the organic chlorine compounds is assumed to be represented by the currently observed average CH_3Cl volume mixing ratios of 0.6 ppbv. For most scenarios, estimated past emission rates of the industrial chlorocarbon gases ($CFCl_3$, CF_2Cl_2, CH_3CCl_3, and CCl_4) were used (Cunnold et al. 1983a, b; Prinn et al. 1983; Simmonds et al. 1983). For the future, in agreement with international proposals (UNEP 1987), starting in 1994 a 20% cut in the worldwide production of the major industrial chlorine-containing organics has been assumed, followed by a further production cut by 30% in 1999. Production of bromine-containing organics should stay on present levels, while no restrictions are assumed for CH_3 CCl_3 and CHF_2Cl. We would like to point out that our assumptions may be optimistic, as a 50% reduction in CFC emissions by the year 2000 only applies to the principal producing countries in the West. Consequently, the stratospheric ozone changes, which we calculate, may well underestimate future developments.

CALCULATED CHANGES IN TEMPERATURES AND OZONE

With the aid of a one-dimensional model which simulates radiative and photochemical processes, as well as vertical transport of trace gases, we have calculated both the vertical distributions of temperatures and chemical constituents in the atmosphere, whereby to a first approximation the results may be considered to represent hemispheric averages at various height levels between the Earth's surface and 60 km. The adopted model is time-dependent so that changes in atmospheric emissions are taken into account. An important feature of the model is also that heat transfer into the ocean surface mixed layer and from there into the deep ocean has been considered. A new tropospheric lapse rate parameterization takes convective and baroclinic processes into account (for a detailed description see Brühl and Crutzen 1988). Due to the limitations of one-dimensional models, cloud feedback is not included. Our model calculates profiles of all important chemical constituents and temperatures that agree favorably with observations.

Vertical profiles of percentage ozone and temperature changes that are calculated for the assumed scenario of trace gas increases are shown in Fig. 1. Especially for the upper stratosphere near 35–50 km, large ozone depletions by several tens of percents and concomitant temperature reductions by some 20°C are calculated to happen by the middle of the next century. The ozone depletions are mainly due to increasing concentrations of chlorine radicals in the stratosphere, while the calculated temperature reductions are caused both by less absorption of ultraviolet radiation due to diminished ozone concentrations and by enhanced infrared radiation emission to space by growing concentrations of carbon dioxide.

In the lower troposphere the opposite results are calculated. Due to enhanced concentrations of CH_4, CO, and NO_x, average northern hemispheric ozone concentrations are calculated to have increased by about 70% since preindustrial times with a possibility of a further increase by about

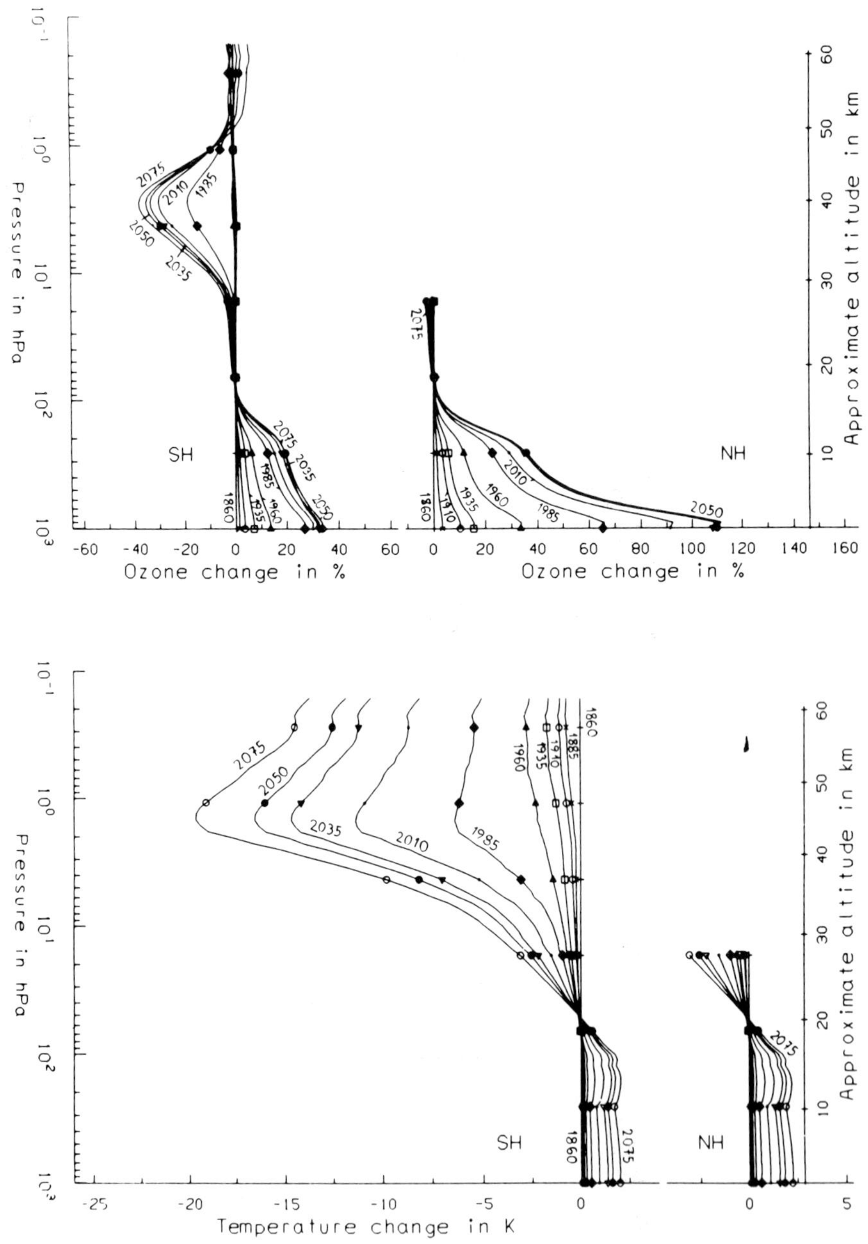

Fig. 1—Calculated changes of average temperature and ozone height profiles in the Northern and Southern Hemispheres, adopting the emission scenario of the present paper.

50% by the middle of the next century. Strong increases in the concentrations of surface ozone over the past century have indeed been shown to have occurred (Levy 1907; Volz and Kley 1988). Current upward trends at background stations in the Northern Hemisphere are close to 1% per year (Oltmans 1985).

With the adopted scenarios, surface temperature increases are calculated to have been about 0.7°C over the past hundred years due to the growth in greenhouse gas concentrations (see Fig. 2). A further increase in average temperatures at the Earth's surface by about 1°C, until the middle of the next century, and larger subsequent heating are estimated. As shown in Fig. 2, the combined contribution of the greenhouse gases other than CO_2 can be about as large as that of CO_2 alone (Ramanathan et al. 1985). We also note that surface temperature increases in the Northern Hemisphere may be larger than in the Southern Hemisphere due to less heat transfer into the oceans and a larger growth in tropospheric ozone.

As shown in Fig. 3, calculated total ozone changes indicate the possibility of depletions by about 5% for the year 2075 in the Southern Hemisphere, but less in the Northern Hemisphere. The difference reflects the anticipated

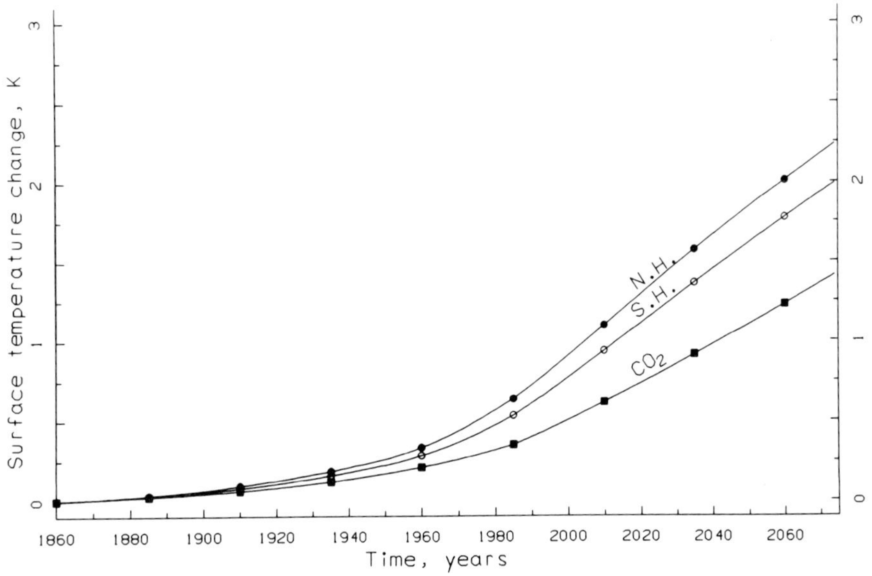

Fig. 2—Estimated temperature increases at the Earth's surface for the emission scenario adopted in this paper. Lower curve shows the "greenhouse effect" by CO_2 alone. Upper curves show total trace gas heating effects in the two hemispheres, which are lower in the Southern Hemisphere due to more heat transfer to oceans and less tropospheric ozone.

increase in tropospheric ozone in the Northern Hemisphere due to expanding emissions of NO_x and CH_4. Because the average atmospheric lifetime of NO_x is only a few days, we must caution, however, that the use of global or hemispheric averages of NO_x emissions in a one-dimensional model clearly overemphasizes the domain of influence of the NO_x emissions, so that the average tropospheric ozone concentration increases in the Northern Hemisphere may be overestimated and total ozone depletions underestimated. Regarding the stratospheric contribution to the total ozone depletion, the result obtained for the Southern Hemisphere indicates that these will be responsible for a total ozone loss of about 4% by the middle of the next century (2050). We should mention that not only CFC emissions but also those of N_2O are responsible for substantial ozone depletions. In the current scenario, N_2O volume mixing ratios are calculated to be about 400 ppb by the middle of the next century.

We must also warn that model calculations of ozone depletions may considerably underestimate future events. Currently measured total ozone depletions are substantially larger than model predictions, especially during September to November at high latitudes in the Southern Hemisphere (Farman et al. 1985; Stolarski et al. 1986), probably due to a combination of special chemical events involving reactions on polar stratospheric haze

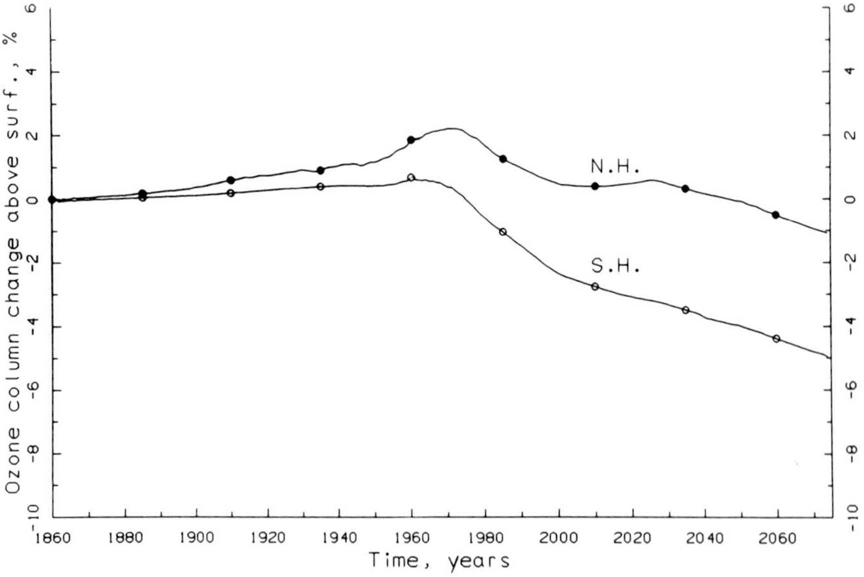

Fig. 3—Estimated percentage of total ozone changes in the two hemispheres, calculated with the emission scenario adopted in this paper.

and cloud particles (Crutzen and Arnold 1986; Molina et al. 1987; Solomon et al. 1986; Tolbert et al. 1987; Wofsy et al. 1988). Reactions of this kind have not yet been included in any predictive models of global stratospheric ozone depletions.

CONCLUSION

As exemplified in this paper by one particular scenario, the global effects of human activities are now so large that major changes in climate and in the global ozone distribution may be foreseen. Major ozone changes have in fact already occurred.

Predictions of the future effects of human activities are still uncertain, e.g., because of lack of knowledge about important climate determining factors, such as cloud feedback, as well as quantitive information about biospheric trace gas or aerosol sources. Analyses of air composition in ice cores will be of the utmost importance, especially in order to understand the natural source strengths. Unfortunately for a number of important trace gases, such as O_3, CO, and N_2O, current trends of concentrations are not well characterized.

Acknowledgements. This work is supported by the Umweltbundesamt, Berlin (German Environmental Agency).

REFERENCES

Andreae, M. O.; Browell, E. V.; Garstang, M. A.; Gregory, G. L.; Harriss, R. C.; Hill, G. F.; Jacob, D. J.; Pereira, M. C.; Sachse, G. W.; Setzer, A. W.; Silva Dias, P. L.; Talbot, R. W.; Torres, A. L.; and Wofsy, S. C. 1988. Biomass burning and associated haze layers over Amazonia. *J. Geophys. Res.* **93**: 1509–1527.

Barnola, J. M.; Raynaud, D.; Korotkevich, Y. S.; and Lorius, C. 1987. Vostok ice core provides 160,000-year record of atmospheric CO_2. *Nature* **329**: 408–414.

Bingemer, H., and Crutzen, P. J. 1987. The production of methane from solid wastes. *J. Geophys. Res.* **92**: 2181–2187.

Bolle, H. -J.; Seiler, W.; and Bolin, B. 1986. Other greenhouse gases and aerosols. In: The Greenhouse Effect, Climatic Change and Ecosystems, eds. B. Bolin et al., SCOPE 29, pp. 157–203. Chichester, New York: Wiley.

Brühl, C., and Crutzen P. J. 1988. Scenarios of possible changes in atmospheric temperatures and ozone concentration due to man's activities, estimated with a one-dimensional coupled photochemical climate model. *Climate Dyn.* **2**: 173–203.

Chameides, W. L.; Stedman, D. H.; Dickerson, R. R.; Rusch, D. W.; and Cicerone, R. J. 1977. NO_x production in lightning. *J. Atmos. Sci.* **34**: 143–149.

Cicerone, R. J.; Shetter, J. D.; and Delwiche, C. C. 1983. Seasonal variation of methane from a Californian rice paddy. *J. Geophys. Res.* **88**: 11022–11024.

Clark, W. C., ed. 1982. Carbon Dioxide Review 1982. Oxford: Clarendon.

Craig, H., and Chou, C. C. 1982. Methane: the record in polar ice cores. *Geophys. Res. Lett.* **9**: 1221–1224.

Crutzen, P. J. 1987. The role of the tropics in atmospheric chemistry. In: Geophysiology of Amazonia, ed. R. Dickinson, pp. 107–130. New York: Wiley.

Crutzen, P. J., and Arnold, F. 1986. Nitric acid cloud formation in the cold Antarctic stratosphere: a major cause for the springtime "ozone hole." *Nature* **324**: 651–655.

Crutzen, P. J.; Delany, A. C.; Greenberg, J.; Haagenson, P.; Heidt, L.; Leub, R.; Pollock, W.; Seiler, W.; Wartburg, A.; and Zimmerman, P. 1985. Tropospheric chemical composition measurements in Brazil during dry season. *J. Atmos. Chem.* **2**: 233–256.

Crutzen, P. J., and Schmailzl, U. 1983. Chemical budgets of the stratosphere. *Planet. Space Sci.* **32**: 1009–1032.

Cunnold, D. M.; Prinn, R. G.; Rasmussen, R. A.; Simmonds, P. G.; Alyea, F. N.; Cardelino, C. A.; and Crawford, A. J. 1983a. The atmospheric lifetime experiment – 4. Results for CF_2Cl_2 based on three years data. *J. Geophys. Res.* **88**: 8401–8414.

Cunnold, D. M.; Prinn, R. G.; Rasmussen, R. A.; Simmonds, P. G.; Alyea, F. N.; Cardelino, C. A.; Crawford, A. J.; Fraser, P. J.; and Rosen, R. D. 1983b. The atmospheric lifetime experiment – 3. Lifetime methodology and application to three years of $CFCl_3$ data. *J. Geophys. Res.* **88**: 8379–8400.

Dianov-Klokov, V. I., and Yurganov, L. N. 1981. A spectroscopic study of the global space-time distribution of atmospheric CO. *Tellus* **33**: 262–273.

Ehhalt, D. H., and Drummond, J. W. 1982. The tropospheric cycle of NO_x. In: Chemistry of Unpolluted and Polluted Atmospheres, eds. H.-W. Georgii and W. Jaeschke, pp. 219–251. Dordrecht: Reidel.

FAO. 1984. 1983 Fertilizer Yearbook 33. Rome: Food and Agricultural Organization of the United Nations.

Farman, J. C.; Gardiner, B. G.; and Shanklin, J. D. 1985. Large losses of total ozone in Antarctica reveal seasonal ClO_x/NO_x interaction. *Nature* **315**: 207–210.

Galbally, I. E., and Roy, C. R. 1978. Loss of fixed nitrogen from soils by nitric oxide exhalation. *Nature* **275**: 734–735.

Hao, W. M.; Wofsy, S. C.; McElroy, M. B.; Beer, J. M.; and Togan, M. A. 1987. Sources of atmospheric nitrous oxide from combustion. *J. Geophys. Res.* **92** (D3): 3098–3104.

Holzapfel-Pschorn, A., and Seiler, W. 1986. Methane emission during a cultivation period from an Italian rice paddy. *J. Geophys. Res.* **91**: 11803–11814.

Kavanaugh, M. 1987. Estimates of future CO, N_2O and NO_x emissions from energy combustion. *Atmos. Envir.* **21**: 463–468.

Keeling, C. D.; Carter, A. F.; and Mook, W. G. 1984. Seasonal, latitudinal and secular variations in the abundance and isotopic ratios of atmospheric CO_2. *J. Geophys. Res.* **89**: 4615–4628.

Keller, M.; Kaplan, W. A.; and Wofsy, S. C. 1986. Emissions of N_2O, CH_4 and CO from tropical forest soils. *J. Geophys. Res.* **91**: 11791–11802.

Levy, A. 1907. Analyse de l'air atmosphérique Ozone. *Annales d'Observatoire Municipal de Montsouris* **8**: 289–291.

Logan, J. A. 1983. Nitrogen oxides in the troposphere: global and regional budgets. *J. Geophys. Res.* **88**: 10785–10807.

Mintzer, I. M. 1987. A Matter of Degrees: The Potential for Controlling the Greenhouse Effect. Research Report No. 5. Washington: World Resources Institute.

Molina, M. J.; Tso, T. -L.; Molina, L. T.; and Wang, F. C. -Y. 1987. Antarctic stratospheric chemistry of chlorine nitrate, hydrogen chloride, and ice: release of active chlorine. *Science* **238**: 1253–1257.

Neftel, A.; Moor, E.; Oeschger, H.; and Stauffer, B. 1985. Evidence from polar ice cores for the increase in atmospheric CO_2 in the past two centuries. *Nature* **315**: 45–47.

Oeschger, H., and Siegenthaler, U. 1988. How did the atmospheric CO_2 change? In: The

Changing Atmosphere, eds. F. S. Rowland and I. S. A. Isaksen, pp. 5–23. Dahlem Konferenzen. Chichester: Wiley.

Oltmans, S. J. 1985. Tropospheric ozone at four remote observatories. In: Atmospheric Ozone, eds. C. S. Zerefos and A. Ghazi, pp. 730–734. Dordrecht: Reidel.

Pearman, G. I.; Etheridge, D.; deSilva, F.; and Fraser, P. J. 1986. Evidence of changing concentrations of atmospheric CO_2, N_2O and CH_4 from air bubbles in Antarctic ice. *Nature* **320**: 248–250.

Pierotti, D., and Rasmussen, R. A. 1976. Combustion as a source of nitrous oxide. *Geophys. Res. Lett.* **3**: 265–267.

Prinn, R. G.; Rasmussen, R. A.; Simmonds, P. G.; Alyea, F. N.; Cunnold, D. M.; Lane, B. C.; Cardelino, C. A.; and Crawford, A. J. 1983. The Atmospheric Lifetime Experiment – 5. Results for CH_3CCl_3 based on three years data. *J. Geophys. Res.* **88**: 8415–8426.

Ramanathan, V.; Cicerone, R. J.; Singh, H. B.; and Kiehl, J. T. 1985. Trace gas trends and their potential role in climate change. *J. Geophys. Res.* **90**: 5547–5566.

Rasmussen, R. A., and Khalil, M. A. K. 1984. Atmospheric methane in the recent and ancient atmospheres: concentrations, trends and interhemispheric gradient. *J. Geophys. Res.* **89**: 11599–11605.

Rinsland, C. P., and Levine, J. S. 1985. Free tropospheric carbon monoxide concentrations in 1950 and 1951 deduced from infrared total column amount measurements. *Nature* **318**: 250–254.

Schnell, R. C. 1986. Geophysical Monitoring for Climatic Change, No. 14. Boulder, CO: Air Resources Laboratory, NOAA.

Simmonds, P. G.; Alyea, F. N.; Cardelino, C. A.; Crawford, A. J.; Cunnold, D. M.; Lane, B. C.; Lovelock, J. E.; Prinn, R. G.; and Rasmussen, R. A. 1983. The Atmospheric Lifetime Experiment – 6. Results for carbon tetrachloride based on three years data. *J. Geophys. Res.* **88**: 8427–8441.

Solomon, S.; Garcia, R. R.; Rowland, F. S.; and Wuebbles, D. J. 1986. On the depletion of Antarctic ozone. *Nature* **321**: 755–758.

Stauffer, B.; Fischer, G.; Neftel, A.; and Oeschger, H. 1985. Increase of atmospheric methane recorded in Antarctic ice core. *Science* **229**: 1386–1388.

Stolarski, R. S.; Krueger, A. J.; Schoeberl, M. R.; McPeters, R. D.; Newman, P. A.; and Alpert, J. C. 1986. Nimbus 7 satellite measurements of the springtime Antarctic ozone decrease. *Nature* **322**: 808–811.

Tolbert, M. A.; Rossi, M. J.; Malhotra, R.; and Golden, D. M. 1987. Reaction of chlorine nitrate with hydrogen chloride and water at Antarctic stratospheric temperatures. *Science* **238**: 1258–1260.

Turman, B. N., and Edgar, B. C. 1982. Global lightning distributions at dawn and dusk. *J. Geophys. Res.* **87**: 1191–1206.

UN. 1984. Demographic Yearbook 1982. New York: United Nations.

UNEP. 1987. Montreal Protocol on Substances that Deplete the Ozone Layer. Nairobi, Kenya: United Nations Environment Program.

Volz, A., and Kley D. 1988. Evaluation of the Montsouris series of ozone measurements made in the 19th century. *Nature* **332**: 240–242.

Weiss, R. F., and Craig, H. 1976. Production of nitrous oxide by combustion. *Geophys. Res. Lett.* **3**: 751–753.

WMO. 1986. Atmospheric ozone 1985. In: WMO Global Ozone Research and Monitoring Project, Report No. 16. Geneva: World Meteorological Organization.

Wofsy, S. C.; Molina, M. J.; Salawitch, R. J.; Fox, L. E.; and McElroy, M. B. 1988. Interactions between HCl, NO_x and H_2O ice in the Antarctic stratosphere: implications for ozone. *J. Geophys. Res.* **93**: 2442–2450.

Zardini, D.; Raynaud, D.; Scharffe, D.; and Seiler, W. 1987. N_2O measurements of air extracted from Antarctic ice cores. Presented at Symposium on Ice Core Analysis, Bern.

Standing, left to right:
David Peel, Thomas Class, Diane Zardini, Terry Hughes, Henrik Clausen, Jochen Rudolph, Ken Rahn
Seated, left to right:
Paul Crutzen, Robert Charlson, Graeme Pearman, Ulrich Siegenthaler

The Environmental Record in Glaciers and Ice Sheets
eds. H. Oeschger and C.C. Langway, Jr., pp. 269–286
John Wiley & Sons Limited

Group Report
What Anthropogenic Impacts Are Recorded in Glaciers?

G.I. Pearman, Rapporteur
R.J. Charlson
T. Class
H.B. Clausen
P.J. Crutzen
T. Hughes
D.A. Peel
K.A. Rahn
J. Rudolph
U. Siegenthaler
D.S. Zardini

INTRODUCTION

The remarkable records of temperature and carbon dioxide recovered recently from the Vostok core in Antarctica have served as a reminder of the powerful tool we have available in ice core analysis for understanding aspects of the past global history. This applies especially to the assessment of anthropogenic impacts. Already, as the background papers show, the scientific community has embarked on a program to take full advantage of the glacial record. Major contributions have been made in glaciology, paleoclimate, anthropogenic pollution, and biogeochemistry. Our discussions at this Dahlem Workshop have shown that the range of possible future applications is enormous. What has been done so far is only the beginning of what promises to be a most productive branch of geophysical science. In this report we did not aim to review the current studies, but rather to explore the potential for future work and to make some suggestions as to the programs which might be encouraged.

RELATING THE GLACIAL RECORD TO REGIONAL AND GLOBAL CHANGES

The group considered that an examination of the mechanisms involved in the incorporation of trace atmospheric constituents into glaciers was overdue. Questions exist concerning the processes involved in the incorporation of atmospheric gases and aerosols into the firn, the transformation of these to ice, and the long-term storage. Such problems were discussed by Group 1 (see White et al., this volume). However, at least two additional issues need to be addressed. These are:-

1. To what extent does the general planetary atmospheric circulation, in its role of transporting materials to the sites of glaciers, give a biased view of global patterns and changes?
2. In what ways do the mechanism of precipitation formation bias the collection of species to be eventually stored in the ice?

Long-range Transport and the Glacial Record

By virtue of the remote location of glaciers, particularly those of polar regions, much of their chemical record is the product of long-range transport. One might assume that at distances of 5,000–15,000 km from anthropogenic sources, the atmosphere would be well mixed. However, this is not the case. The global atmosphere is stratified and somewhat structured in its circulation, leading to the expectation that glaciers may preferentially sample air contaminated or influenced more by certain regions than others. The discussion below is aimed at encouraging investigations to assess the extent to which this is true, as any lack of representivity, and the possibility of changes in the selectivity with subtle changes in the general circulation, will be important in interpreting glacial records.

Vertical structure in the atmosphere. It is a meteorological fact that the atmosphere is stratified. This is particularly so at high latitudes where strong stability inhibits vertical mixing of the atmosphere. Furthermore, transport in general will be more effective in the mid to upper troposphere than in the lower troposphere because of the higher wind speeds, less scavenging by precipitation, and often lower concentrations of reactants. In the stratosphere, the chemical stability for some species is reduced where certain photochemical reactions proceed much faster than in the troposphere.

It has been long recognized that certain substances are concentrated at specific levels in the atmosphere. Hence, we talk about structures such as the ozone layer and the Junge aerosol layer. Based on particle counts and limited aerosol chemistry, Junge and colleagues developed the concept that the troposphere, above about 5 km, is chemically homogeneous with respect to trace constituents; however, more recent measurements, at least of some gaseous species, suggest that such a concept belies the real structure in the atmosphere. Modern atmospheric chemistry has exposed the complexity of chemical weather, the chemical analogue of physical weather. The chemical composition is temporally and spatially as variable as is the physical atmosphere.

Surface materials get into the upper layers of the troposphere more effectively in some regions of the world than others. The deep convective mixing at low latitudes, and to a lesser extent specific energetic events such as dust storms, are the usual mechanisms while volcanic eruptions and nuclear weapons tests are very effective albeit less frequent.

Many pollutants are released into the atmosphere in regions where insufficient energy is available to induce rapid and deep vertical mixing. Instead, the material is advected more or less horizontally. Studies of Arctic haze have suggested that because air parcels tend to be advected isentropically, higher layers over the Arctic, which have higher potential temperatures than surface layers, have come from warmer surface regions to the south.

While chemical layering is perhaps best known for aerosols, there is growing evidence that this exists for reactive gases as well.

Origin of constituents in precipitation. There exists considerable evidence that precipitation preferentially samples the atmosphere at altitudes 1–5 km above the local surface. Examples of this tendency are known for anthropogenic source regions and remote areas. In Narragansett, Rhode Island, precipitation tends to represent sources upwind (to the west) while aerosols are more locally representative. At least three sites in the U.S. Rocky Mountains indicate that the precipitation there contains a marked marine component, presumably from the Pacific, whereas aerosol measurements at the sites do not. In each case the extra component in precipitation represents upwind sources which are presumably contained in layers above the sampling site. The preferential sampling of layers aloft is presumably linked with the major role of nucleation and precipitation formation (see section on *Precipitation Processes*).

The importance of mean or extreme atmospheric circulation. Precipitation in polar regions requires meteorological conditions which are unrepresentative of an entire season or year. Generally the moisture in precipitation comes from lower latitude areas. Furthermore, the conditions which led to the particular circulation pattern responsible for the precipitation may influence the source strength (particularly of natural sources) so that they may represent extreme events rather than some average condition. The magnitude of this meteorological discrimination is not yet known.

General comments concerning glacial representativeness. For the reasons given above, polar glaciers may well preferentially sample, via air aloft, sources remotely sited in regions of strong vertical transport and perhaps anomalous rather than average circulation conditions.

The clear volcanic and nuclear test debris signal in both Arctic and Antarctic glaciers may be an indication of the association between the polar glacial record and the composition of higher altitude air. An extreme example of how such biases might occur is the recording of Arctic haze by Greenland glaciers. In the Arctic atmosphere in winter, most of the haze is below 2–3 km. Most of this material comes from the northern parts of its source region, including the Soviet Union, and is found below 1–2 km. The Arctic haze from more southerly sources in Central Europe and possibly North America and Eastern Asia tends to be found above 1–2 km in the Arctic. The Greenland ice sheet has an altitude of 1.5–3.5 km and will thus most probably receive precipitation representative of altitudes above 2.5–4.5 km. It is possible, therefore, that Arctic precipitation will be more representative of pollution of more southerly origin.

On the other hand, the process of dry deposition will incorporate material from immediately over the ice and thus be more locally representative. Conclusions drawn from the ice may well need to take into consideration the different degrees of representativeness of wet and dry deposition.

The extent to which these dynamical effects influence the representativeness of glacial ice in general and pollution sources in particular needs to be determined. It is likely to be important in understanding the chemical composition and trends recorded in glaciers.

Precipitation Processes

Many of the chemical constituents in the glacial records will have been originally transported to the glacier as aerosols or impurities with the precipitation. The processes, particularly those involved in the microphysics of precipitation formation, will markedly influence the final composition of the snow. In other words, these processes will selectively determine the concentration of the snow and ice record.

As can be seen in Fig. 1, aerosol substances are advected in and out of the polar regions and the amount deposited by wet and dry removal is reflected in the relative magnitude of these two quantities. In Antarctica the average residence time of air may be as short (days) as compared to that of aerosol particles (10–100 days), such that the ice record represents an average over a larger area of the high-latitude region of the Southern Hemisphere.

The presently recognized pathways for incorporation of materials into the snow surface (and eventually ice) are numerous as shown schematically in Fig. 1. There is presently no agreement as to which are dominant in any given location, season, or meteorological condition. Furthermore, it is important to recognize that below −40°C, the liquid phase processes may be entirely absent. A major uncertainty exists in the case of clouds which

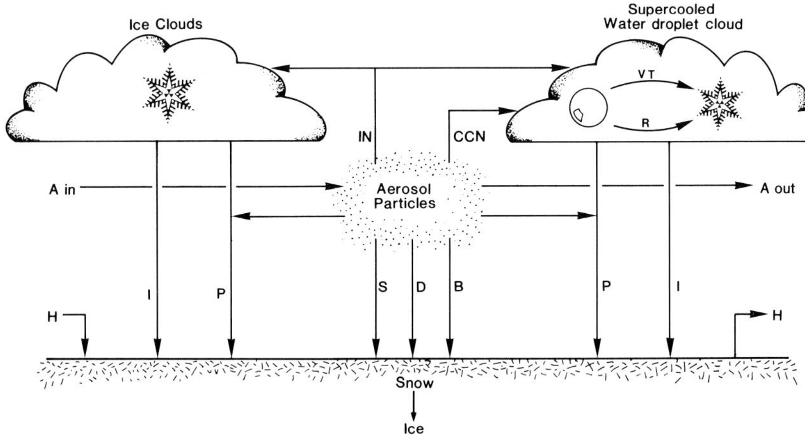

Fig. 1—Schematic representation of the processes delivering aerosol substances to glacial ice. Note that depending on the relative amount of VT and R, varying amounts of CCN substances occur in the falling snow. Key: A_{in}, advection into region; A_{out}, advection out; B, Brownian motion; CCN, cloud nucleation; D, impaction; H, wind-blown snow; I, impaction; IN, ice nucleation; P, precipitation; R, riming (collection of supercooled droplets by snow flakes); S, sedimentation; VT, vapor transfer from supercooled droplets to snow flake.

contain both ice and supercooled liquid. If a snow flake falls and collects supercooled droplets (by riming), it will contain the solute of the droplets which were derived from the cloud condensation nuclei (CCN). If the ice particle grows by deposition of vapor from the liquid phase, it will not contain the CCN substance. As a result, different snow (and therefore glacial ice) compositions can result from the same atmospheric composition. Scott (1981) showed a difference in sulfate concentrations in snow of a factor up to ca. 50, depending on the presence of riming (rimed snow had the highest concentrations). This ambiguity limits our ability to interpret ice core records for aerosol substances such as SO_4^{2-}, Na^+, Cl^-, dust and perhaps NO_3^-.

It is worthwhile to consider the maximum variations likely to be caused by changes in deposition alone. In this way we can identify when changes in concentration in an ice core most probably reflect actual changes in airborne concentrations.

The size of these maximum variations caused by changes in deposition rates will vary from species to species. For example, the efficiency with which the sulfate is scavenged by wet deposition for individual events may vary by a factor of ten or more. The efficiency with which nitric acid is scavenged by wet deposition may vary by even greater amounts. Concentrations in an ice core may be greater during times of low precipitation rate, due to the influence of dry deposition.

When an ice core sample represents an average over several years, the variations caused by changes in deposition rate may be considerably reduced.

This is because seasonal variations in meteorology, which affect deposition as well as year-to-year fluctuations in precipitation, may be averaged out. Long-term changes in atmospheric temperature may affect scavenging by altering the scavaging mechanism. Thus a variation in concentration of a species in an ice core by an order of magnitude or more is likely to reflect a true change in atmospheric concentration. It must be said, however, that the amount of change of airborne concentration cannot be quantified utilizing current knowledge.

There is considerable hope that the unique case of the Antarctic plateau will prove to have simpler mechanisms and will therefore be quantified more easily. The reason for this is that the major causes of variability in scavenging ratio appears to be variable amounts of riming versus vapor deposition as described above. Below ca. −40°C, liquid water ceases to exist in the atmosphere and therefore riming cannot occur. Thus a "dry" process is probably responsible for delivery of aerosol (e.g. SO_4^{2-}) to the Antarctic plateau. This should be a much less variable process. Key experiments are needed to demonstrate that the scavenging efficiency is (*a*) a systematic function of temperature and (*b*) less variable at temperatures below ca. −40°C.

What Can Ice-core Records Tell?

Given the problems described above, of relating the observed chemical composition in glaciers with anthropogenic activites in particular regions, it is worth, for the sake of perspective, to look at some selected studies where despite this, scientific objectives have been achieved.

Aerosols. The well-known and obvious modification by man of the atmospheric sulfur and nitrogen cycles, as well as the emission of traces of metals and other substances that have low vapor pressures, results in the formation of submicrometer aerosol particles which have relatively long lifetimes (up to months) (Shaw, this volume). These substances are removed by precipitation in many regions of the globe so that glacial ice formed from snow contains a permanent record of these anthropogenic pollutants. In the extremely arid areas of the Antarctic plateau, the aerosol may be instead directly deposited on the snow surface because precipitation is so small; however, the record is still present in ice. It is therefore very attractive and useful to learn how to interpret the glacial record in terms of anthropogenic pollutants. Several of the background papers (Shaw, Davidson, Wagenbach, Peel, Clausen and Langway, Delmas and Legrand, all this volume) focus directly on various aspects of the records of anthropogenic species in ice, including Arctic, Antarctic, and temperate glaciers.

Of the aerosol species that do not have gaseous precursors, the heavy metals have a demonstrated human global-scale impact. The data set of reliable values is restricted and only for lead is there a description of time trends through the industrial period. A clear signal in Greenland, where lead concentrations in snow have increased 200–fold during the past 2,000 yrs, contrasts with a much weaker 4– to 10–fold increase in Antarctica. This result accords with expectations from the global distribution of lead emissions and the restricted transport of air masses across the equator and across the Antarctic circumpolar circulation. Preindustrial/agricultural concentrations of lead measured in ancient ice are in line with estimated natural crustal abundances, demonstrating that there are no important natural enrichment mechanisms for this element.

It is worth noting that traces of organochlorine compounds such as DDT(E) and the PCBs, which have no natural sources, have been identified in modern polar snow in both hemispheres.

For some of the species, the particles are effectively dissolved in the ice (e.g., SO_4^{2-}) while insoluble materials are maintained in a colloidal form, which allows more fundamental studies of the role of aerosol physics in the delivery of these materials to the ice.

As the foregoing discussion shows, there are uncertainties in the mechanisms involved in transporting aerosols from the source to the sampling site and in the incorporation of aerosol into snowfall. Nevertheless, several clear impacts have been positively identified in glacial ice deposited during the period for which we have accurate knowledge of anthropogenic emission rates and their geographical distribution.

1. The record of radioactivity in parallel with nuclear weapons testing. Time series from the polar ice sheets reflect the changing yield of successive weapons test programs. The radionuclide composition also follows the known development of weapons type.
2. Increases in NO_3^- and SO_4^{2-} in the Arctic during the past 200 yrs broadly follow the known inventory of emissions. Increases in NO_3^- were not detectable before the 1950s whereas SO_4^{2-} has increased progressively during the last century. Qualitatively known increases in the combustion of coal and oil, and the rapid growth of automobile emissions of NO_3^- have occurred since the 1950s.
3. Increases of Pb emissions are clearly reflected in Arctic snows and to a smaller extent in the Antarctic. Most of the increase has occurred since the 1950s as expected from the rapid increase of emissions from leaded fuels in motor vehicles.

Atmospheric trace gases. At the beginning of this report, and in background papers by Khalil and Rasmussen, Crutzen and Brühl, and McElroy (all this volume), the recent secular changes of the key greenhouse gases CO_2, CH_4, and N_2O were described. Their temporal changes during glacial and interglacial conditions together with anthropogenic effects over the last two centuries are now clearly defined. In the case of CO_2 the stable carbon isotopic change has also been measured leading to a strong constraint on estimates of the magnitude and timing of significant biospheric contributions to the CO_2 increase. However, of special interest are the results of comparisons of the concentrations of CO_2 and CH_4 observed in the air occluded in the ice cores with the results from modern atmospheric monitoring programs. For example, over the 20 year period 1961–1980, the CO_2 concentration observed in the ice in a recent Australian study on an Antarctic core (Pearman, personal communication) and that recorded for the same years at the South Pole were indistinguishable. Similarly for CH_4, over the period 1978–1982, the concentration observed in the ice (Pearman 1988; see Fig. 2) and that recorded at the Cape Grim Observatory, Tasmania, were the same within the experimental precision.

Nitrate and sulfate. Particularly important findings from the Greenland ice cores include the marked increases in SO_4^{2-} and NO_3^- depositions, by factors of 3–4 and almost 2, respectively, that have been detected since the beginning of the industrial era.

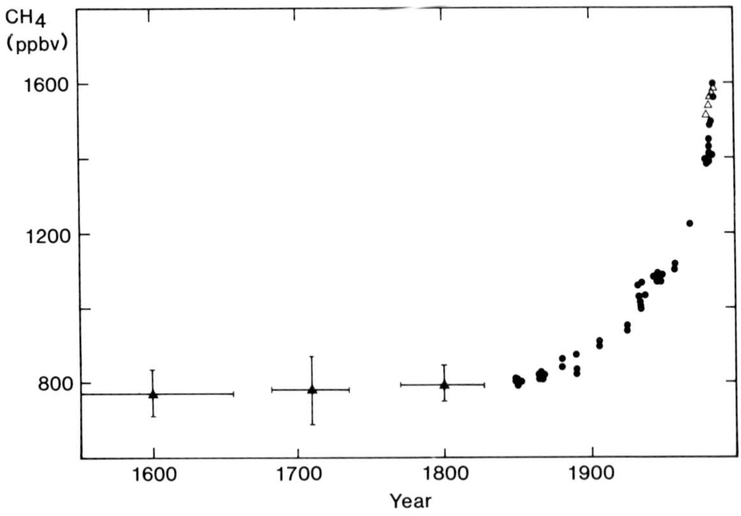

*Fig.*2—Historical changes in atmospheric methane as deduced from air extracted from bubbles trapped in Antarctic ice. The triangles represent the direct observations of atmospheric methane at the Cape Grim Observatory, Tasmania. After Pearman (1988), unpublished data of D. Etheridge (Australian Antarctic Division), G.I. Pearman and P.J. Fraser (CSIRO). Horizontal bars indicate the range of ages of samples; vertical bars are standard deviations of concentration measurements for groups of pooled data.

Currently estimated sources of sulfur consist largely of an anthropogenic input of 100 Tg S/year and an oceanic emission of about 40–50 Tg S/year. For the Northern Hemisphere this implies an increase in emissions by as much as a factor of 6–7, which is only in part recorded in the ice cores. This may be due to a combination of the following factors: significant natural emissions occur also from land areas in the Northern Hemisphere ($\approx$ 25 Tg S/year) and a relatively larger influence of oceanic S emissions due to the oceanic location of Greenland.

Regarding $NO_x(NO + NO_2)$, the currently estimated budget is more complicated, as summarized by Crutzen and Brühl (this volume). Preindustrial emissions in the Northern Hemisphere were maybe of the order of 12 Tg N/year, consisting of 8 Tg N/year from soils and 4 TG N/year from lightning. The latter sources are mostly deposited throughout the atmosphere. Both contributions arise substantially from the tropics and subtropics. Anthropogenic emissions have added about 20 Tg N/year since preindustrial times, so that total emissions in the Northern Hemisphere are estimated to have increased by a factor of about 3. However, the anthropogenic NO_x emissions originate from regions much closer to the Greenland glaciers than the natural emissions. Nevertheless, the NO_3^- depositions in the ice cores increased by less than a factor of 2. This discrepancy may suggest that background NO_x and HNO_3 are influenced by the lightning source and that long-range transport of lightning and possibly stratosphere NO_x may be a very important phenomenon.

Concentration measurements of NO_3^- in Antarctic snow, which show generally lower values by a factor of 2–3, may be taken to support this viewpoint. An open question is, however, whether NO_3^- deposition may not be influenced by the stratospheric NO_x source. This is of particular interest as the condensation of HNO_3 is probably occurring on polar stratospheric cloud particles during winter and early spring. These particles may settle out from the stratosphere.

NONPOLAR GLACIERS

Polar glaciers, particularly those in Antarctica, do not necessarily respond sensitively to anthropogenic changes in the atmospheric levels of short-lived substances, e.g., substances being deposited by rain and originating from northern midlatitudes. Such a criticism may also be made with respect to the use of polar glaciers for interpreting other changes in the Earth's surface at lower latitudes. Such changes may be more sensitively recorded in midlatitude, cold and high altitude mountain glaciers.

Such glaciers may be situated in the middle of the major anthropogenic source areas, yet are sufficiently isolated due to poor vertical transport that the records are regionally representative. As with the polar glaciers, the actual source area represented in the composition of such alpine glaciers will be difficult to determine and weight.

A specific problem with alpine glaciers is the contribution of long- and short-distance transport of dust. Sampling of dust layers in the glaciers is to be avoided since they may cause interference for the present purpose. They are, however, possibly useful in paleoclimatic studies. Furthermore, particularly at an exposed site, snow deposition may be irregular and the seasonal variations may not represent such variations of the source and could lead to a biasing of both long- and short-term averages. Finally, dating of cores of such glaciers may be difficult. By suitable dating methods and with a flow model which defines the upstream location of snow deposition, the chronology may nevertheless be established for the last century.

Despite the many problems which may make the interpretation of nonpolar glacial records difficult, the group emphasized that with a combination of careful selection of sites and inventive procedures they remain optimistic about the future of such studies. The difficulties perceived must be seen in the perspective of the potential gains in knowledge that may be obtained.

More specifically, it should be added that, e.g., for SO_4^{2-} and NO_3^- a good record of the anthropogenic rise during the last century has been established from data from the Swiss Alps (Wagenbach et al. 1988).

GLACIAL RECORDS AND REACTIVE ATMOSPHERIC CHEMISTRY

In the past decade or so, the field of atmospheric chemistry has developed enormously. It is clear from these developments that the atmosphere plays a key role in the oxidation of reduced gases from biological sources. A key species in this process is the hydroxyl radical (OH). Methane (CH_4) is one of the main biotic species to be oxidized. Apart from its role in this reactive chemistry, it is quite a significant greenhouse gas. The fact that CH_4 is increasing in the atmosphere has drawn attention to the sensitivity of such gases to global change, either due to increases in their sources or due to a reduction in the oxidizing capacity of the atmosphere. Either way, such changes are intimately connected to a range of species that take part in the reactive cheimistry (see Crutzen and Brühl, this volume).

There are a number of chemical compounds which can provide important information about the reactive chemistry of the atmosphere and how this may have changed in response to anthropogenic and natural factors. It would be most useful to this area of science if the history of these compounds or some proxy parameter could be recovered from glacial records. The compounds of importance are as follows:

1. O_3: This is the most important gas in the reactive chemistry of the

atmosphere. The feasibility of O_3 measurements and the chemical activity of O_3 in ice cores should be explored.

2. H_2O_2: This species has already been measured in ice cores, is an important temporary odd hydrogen reservoir gas, and as such is a sensitive indicator of the oxidative capacity of the atmosphere.
3. CH_4, CO, H_2, CH_2O: These gases either play important roles in the background odd hydrogen chemistry of the troposphere (CH_4, CO, CH_2O) or are products of photochemical processes (CO, H_2, CH_2O). Of these CH_4 has already been measured, indicating a large atmospheric increase during the past 300 years and similar variations with glacial and interglacial periods. For CO, H_2, and especially for CH_2O it must be ascertained whether ice cores are indeed representative of atmospheric conditions. Preliminary work indicates a substantial enrichment of CO in ice cores vis-a-vis the atmosphere. For H_2, molecular diffusion in the ice may be a problem.
4. CH_3Cl: Because this gas is largely removed from the atmosphere by reactions with OH and its source may be dominated by oceanic emissions, it may be a good indicator of changing OH concentrations (or oxidative capacity of the atmosphere). A problem, however, may be that biomass burning could likewise be a significant source of CH_3Cl.
5. C_2H_6: This gas is the second most abundant hydrocarbon gas in the atmosphere and is of significant importance due to its potential as a precursor for peroxyacetylnitrate (PAN), acetaldehyde, and acetic acid as well as its value as a diagnostic for changes in the chemistry of the atmosphere.
6. COS: This gas is the most important carrier of sulfur to the stratosphere during nonvolcanic periods and is responsible for the formation of the stratospheric sulfate layer. The stability of COS in ice should be ascertained.

Other trace gases may provide interesting information on the photochemical functioning of the atmosphere, but in many cases concentrations are so small that development of more sensitive techniques is required.

LOOKING TO THE FUTURE

Scientific Questions

Long-lived gases. Analyses of polar ice cores have demonstrated, in an impressive way, the anthropogenic increase in the atmosphere of CO_2, CH_4, and N_2O. Scientific questions related to these gases are centered on the consequences for climate and the environment, the causes of the observed increases, and the changes in sources and sinks. For estimating a net source strength (emission rate) as a function of time, one has to know not only c(t) (the concentration in the atmosphere at time t), but also dc/dt (the rate of change with time). This means that the precision of the concentration measurements and of the dating of the trapped air has to be very good.

While concentration changes can only tell about changes of the total atmospheric gas amounts, isotopic data can help to discriminate between different sources, for instance, ^{14}C in CO_2 and CH_4 between fossil and nonfossil carbon, or $^{13}C/^{12}C$ between oceanic and biospheric or fossil CO_2. The isotopic ratios for CH_4 and N_2O (cf. McElroy, this volume), similarly contain information about their sources.

To obtain a net anthropogenic signal, it is necessary to know the natural baseline concentrations and their variability. The existing ice core data show that the atmospheric CO_2 level has in the past 160,000 yrs never been as high as at present, which clearly counters the sometimes suggested possibility that the recent increase might just be a natural fluctuation. Measurements made on a South Pole ice core indicate that between A.D. 1200 and 1300 the CO_2 concentration may have increased by ca. 10 ppmv (Siegenthaler et al. 1988); a detailed assessment of such variations of gas concentration is obviously of great vaue for better understanding the global cycles and therefore being able to predict future changes.

The comparison of ^{10}Be results from ice cores with tree-ring ^{14}C data has revealed that simultaneous and rapid (100–200 yrs) variations in both isotopes over the last 5000 yrs must be due to a common cause, i.e., changes of solar activity (Beer et al. 1988; Raisbeck and Yiou 1988; Dansgaard and Oeschger, this volume; Lorius et al., this volume). Accepting this, a ^{10}Be-^{14}C comparison can, inversely, help to check on the models of the global carbon cycle used to study the anthropogenic perturbations.

Fossil-fuel burning must have led to a slight decrease of the O_2/N_2 ratio in the atmosphere. While a determination of this decrease is not easy analytically, it would yield information complementary to the CO_2 data. Furthermore, in view of the overall importance of molecular oxygen for life and for the whole geochemistry of the earth/atmosphere system, the oxygen cycle certainly deserves enhanced attention. This also includes studies of variations of the O_2/N_2 ratio and of the isotopic composition of O_2 over glacial/interglacial cycles.

Reactive gases. In the section on *Glacial Records* a list of gases involved in reactive chemistry or symptomatic of the state of this chemistry was developed. They included O_3, H_2O_2, CH_4, CH_2O, CO, H_2, CH_3Cl, C_2H_6, and COS. Other measurements mentioned during our discussion of future measurement programs included CS_2, MSA (methanesulfonic acid), and DMS (dimethyl sulfide) as important components in the sulfur cycle; HO_2 and organic species such as chloroform, propane and formaldehyde; isotopic measurements including ^{15}N and ^{36}Cl; and natural halogens.

Ecosystems. Associated with anthropogenic changes there have been major ecosystem modifications. Some of these are reasonably well understood; others are not. Measurement programs aimed at obtaining information from the glaciers concerning these changes should be pursued. Pollen analyses and the presence of elemental carbon (soot) as an indicator of burning of biomass are possibilities, but others may exist. Changes of DMS need to be assessed (Charlson et al. 1987).

Multiphase aspects. Multiphase atmospheric chemistry describes the behavior of the condensed phases of the atmosphere (i.e., aerosol particle substances, aqueous solution droplets, and ice particles). Regarding the ice core records, it is of interest to be able to quantitatively relate constituents in the ice to that of the air. Thus we can list several key scientific questions that are suitable for direct experimental study.

1. What are the scavenging ratios and the factors controlling them at sites where cores can be retrieved?
2. Does the scavenging ratio follow the temperature dependence described above in the section on *Precipitation Processes*?
3. Will size-dependent sampling of aerosol particles in polar regions reveal the general presence of the bimodal aerosol mass concentration as found universally at lower altitudes?
4. Will laboratory experiments reveal the exclusion of H_2SO_4 solution from forming ice crystals during cloud formation as the source of concentrated sulfate solution at grain boundaries in glacial ice?
5. What aerosol species besides the ones emphasized here (SO_4^{2-}, NO_3^-, Pb, elemental carbon, Na^+, Cl^-, soil dust, ^{10}Be, and other nongaseous radionuclides) may be studied in ice cores?

We stress that the emphasis in the programs should be on quantifying and understanding processes as the means by which ice core chemistry can best be interpreted for substances from the atmospheric aerosol.

Polar pollution. Gaseous and particulate pollutants released in regions where transport is likely to carry them to glaciers may be particularly important to glacial records for at least two reasons. First, the proximity of these sources to polar glaciers tends to increase their relative effect, selective sampling due to dynamics and precipitation processes aside. Second, the stability and dryness of polar atmospheres, as well as the lack of sunlight and low concentrations of pollutants, decrease the removal of particles and soluble or reactive gases to rates well below those of midlatitudes.

For these reasons, it is important to establish the baseline of current polar pollution now, as it is sure to increase in the future as human activity in and near polar regions increases. Fortunately the interest in Arctic haze since the mid-1970s has provided the Arctic baseline for many species. This work should continue. Fewer measurements are available for the Antarctic, however. With the Antarctic Treaty coming up for review, and the types of human activity presently there subject to change, it is important to measure more constituents at more sites in and around Antarctica. New observatories, or attention to the programs of existing observatories in the polar regions, should be encouraged in the context of understanding the glacial records.

Atmospheric Data

In attempting to correlate the concentrations of trace species in polar ice cores with their atmospheric abundances, one becomes aware that for a large number of important atmospheric trace components, there is a dearth of measurements in the polar atmosphere. Apart from the long-lived species such as CO_2, N_2O and CH_4 and the chlorofluorocarbons, which are almost uniformly distributed in the two hemispheres, the atmospheric concentrations of reactive trace gases in the Arctic and especially the Antarctic regions are not very well established, often completely unknown. Likewise, there is a serious lack of precipitation chemistry stations at background locations.

There are some data available on the composition of the polar aerosol. Many of these studies were made in connection with the Arctic haze phenomenon.

As far as the nonmethane hydrocarbons are concerned, the atmospheric mixing ratios at high latitudes are only reasonably well known for ethane. For all the other nonmethane hydrocarbons, their abundance in polar regions is very uncertain. Since several of these species have significant biogenic sources and are highly reactive, they may have a considerable impact on the formation of aldehydes or ketones, carboxylic acid, peroxides, etc. This is especially important if one tries to determine changes in the photochemistry of the atmosphere, e.g., the concentration of OH, radicals from measurement of species such as H_2O_2, or CH_2O in ice cores.

In addition to this almost complete lack of measurements of these species in the polar troposphere, one of the most important factors for the formation of H_2O_2, O_3, etc. from hydrocarbon oxidation, namely the concentration distribution of NO_x, HNO_3 and PAN in the polar troposphere, is essentially unknown.

Even for the purpose of establishing long-time trends for trace gases with medium atmospheric lifetimes, e.g., CH_3Cl and COS, the present-day global distribution and seasonal variability is not always well enough known to deduce directly global changes from possible changes in ice cores.

For the highly soluble species such as formic acid, acetic acid, mineral acids, formaldehyde, etc., the present-day data bases of measurements in precipitation for "background areas" are still rather limited and again not sufficient to establish whether polar ice cores can be taken to indicate changes on a global scale. Especially important are simultaneous measurements of the highly soluble trace gases in the gas phase, the aerosol phase, in precipitation, and in snow on the ground in polar regions.

What Technical Advances Are Needed?

Present knowledge of the anthropogenic impact on glaciers is limited to a few stable trace gases (CO_2, CH_4, and N_2O) and soluble chemical species (SO_4^{2-}, NO_3^-, and the heavy metal elements). Studies on the gaseous components are in many respects better developed than those made on the soluble aerosol constituents. These species have a short residence time in the atmosphere. Consequently they are sensitive to global transport processes. However, the assessment of a human impact from time-series investigations is complicated by the fact that natural sources are significant contributors to global flux rates of the species studied so far. The problem is especially severe in areas, such as Antarctica, that are very remote from the pollution source regions. Technical advances are needed in several areas.

Improved anthropogenic tracers. We need to identify new tracers that are of unambiguous anthropogenic origin:

1. Multi-element tracers. This approach has been used successfully to interpret the Arctic aerosol. Application to snow and ice studies is currently limited by the low concentrations encountered combined with lengthy analysis times which restrict a statistical analysis of the data. Suitable systems applicable to other geographical areas should be developed.
2. Isotope ratios or chemical speciation. Isotope ratios of either gaseous or solid elements can help to characterize a source area of source type, e.g., $^{206}Pb/^{207}Pb$ has been found to vary widely between different lead deposits, making it suitable as a tracer of atmospheric pathways. The same is true for gases (CH_4, CO_2, N_2O). A related approach may be possible if the particular chemical species contributing to a total elemental concentration in ice can be identified.
3. New species of exclusive anthropogenic origin. Species should be identified that have no known natural origin. In addition, they should be sufficiently stable to resist decomposition during long-range transport and following deposition. In this way it will be possible both to establish that the analytical technique is free from contamination in studies of ancient ice and to trace anthropogenic trends from an unambigious baseline. Combined CG/MS analysis is now eminently suitable for ultra-trace studies of organic species yet has seen hardly any application to

glacier ice studies. Several groups of organic compounds such as the organochlorines, e.g., the pesticides, chlorofluorocarbons and PCBs, could provide useful starting points.

New analytical techniques. In order to make significant progress in understanding long-range transport processes it is clear that there will be a rapidly growing demand for more highly resolved (to sub-seasonal level) time series of a wider range of species. Moreover, in order to reduce ambiguities it will be important to make the studies in parallel on the same sections of ice core. Drilling should be regarded as an integral part of the analytical process, representing the first stage of sample preparation.

1. Improvements in drilling technology

   a. Improved drill design to ensure consistent core quality (no hair-line fracturing) through period of interest.
   b. Research on materials for drill components to ensure compatability with chemical analyses on the ice.
   c. Research on drilling fluids to ensure chemical compatibility.
   d. Larger diameter cores to enable more complete core clean-up and adequate sample for multiple analyses.

N.B. If (a) can be solved then the pressure on (b) and (c) could probably be relaxed.

2. Improvements in analytical technique.
The following features are required generally:

   a. A need for rapid analysis with multi-element or multi-species capability for compatability with high resolution analysis of ice cores.
   b. A need for simplicity. They should require minimal manipulations to reduce contamination problem.
   c. Low sample volume. Necessary for trace-species analysis to enable cross-core profiling as contamination control. In the case of gas studies it is important to study the composition and age of different gases in the firn and to elucidate the mechanism by which they become trapped in the ice.

N.B. Several techniques that have seen little application to ice core analysis show considerable promise. They include continuous-flow injection analysis, laser-resonance mass spectrometry, ICR mass spectrometry, and combined GC/MS.

Evidence on atmospheric processes involved. We need firmer evidence of processes involved in transport of pollution from the source region through to the point of deposition. Transfer functions during this final stage are crucial to the interpretation of ice core data. Evidence so far suggests that these vary between different chemical species and that they are sensitive to the geographical situation and to the particular weather conditions at the time of deposition.

Having established the aerosol composition at the remote deposition site this must be connected through to data from similar widely dispersed sites in order to evaluate transport mechanisms. The following issues need attention:

1. Transfer functions at sample site. Continuous aerosol/snowfall collections are needed at remote sites, on a year-round basis. Preferably such sites will be remote also from manned stations. Reliable, automatic systems would be desirable. Means for extending such studies through a vertical column in the atmosphere to cloud level should be explored.
2. Large-scale transport. Remote sensing techniques should be exploited to enable an understanding of the dynamical history of air masses approaching the sampling areas. Reverse trajectory analysis would be valuable for interpreting continuously sampled aerosol data. Possibilities for tracking atmospheric constituents, such as total dust loadings, should be explored.
3. Atmospheric chemistry. Several anthropogenic species may become involved in chemical reactions either in the troposphere or in the stratosphere. We need to identify those sensitive species, particularly those that can be studied with existing technology. We need to study species that can be identified clearly as having a stratospheric origin or a tropospheric origin in order to trace the exchange of air at different levels in the atmosphere.

Efficient data and sample exchange. There is a growing need for comprehensive storage of data from all geographical sectors, and for the data to be available to all groups. This is partially satisfied by the World Data Centers, and with the quantity of data available to date, the publication network may still be adequate. Establishing a comprehensive data base will not be straightforward because data subject to varying uncertainties would be incorporated. Indeed this fact is often not recognized until many years after the original data have been first published. In some areas of research, especially in trace gas analysis, an effective system of sample exchange has been developed, supported by analytical standards, such as those at NBS. This procedure should be extended to analyses of the ultra-trace soluble constituents of ice.

REFERENCES

Beer, J.; Siegenthaler, U.; Bonani, G.; Finkel, R.C.; Oeschger, H.; Suter, M; and Wolfli, W. 1988. Information on past solar activity and geomagnetism from ^{10}Be in the Camp Century ice-core. *Nature* **331**: 675–679.

Charlson, R.J.; Lovelock, J.E.; Andreae, M.O.; and Warren S.G. 1987. Oceanic phytoplankton, atmospheric sulfur and albedo and climate. *Nature* **326**: 665–661.

Pearman, G.I. 1988. Greenhouse gases: evidence for atmospheric changes as anthropogenic causes. In: Greenhouse, Proceedings of the Conference Greenhouse 87: Planning for Climate Change, Monash University, Melbourne, ed. G.I. Pearman. CSIRO/Brill, Melbourne/Leiden, 752 pp.

Raisbeck, G.M. and Yiou, F. 1988. ^{10}Be as a proxy indicator of variations in solar activity and geomagnetic field intensity during the last 10,000 years. In: Secular Solar and Geomagnetic Variations in the Last 10,000 Years, eds F.R. Stephenson and A.W. Wofendale. Dordrecht: Reidel, in press.

Scott, B.C. 1981. Sulfate washout ratios in winter storms. *J. Ap. Meteor.* **20**: 619–625.

Siegenthaler, U.; Friedli, H.; Loetsher, H.; Moore, E.; Neftel, A.; Oeschger, H.; and Stauffer, B. 1988. Stable isotope ratios and concentrations of CO_2 in air from polar ice cores. *Ann. Glaciol.* **10**: 151–156.

Wagenbach, D.; Munnich, K.O.; Schotterer, U.; and Oeschger, H. 1988. The anthropogenic impact on snow chemistry at Colle Gniffetti, Swiss Alps. *Ann. Glaciol.* **19**: 183–187.

The Environmental Record in Glaciers and Ice Sheets
eds. H. Oeschger and C.C. Langway, Jr., pp. 287–318
John Wiley & Sons Limited

Past Environmental Long-term Records from the Arctic

W. Dansgaard[1] and H. Oeschger[2]

[1]Geophysical Institute, University of Copenhagen,
2200 Copenhagen N, Denmark
[2]Physikalisches Institut, Universität Bern
3012 Bern, Switzerland

Abstract. Long time series of environmental parameters have been established by measurements along boreholes and by analyses of deep ice cores from Greenland. The temporal deformation of boreholes gives information about the mechanical properties of the ice, which is important to ice flow modeling. The ice deposited under glacial conditions is much softer than postglacial ice, which suggests that the ice sheet was considerably thinner than today. However, this problem is still open to discussion.

Detailed $\delta^{18}0$ measurements along ice cores are the basis for absolute dating of up to 8000 yr old ice and, in addition, they reveal surface temperature changes over the last 100,000 years or more. One of the most interesting features of the δ profiles is the indication of frequent and abrupt climatic shifts between mild and very cold phases of the glaciation. All the environmental parameters shifted more or less simultaneously. The shifts were probably associated with retreats and advances of the North Atlantic sea ice cover, as was the case during the last glacial climatic oscillation, Bølling-Allerød-Younger Dryas, which stands out clearly in Greenland but is less pronounced, if present at all, in Antarctica. The deuterium excess, $d = \delta D - 8\delta^{18}O$, in recent and glacial deposits gives evidence for a dominating subtropical moisture source for the precipitation at high elevations on the Greenland ice sheet.

The air bubbles in polar ice reflect the ancient atmospheric gas composition. Of special interest is the question of the preindustrial concentrations and the natural variability of gases such as CO_2, CH_4, and N_2O, which are now rising due to anthropogenic emissions. CO_2 and CH_4 show a decrease from about 60,000 yrs. B.P. till about 20,000 yrs B.P., followed by a rapid increase around 15,000 yrs B.P. to essentially constant Holocene concentrations till the beginning of industrialization. During the last glacial period, CO_2 and CH_4 also showed rapid variations superimposed on the long-term trend.

Measurements of the cosmic ray produced radioisotope, ^{10}Be, by accelerator mass spectrometry show fluctuations which reflect production rate changes and/or changes in atmospheric transport and deposition. During the last glaciation the concentrations were higher by a factor of 2 to 4, which can be explained by smaller accumulation rates. During periods of a quiet sun, e.g. from A.D. 1640 to A.D. 1710, the ^{10}Be concentrations were higher by a factor of ca. 1.5, reflecting a reduced solar magnetic shielding of cosmic radiation. These short-term (200 yrs) ^{10}Be fluctuations show a high correlation with the short-term ^{14}C fluctuations found in tree-rings.

Generally, a broad spectrum of environmental parameters are recorded in ice cores, which has already contributed substantially to the understanding of Earth system processes.

INTRODUCTION

By long-term records we mean those that reach beyond the Pleistocene-Holocene transition 10,700 yrs ago (Hammer et al. 1986), i.e., into the last glaciation, Würm/Wisconsin. In practice, this excludes temperate glaciers, i.e., ice masses at the pressure melting point, because the geothermal heat flux removes the bottom layers by melting, before they reach an appreciable age. Hence, we shall concentrate on discussing cold glaciers with ice temperatures below the pressure melting point.

In cold glaciers, some melting may occur on the surface in the summertime, but the meltwater percolates into the cold underlying snow strata and refreezes, leaving a distinct layer of bubble-free ice in an ice core. This process complicates the interpretation of several types of ice core analyses. For example, a wet surface layer of snow absorbs CO_2 from the atmosphere causing elevated CO_2 content in the refrozen melt layer. Hence, the most favorable areas for deep ice core drilling and analyses are those where summer melting rarely or never occurs. Under the present climatic conditions, such areas in the Northern Hemisphere are limited to the high elevation parts of Central and North Greenland, and perhaps the northernmost part of the Canadian Arctic. During the last glaciation, however, the melt-free area was much larger, and any ice mass from the glaciation still in existence, in South Greenland or in minor ice masses, was certainly deposited on a surface without significant summer melting.

As the layers sink into an ice sheet they are stretched horizontally (Fig. 1): the thickness of the annual layers approaches zero close to the bedrock, provided no bottom melting takes place. In any case, the vast majority of annual layers are situated close to the bedrock. This is why it is so important for deep ice core drilling to be extended all the way to the bottom.

ENVIRONMENTAL PARAMETERS

Deep ice cores offer the possibility of establishing a broad selection of time series of environmental parameters. Many of them have been treated in the preceding chapters of this volume and we shall therefore concentrate on profiles of borehole data (particularly on temperature), stable isotopes, CO_2, and cosmogenic radioisotopes.

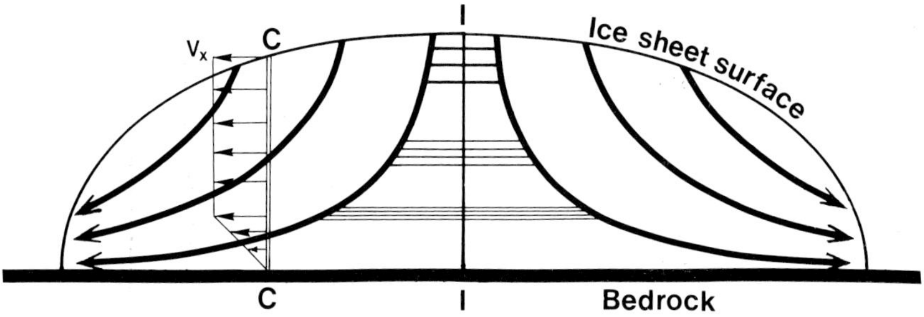

Fig. 1—Schematic ice flow pattern in a vertical cross-section of an ice sheet resting on a horizontal bedrock. The further inland an ice particle is deposited on the surface, the closer to the bedrock it will travel (as indicated by the long arrows). It ultimately reaches the margin where it melts away, or is extruded as part of an iceberg. Annual ice layers are gradually stretched and therefore thinned as they approach the bedrock (cf. the symbolized layers close to the ice divide I-I). In an ice core C-C drilled at some distance from I-I, the deeper layers originate further inland. The arrows along C-C show the horizontal ice velocity components v_x. Close to the bedrock, v_x is zero if the temperature of the deepest ice is below the pressure melting point.

Bore hole Measurements

Deep boreholes have to be filled with a fluid of the same density as the ice in order to counterbalance the increasing pressure downwards. Repeated measurements of the changing geometry of the hole by a submerged logging device gives the deformation rate of the plastic ice (Fig. 2), and thereby its mechanical properties (Gundestrup and Hansen 1984; Dahl-Jensen 1985), which are essential to realistic ice flow modeling (see Reeh, this volume). At great depths, such measurements are particularly important for modeling the extension and shape of the Arctic ice sheets under glacial conditions (Reeh 1984).

At a given temperature, ice deposited in the Northern Hemisphere during the glaciation is up to 5 times softer than Holocene ice (Gundestrup and Hansen 1984; Paterson 1981; Fisher and Koerner 1986), probably due to very high concentrations of continental dust (Dahl-Jensen 1985; Fisher and Koerner 1986). This leads to faster ice flow and, combined with lower accumulation rate, to smaller ice sheet thickness under glacial conditions if it were not for the compensating effect of lower temperatures, which makes the ice stiffer. This is further discussed in the section *The entire glacial to postglacial transition*.

The same logging device also measures the temperature profile along the hole (see thick curve in Fig. 3), measured along the 2038 m deep hole to bedrock at Dye 3 in Southeast Greenland (65° N, 44° W) (Gundestrup and Hansen 1984). The remarkable temperature increase in the bottom 600 m of the ice profile is due to the geothermal heat flux.

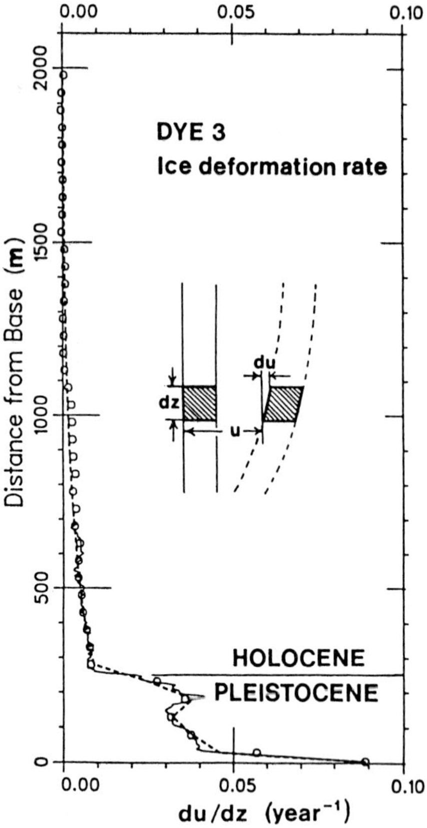

Fig. 2—Measured ice deformation rate du/dz (see insert) along the Dye 3 deep core. Ice deposited during the last glaciation is 4 to 5 times "softer" than postglacial ice (from Gundestrup and Hansen 1984).

A combined ice flow and heat transfer model has been developed (Dahl-Jensen and Johnsen 1986) for calculating the temperature profile (thin curve in Fig. 3) as it would have been under steady state conditions, i.e., if the surface temperature and the accumulation rate had remained unchanged throughout the ages. The temperature minimum close to 800 m depth is due to the deeper ice evolving from higher and, therefore, colder parts of the ice sheet (cf. Fig. 1). Furthermore, under steady state conditions the bedrock temperature would be −8.2° C (not shown in Fig. 3), instead of the actual 13.2° C.

The model is also able to calculate the temperature profile under given nonsteady state conditions. Complete fit with the entire measured profile within the experimental error of 0.03° C is obtained with the assumptions listed in Table 1. Most interesting in this context is that during the glaciation,

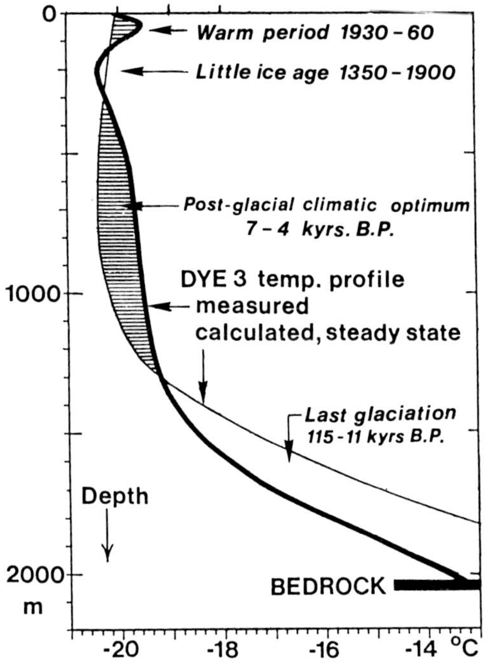

Fig. 3—*Thin curve*: Temperature profile along the Dye 3 deep core, calculated with the assumption of steady state conditions, i.e., unchanged temperature and accumulation backwards in time. The deepest layers are warmest due to the geothermal heat flux from below. *Heavy curve*: Measured temperature profile. The positive (hatched) deviations from the thin curve are due to warmer than present climatic conditions 1930–1960 and during the postglacial climatic optimum. The negative deviations reflect the "little ice age" A.D. 1350–1900 and the Würm/Wisconsin glaciation prior to 10,700 yrs B.P. (after Dahl-Jensen and Johnsen 1986).

the surface temperature in South Greenland was 12±2° C lower than today. The surface elevation was hardly much different from the present because the 100 m lower sea level was compensated, at least in part, by isostatic rebound, and, as mentioned above, the thickness of the central parts of the ice sheet was probably close to the present value. Hence, the climate on the South Greenland ice sheet was probably 12±2° C colder than now, an estimate essentially in agreement with the CLIMAP reconstruction (CLIMAP 1976) of a maximum sea-surface temperature anomaly of 14° C in the North Atlantic Ocean during the last glacial maximum.

Stable Isotopes

The occurrences of the three most important isotopic components of water are related in nature approximately as

Table 1

Period	Surface temperature ° C	Accumulation rate % of present value
Last glaciation	−32±2	50±25
6000–3000 yrs B.P.	−18.8±0.5	100
"Little ice age" A.D. 1400–1900	−20.9±0.5	100
A.D. 1930–1960	−18.4±1	100
Present	−20.1	100

$$H_2{}^{16}O : HD^{16}O : H_2{}^{18}O = 997680 : 320 : 2000. \quad (1)$$

The concentrations, c, of the two heavy components in a water sample are always given as relative deviations from the concentrations c_o in Standard Mean Ocean Water (SMOW):

$$\delta = \frac{c - c_o}{c_o} \cdot 1000‰ \text{ (per mil)}. \quad (2)$$

The measuring accuracy with the conventional mass spectrometer technique is better than 1‰ and 0.1‰ for δD and $\delta^{18}O$, respectively. In both cases this corresponds to less than 1/600 of the variability in natural waters.

One important reason for this variability is that the vapor pressure of the light component $H_2{}^{16}O$ is ca. 8% higher than that of HDO and ca. 1% higher than that of $H_2{}^{18}O$ (Jouzel and Merlivat, 1984). Therefore, the isotopic composition of vapor in equilibrium with SMOW will be

$$(\delta D, \delta^{18}O) = (-80‰, -10‰), \quad (3)$$

see Fig. 4. For the same reason, cooling of such vapor will result in a first stage condensate (precipitation) of the same composition as SMOW. Further cooling and immediate removal of the condensate (Rayleigh condensation) will cause gradual depletion of the heavy components in the remaining vapor, and consequently in the precipitation. In other words, the lower the condensation temperature, the lower the δs of the precipitation.

This very simplified model is only a first attempt to explain the linear relationship between mean annual surface temperature T and mean $\delta^{18}O$ of the snow pack on the Greenland ice sheet (Fig. 5; Johnsen et al. 1988)

$$\delta^{18}O = 0.67\,T - 13.7‰, \quad (4)$$

and of the linear relationship between δD and $\delta^{18}O$ in precipitation (cf. the Meteoric Water Line with slope 8 in Fig. 4; Craig 1961; Dansgaard 1964).

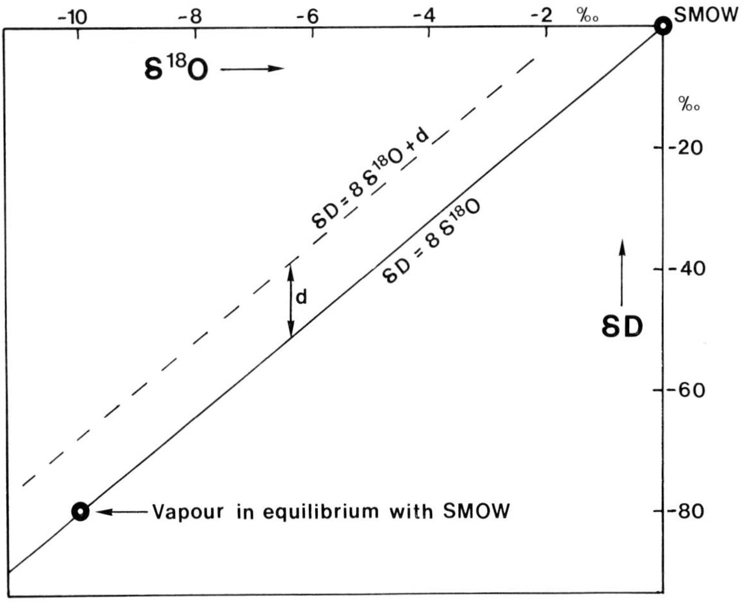

Fig. 4—δD versus $\delta^{18}O$ diagram. The zero point is defined as the isotopic composition of SMOW. The deuterium excess is defined as d = δD − 8 $\delta^{18}O$. The dashed line with d = 10‰ is the Meteoric Water Line showing the normal compositions of natural waters.

The full explanation includes kinetic effects, both in nonequilibrium evaporation from the oceans (Merlivat and Jouzel 1979) and in nonequilibrium sublimation from super-cooled vapor to ice crystals at late stages of the cooling process (Jouzel and Merlivat 1984). Furthermore, it is necessary to account for the temperature dependence of the isotopic fractionation during phase shifts (Majoube 1971).

Attempts to use general circulation models to reproduce the global distribution of δ in precipitation have been successful in many respects, but so far these models have been unable to reproduce important features such as the seasonal δ variation in Greenland ice sheet precipitation (Jouzel et al. 1987). This may be because such precipitation is formed by a special mechanism, not taken into account by the models. We shall return to this problem later.

Long-term Greenland δ records. Figure 6 shows the δ profiles from approximately 300 m of the deepest parts of the ice cores through the ice sheet at Dye 3, S.E. Greenland and Camp Century, N.W. Greenland. The ice deposited during the last glaciation is situated from 1786 m to 2015 m depth in the Dye 3 core, and from 1157 m to 1343 m depth in the Camp Century core, as indicated by the generally more negative δs.

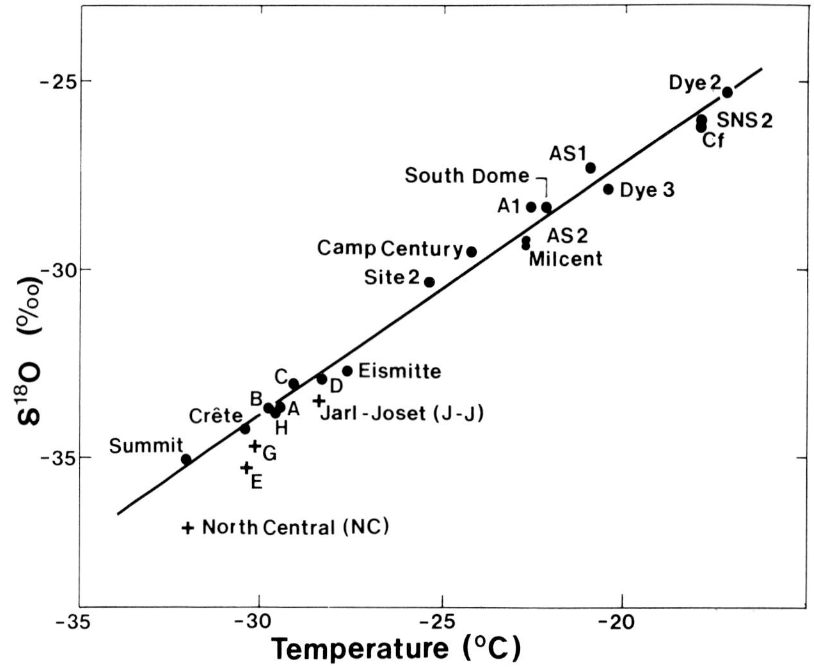

Fig. 5—Relationship between recent mean annual values of $\delta^{18}O$ and temperature of the snow pack at Greenland ice sheet stations. The crosses refer to Northeast Greenland (from Johnsen et al. 1988).

The numbered double arrows point at layers of equal age in the two cores, to judge from similarities between details in the records. Arrows 1 and 2 are questionable, however, because at 1950 m depth the Dye 3 core contains a visible layer of solid particles, apparently from the bedrock, which suggests that the deepest 87 m of the Dye 3 record may not represent a continuous time series. In that case, the reason is most probably the hilly bedrock topography upstream from Dye 3 (Overgaard and Gundestrup 1985), which complicates the flow of the deepest ice.

Dating of the old ice in Greenland. Dating of the layers indicated by arrows 9 and 10 in Fig. 6 was accomplished by counting annual layers downwards from the surface (Hammer et al. 1986), continuously through the last 5900 years. The dating prior to the Bølling-Allerød-Younger Dryas, however, is unsatisfactory, i.e., below arrow 8 in Fig. 6. This is due to:

1. Seasonal δ variations,which are a powerful tool for absolute dating of Holocene ice by annual layer identification, have been obliterated by diffusion in older ice (Johnsen 1977).

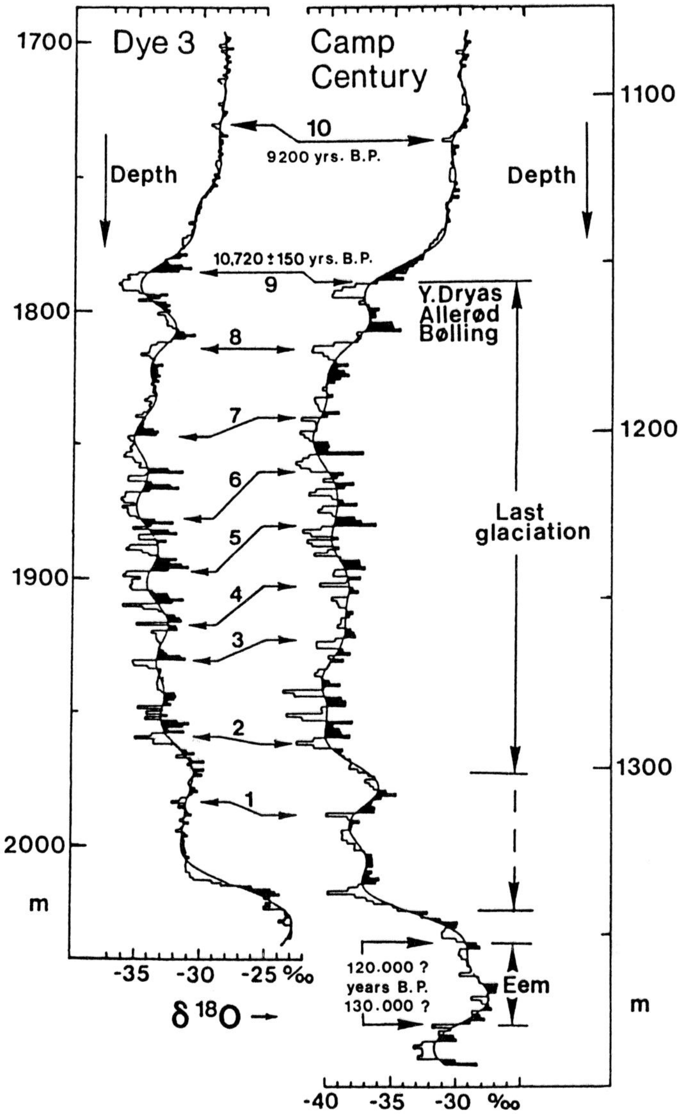

Fig. 6—$\delta^{18}O$ profiles along the deepest (300 m) of the ice cores from Dye 3 and Camp Century. The numbered arrows indicate layers of simultaneous deposition (from Dansgaard et al. 1982).

2. Seasonal acidity variations are suppressed in most of the Pleistocene ice in Greenland. This ice is alkaline due to the then high alkaline aerosol load that obviously neutralized the atmospheric acids, at least at high northern latitudes (Hammer et al. 1985).

3. Ice flow modeling has so far not been able to provide a reliable time scale through the Pleistocene ice in the two existing deep Greenland ice cores because, in contrast to the Vostok core, they were both drilled far from the main ice divide and in areas where the Pleistocene ice is situated only some 200 m above the bedrock.
4. Until recently, radiocarbon dating of glacier ice (using the atmospheric CO_2 entrapped in the air bubbles) required more ice than was available in the deep ice cores. In the future, the new accelerator mass spectrometer technique will make radiocarbon dating possible, but hardly beyond 30,000 years B.P. (see section on *Cosmogenic Radioisotopes*).
5. Indirect dating by reconciliation of δ profiles along ice and sea sediment cores has been attempted (see Fig. 7; Dansgaard et al. 1982) with the assumptions that (*a*) the dating of the sea sediment cores is correct; (*b*) the major global climatic changes are reflected the same way in continental and marine records (which seems to be the case in Antarctica, see Lorius et al., this volume), and (*c*) possible phase differences are, at most, of the same order as the dating accuracy of the marine record (5% through the last glacial cycle?). This technique involves personal judgement, which can be questioned. For example, is the δ maximum between arrows 1 and 2 in Fig. 6 really identical with Emiliani stage 5a, as suggested in Fig. 7? The identification of ice from the Eemian/Sangamon interglacial just below 1350 m depth in the Camp Century core (see Fig. 6) seems though not to be controversial.

In summary, development of an independent dating method is highly desirable. For the state of the art of radioactive dating, see Stauffer (this volume). Ice flow modeling may be applicable on deep ice cores from areas of relatively simple ice flow pattern, such as Central Greenland. Other possibilities lie in further development of the stratigraphic methods based on seasonal variations in dust content and alkalinity, see Peel (this volume).

The Younger Dryas to Pre-boreal (YD/PB) boundary. Arrow 9 in Fig. 6 is extremely well defined in both ice cores (10.720 ± 150 yrs B.P.). In the Dye 3 core, $\delta^{18}O$ shifts from −35.2‰ in YD to −30.5 in PB, within a one meter increment. At the same time the annual layer thickness increases from approximately 2.1 to 3.4 cm (Hammer et al. 1986), indicating a transition period of the order of only 40 years.

During this short time interval the environmental conditions changed drastically in the North Atlantic region. In Greenland, the mean air temperature rose 7° C, as estimated from Fig. 6 and Eq. 1; the accumulation rate increased approximately 60% to judge from the rapidly increasing annual layer thickness; and the dust content, the acidity, and the chemical

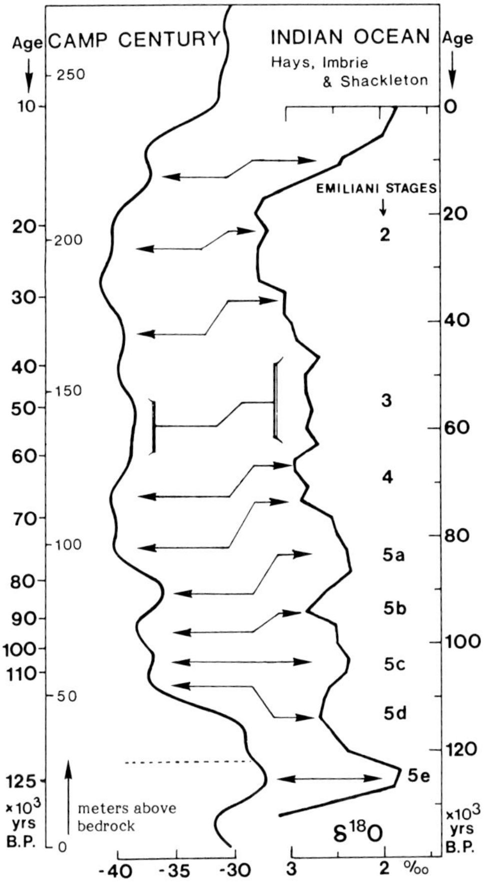

Fig. 7—Smoothed version of the deepest 270 m of the Camp Century $\delta^{18}O$ profile compared with the upper 4 m of a planktonic foraminiferal $\delta^{18}O$ profile from the South Indian Ocean. The linear time scale along the deep-sea sediment core is transferred to the Camp Century ice core in accordance with the correlation suggested by the arrows, assuming only minor time lag between climatic conditions in Greenland and global amounts of continental ice (from Dansgaard et al. 1982).

composition of the deposited ice changed (Hammer et al. 1985; Herron and Langway 1985). In the North Atlantic Ocean, the winter ice cover boundary was pushed far northwards (Ruddiman and McIntyre 1981) to the eastern part from Biscay into the Norwegian Sea. Also in Europe, a rapid change of the flora is seen in all pollen diagrams.

The Bølling-Allerød-Younger Dryas oscillation (BA-YD). The YD/PB boundary marks the end of the BA-YD, the last climatic oscillation during the glaciation. The Greenland δ record of it matches many of the ^{14}C dated

European records, even in detail (Oeschger et al. 1984); see for example Fig. 8. In North America, however, the BA-YD has only been recorded at a few locations in the northeasternmost part (Ruddiman and McIntyre 1981). In view of the general atmospheric circulation pattern, this suggests that the BA-YD was a regional North Atlantic, rather than a global, phenomenon. It also raises the question of whether a similar yet less pronounced feature recorded in Antarctic ice cores (cf. Lorius et al., this volume) could be identical with BA-YD. ^{14}C dating of the Antarctic cores may give the answer.

The entire glacial to postglacial transition. The total δ shift during the entire Pleistocene to Holocene transition amounts to ca. 7‰ in the Dye 3 core and ca. 11‰ in the Camp Century core. According to Eq. 1, this corresponds to a warming of ca. 10.5 and 16.4° C, respectively. Since Eq. 1 primarily describes the present geographic $\delta^{18}O$ to temperature relationship, it is not evident why it should be valid for temporal changes of $\delta^{18}O$ and temperature. Nevertheless, 10.5° C warming of the Dye 3 area agrees essentially with the 12 ± 2° C, calculated independently from the borehole experiments. It is also interesting that application of the present accumulation rate to $\delta^{18}O$ relationship (8% per ‰ change in $\delta^{18}O$; Clausen et al. 1988) on the 7‰

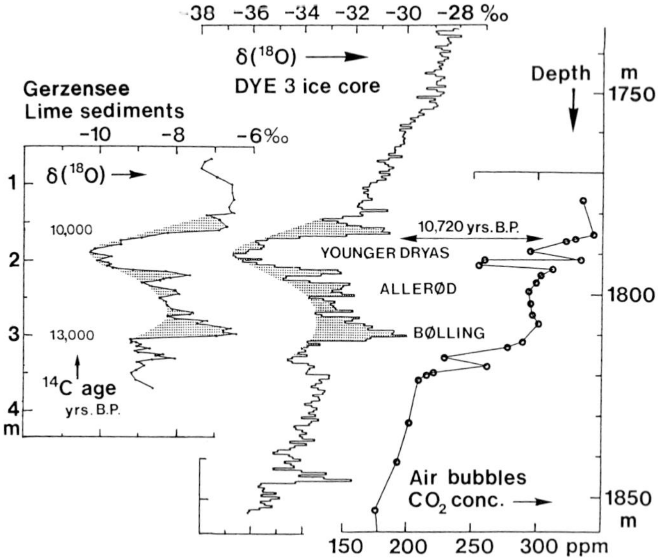

Fig. 8—*Middle*: Detailed $\delta^{18}O$ profile along a 120 m increment of the Dye 3 deep core containing ice deposited during the entire Pleistocene-to-Holocene transition. *Right*: CO_2 concentration in air bubbles (see Fig. 11 for further details). *Left*: $\delta^{18}O$ of lime sediments in the Swiss Lake Gerzen.

$\delta^{18}O$ shift in the Dye 3 core suggests a ca. 56% lower accumulation rate during the glaciation than now. This is again in agreement, yet is not very significant, with the borehole calculation (see Table 1).

The apparent stronger warming trend in the Camp Century area agrees with data from an Ellesmere Island ice core (Fisher et al. 1983) substantiating our interpretation that global climatic changes are generally more pronounced in the higher latitudes (e.g. the general warming in the 1920s was 1.6° C in South Greenland, but 3.8° C in North Greenland). However, in an ice core from the Canadian Devon Island, at almost the same latitude as Camp Century, the $\delta^{18}O$ shift was only 8‰ (Paterson et al. 1977), i.e. the same as at Dye 3.

It has been argued (Flint 1971) that the "too low" glacial δ values at Camp Century might be due to the possible existence of an icebridge between Northwest Greenland and Arctic Canada and, consequently, much higher surface elevations of the Camp Century catchment area. Another hypothesis is that the Würm/Wisconsin ice at Camp Century has traveled from high elevation areas in central North Greenland (Budd and Young 1983).

Both viewpoints are supported by the atmospheric air content in glacial ice being lower than in postglacial ice, corresponding to an elevation difference of 600 m (Raynaud and Lorius 1973) and a $\delta^{18}O$ difference of 4–6‰. However, both viewpoints are contradicted by ice flow modeling (Reeh 1984), which shows that the deep Camp Century ice is of local origin. Furthermore, glacial ice from central North Greenland, sampled close to the margin (cf. Fig. 1), exhibits $\delta^{18}O$ values as low as −47‰ (Reeh et al. 1987), i.e., significantly lower than the lowest δ value in the Camp Century ice core.

A possible explanation of the low air content in deep Camp Century ice is that the firn density ρ, at which the air bubbles were closed off, may have been slightly different from today. A 1–2% higher close-off density would account for the difference in air content (Dansgaard et al. 1973).

In summary, the high δ shift in the Camp Century and Ellesmere Island ice cores most likely reflects a higher temperature shift than in South Greenland, but the smaller δ shift on Devon Island is still an unsolved problem.

Abrupt climatic changes. It appears from Fig. 6 that the Bølling-Allerød-Younger Dryas oscillation was not a unique event. Instead it was the last of a long series of similar events through a major part of the glaciation, during which the δs had a tendency to alternate between two levels of approximately −35.5‰ and −32‰ in the Dye 3 record, and approximately −42‰ and −38‰ in the Camp Century record, corresponding to 5–6° C temperature shifts.

A detailed plot (Fig. 9; Dansgaard et al. 1984) of the δ peaks from 1857 to 1890 m depth in the Dye 3 core reveals their sawtooth shape, with considerably more abrupt δ shifts in cold to mild phases rather than visa versa. All of the δ peaks are provided with a secondary peak or, at least, a "shoulder" on the younger side.

The reason for these features is still unknown. They have also been noticed in North Atlantic sediment cores (Pisias 1973) and in a pollen record from northeastern France (Woillard and Mode 1982). They, however, are not present in Antarctic cores, at least not with the same regularity and high amplitude. This suggests, in analogy with the BA-YD case, that they were connected to a "flip-flop" mechanism in the North Atlantic Ocean,

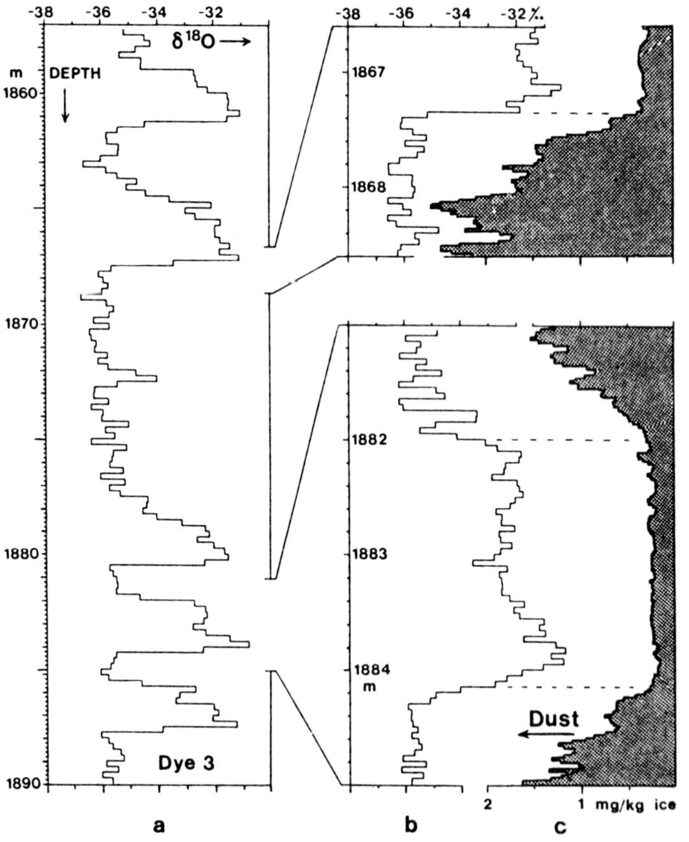

Fig. 9—*a*: Details of the δ oscillations marked to the left of the Dye 3 record in Fig. 6. *b*: Further details of some of the δ shifts. *c*: Concentration of insoluble microparticles. Note the increasing values towards the left (from Dansgaard et al. 1984).

perhaps a "turn-on and -off" of the North Atlantic current (Broecker, 1987). This would also influence the deep water formation and, thereby, the CO_2 content of the atmosphere; yet even if this were true, the cause-and-effect problem remains unsolved.

It should also be noted that the dust concentration changed drastically, but not simultaneously with the δs (see Fig. 8b and c). This provides strong evidence for the core being continuous (Dansgaard et al. 1984). High dust concentrations as found throughout the glaciation but particularly in the coldest phases, indicate dry and stormy conditions in the source areas along the southern margins of the great ice sheets in North America and Europe.

The origin of Arctic precipitation. The isotopic composition of any water sample may be expressed as

$$\delta D = 8\,\delta^{18}O + d. \tag{5}$$

The deuterium excess d is the deviation from the line $\delta D = 8\,\delta^{18}O$ (see Fig. 4) and for a given sample, d depends on the physical conditions in the processes that led to formation of the water (Johnsen et al. 1988), i.e., temperature and humidity during evaporation, the degree of cooling of the vapor prior to condensation/sublimation, and the degree of supersaturation in sublimation processes.

A model has been developed (Johnsen et al. 1988) which takes these processes into account. It follows the vapor from its origin in the ocean until it ends up as snow at high elevation localities on the Greenland ice sheet. The model predicts that the initial mixing ratio, w_{so}, in precipitating air determines the slope of the d versus δ relationship at late stages of the precipitation process, and that the sea surface temperature T_s in the source area of the moisture only influences the d level. The generally high d values in present ice sheet precipitation are compatible only with high values of w_{so} and T_s, which suggests that the subtropical part of the North Atlantic Ocean is a dominating moisture source for ice sheet precipitation.

This is supported experimentally: When the model is run with monthly w_{so} and T_s mean values observed at Ship E (35° N, 48° W), it reproduces the high d level, the amplitude of the seasonal d variations, and the few months phase difference between d and δ on the ice sheet. The model does not produce these features using a local, high latitude moisture source. Thus, precipitation from such sources is confined mainly to low elevation areas.

Detailed isotope analyses of the ice core increments shown in Fig. 9, spanning several abrupt climatic shifts under glacial conditions, yield close to the present-day d values during the cold phases, but lower d values during the mild phases. This suggests that the annual mean of T_s and the humidity (relative to saturation at T_s) in the then moisture source region was nearly the same as it is today, i.e., approximately 21° C and 75% respectively. Such evaporation conditions, now existing in the Ship E area (35° N), characterized a region some 10° latitude further south (CLIMAP 1976).

In the mild phases of the glaciation, however, T_s in the source region seems to have been 2° C *lower* and humidity 10% higher than in the cold phases. This might reflect retreats of the North Atlantic sea ice cover, which gave a larger contribution of moisture evaporated from open sea water of lower T_s.

CO_2 and other Gases

The air bubbles in polar ice constitute physically occluded samples of the atmosphere at the time of gas enclosure. This can be concluded from the concentrations of N_2, O_2 and Ar, which correspond to those of atmospheric air. Also, the CO_2 and CH_4 concentrations in air bubbles in newly formed ice agree with atmospheric concentration data (Neftel et al. 1982; Stauffer et al. 1985). Detailed studies on an ice core from Siple Station, Antarctica, enabled the reconstruction of the *increases* in CO_2 and CH_4 since ca. A.D. 1800 due to anthropogenic emissions.

An important question often raised in this context concerns the possible natural variability of atmospheric constituents like CO_2 and CH_4. It is estimated that the yearly CO_2 exchange fluxes between the atmosphere and biosphere, and between the atmosphere and the oceans, amount to 10% and 15% of the atmospheric CO_2 content, respectively; a slight imbalance in these fluxes would lead to shift in the atmospheric CO_2 concentration. The atmospheric CH_4 concentration is controlled by the flux from the surface of the Earth due to anaerobic decomposition and by the oxidation in the atmosphere. Thus, the atmospheric CH_4 concentration could change either due to a variance in its source strength or a change in the oxidizing capacity of the atmosphere.

Both CO_2 and CH_4 are infrared active gases which contribute to the greenhouse effect. Their changes could lead to shifts in the global temperature. CH_4 is the most abundant reactive trace gas in the atmosphere and thus has a strong impact on tropospheric chemistry. It interacts with the O_3 and the O and OH radicals.

The ice cores from Greenland and Antarctica offer the possibility to reconstruct the concentration of CO_2, CH_4, and other trace gases through the Holocene back into the Wisconsin glaciation. Detailed CH_4 measurements on samples from the preindustrial era back into the glacial period have been started only recently, whereas the first reliable CO_2 concentration values for air extracted from Wisconsin ice were obtained in 1979 (Berner et al.

1980; Delmas et al. 1980). Surprisingly, CO_2 concentrations in the range of 180–200 ppmv were observed compared to Holocene concentrations of about 280 ppmV. In Fig. 10 (Stauffer and Oeschger 1985) the CO_2 concentrations in air samples from the Dye 3 core are plotted together with the $\delta^{18}O$ values. We can distinguish a long-term trend and superimposed rapid variations in the increment between 1860 and 1900 m, corresponding to 30,000 to 40,000 yr B.P. where detailed studies were performed.

Long-term trend. From layers close to the bottom, up to a depth of about 1850 m below the surface (corresponding approximately to an age of

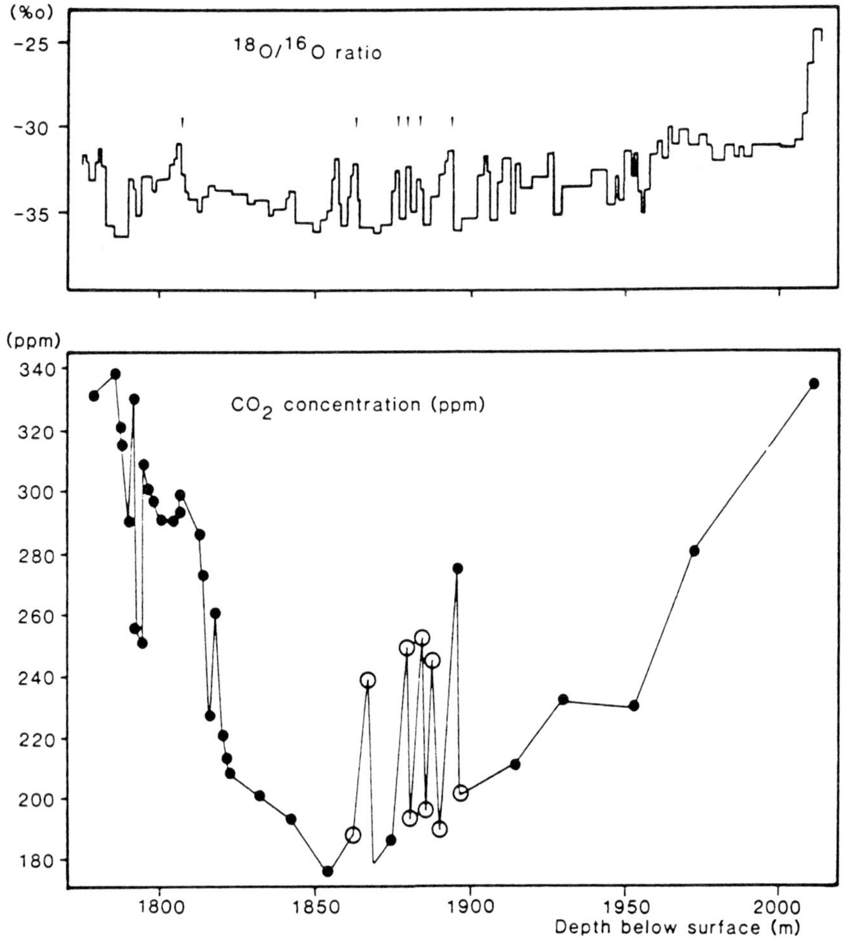

Fig. 10—$\delta^{18}O$ and CO_2 concentration measured on the deep ice core from Dye 3 (from Stauffer and Oeschger 1985).

20,000 yr) the CO_2 concentration is falling from ca. 300 ppmV to a minimum value of 180 ppmV. After this follows a slow increase up to a depth of about 1820 m and a rapid increase to values of about 280 to 300 ppmV between 1812 m and 1792 m. The rapid increase occurs a few meters below the $\delta^{18}O$ transition, which is considered to mark the first warming at the end of the last glaciation; the transition from the Oldest Dryas to the Bølling-Allerød pollen zone.

The core increment between 1786 m and 1794 m consists of ice from the Younger Dryas cold period. Two values between 250 and 260 ppmV indicate a drop of CO_2, but they are followed by a high value of 325 ppmV. Thus, unequivocal evidence does not exist to show that during the Younger Dryas the atmospheric CO_2 concentration really decreased to lower values before its increase to the high Holocene concentrations. In fact, from the analysis of Holocene ice from the Dye 3 core, we know that during a warm climate, melt layers are present in which CO_2 is enriched compared to the other air gases. This leads to CO_2 concentrations in the air occluded in the ice which are well above those of the atmosphere. Based on this we might conclude that the rapid increase at the Oldest Dryas-Bølling boundary has already been influenced by meltlayers and, therefore, one cannot exclude the possibility of a slow CO_2 increase, as indicated in Fig. 11 (Stauffer et al. 1985).

Recently Stauffer et al. (1988) also found a very similar, general trend for CH_4 in the Dye 3 and the Byrd Station cores. A decrease of the CH_4 concentration from 500 ppbv 60,000 yr B.P. to 350 ppbv about 20,00 yr B.P. is followed by a fast increase to Holocene values of 650 ppbv. In the Dye 3 core for the Younger Dryas period, a lower value of 447 ppbv was observed. Though very different in their biogeochemical behavior, the trends of the two gases of low abundance in the atmosphere, CO_2 and CH_4, support the new concept of the Earth as a complex system with numerous physical, chemical and biological interactions, both in regards to long- and short-term behavior.

Rapid Wisconsin CO_2 variations. In the Dye 3 core in an increment, which corresponds to the time interval 30,000 to 40,000 yr B.P., CO_2 variations between low values in the range of 180–200 ppmv and higher values in the range of 240–260 ppmv have been observed (Fig. 12; Stauffer et al. 1984). The low values are found in samples with low $\delta^{18}O$ and high particle and chemical impurity contents, reflecting precipitation from periods of cold climate; the higher CO_2 concentrations are measured in samples reflecting mild climatic conditions. The changes between the two climatic states approximately occur within a century or less. This observation gave rise to the following questions:

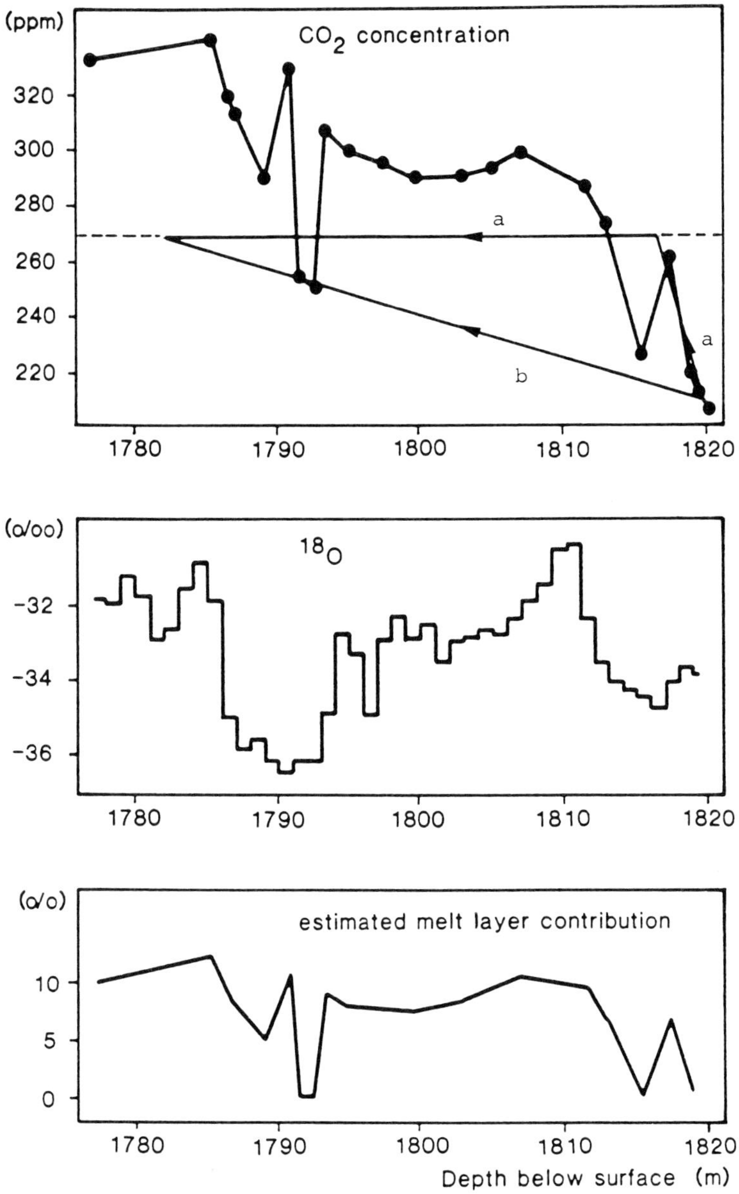

Fig. 11—CO_2 concentrations and $\delta^{18}O$ values in the Dye 3 ice core at the end of the last glaciation. The depth interval 1812–1787 m corresponds to the Bølling-Allerød-Younger Dryas climatic oscillation. Bottom: Estimated meltlayer contribution assuming that atmospheric CO_2 concentration increased according to line *b*. A comparison with $\delta^{18}O$ values shows that the formation of meltlayers is unlikely and that the atmospheric CO_2 concentration more likely increased to line *a* (from Stauffer et al. 1985).

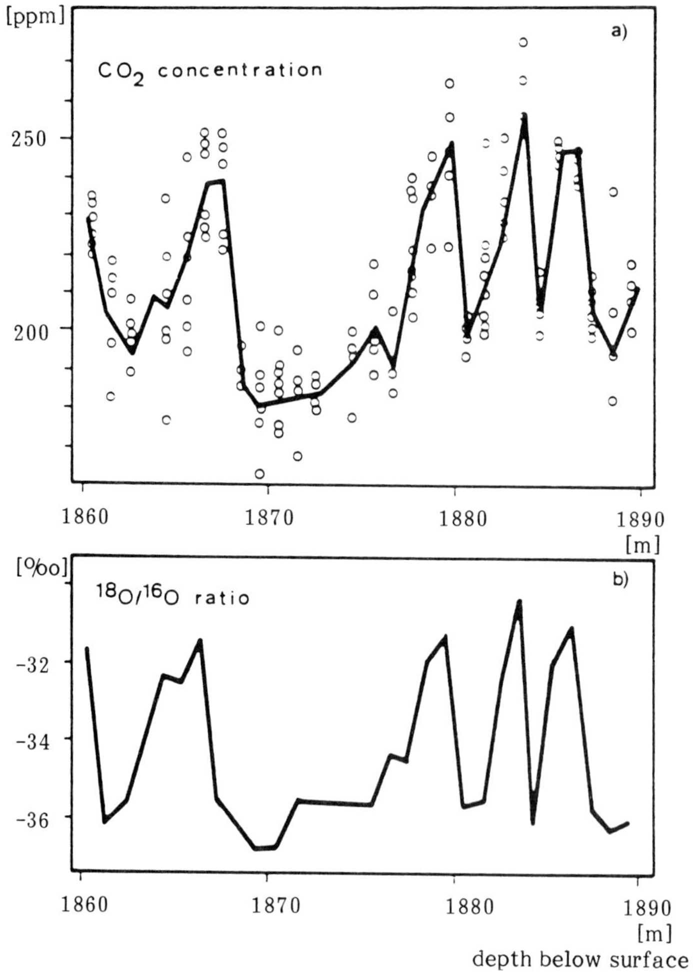

Fig. 12—CO_2 and $\delta^{18}O$ values measured on ice samples from Dye 3. The 30 m increment corresponds to about 10 ka (from Stauffer et al. 1984). (*a*) Circles indicate the results of single measurements of the CO_2 concentration of air extracted from ice samples. The solid line connects the mean values for each depth. (*b*) The solid line connects the $\delta^{18}O$ measurements made on one sample every metre.

1. Do these rapid CO_2 changes observed in the air occluded in the ice indeed reflect atmospheric CO_2 concentration changes, or are they caused by the interaction of the gases with the ice matrix or by the presence of meltlayers in which CO_2 is enriched?
2. What mechanisms in the carbon system could cause such rapid atmospheric CO_2 changes?

Below we try to give an answer to the first question.

As shown in Fig. 6, the $\delta^{18}O$ records in the Greenland Dye 3 and Camp Century ice cores are highly correlated, regarding the rapid changes between cold and mild climate during the last glaciation. It was, therefore, obvious that one should check whether in the Camp Century ice core the rapid Wisconsin $\delta^{18}O$ changes are also accompanied by CO_2 changes. Detailed studies on the Camp Century core confirm the strong coupling betwen CO_2 and $\delta^{18}O$ found first in the Dye 3 core (Staffelbach et al. 1988).

This supports the assumption that the rapid CO_2 changes reflect an atmospheric phenomenon since the physical and chemical properties of the two ice cores are different, and, especially, since the temperature at Camp Century is lower than at Dye 3. Thus, meltlayer contributions to the CO_2 excursions seem less probable. On the other hand, the rapid CO_2 changes should also be present in the Antarctic ice cores, assuming that the atmosphere was well-mixed within a few years, as it is at present.

Thorough investigations of the ice core from Byrd Station, Antarctica, (Fig. 13) revealed CO_2 concentrations in a 180 to 220 ppmv band during the time interval which should overlap with that during which the rapid changes in the Greenland cores (with values up to 240–260 ppmv) were observed (Neftel et al. 1988). The Byrd core values are significantly lower.

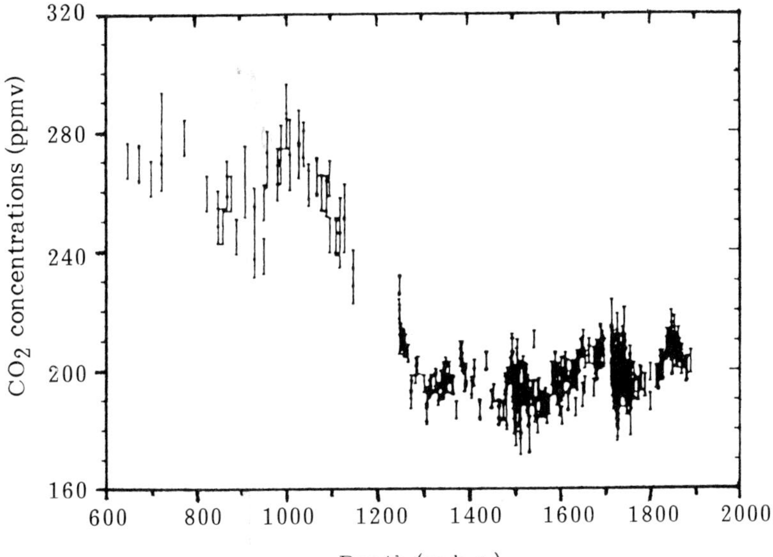

Fig. 13—All measured data points as a function of the depth below surface. The data points are indicated with a bar corresponding to the 1 σ error (from Neftel et al. 1988).

However, clear trends in the CO_2 concentration and periods of a strong scattering of the values can be identified.

At Byrd Station the present annual accumulation is 0.16 m of water equivalent, compared to 0.50 m of water equivalent at Dye 3. This means that at Byrd Station both the difference between the age of the surrounding ice and the mean age of the air and the width of the age distribution of the gases are larger than at Dye 3. Therefore, one might argue that in the Byrd core the CO_2 changes are dampened due to the wider age distribution of the gases in the ice. This effect was modeled by a deconvolution procedure to reinstall the range of possible atmospheric CO_2 concentrations. This procedure leads to a somewhat wider band of possible CO_2 concentrations; in some periods it is 170–230 ppmv. It does not, however, produce values in the 240–260 ppmv range observed in the Greenland ice cores. All this new information leads to the following ambiguous situation:

1. The new Camp Century ice core CO_2 data show rapid Wisconsin CO_2 changes similar to the Dye 3 data in spite of the differences in temperatures and physical and chemical properties of the two ice cores.
2. Similar CO_2 changes do not appear in the Byrd Station ice core from Antarctica, in the core increment thought to represent the corresponding time interval. One must add, however, that until now an unequivocal matching of the time scales of the Greenland and Antarctic ice cores has not been possible.

There are essentially three possibilities to resolve the discrepancy between the Greenland and the Antarctic CO_2 information assuming a globally well-mixed atmospheric CO_2:

1. In the Greenland cores the CO_2 is artificially enriched in certain periods due to a not yet identified mechanism.
2. In the Byrd core the rapid CO_2 variations are strongly dampened due to the longer gas occlusion time.
3. The investigated ice core increments of the Greenland and Antarctica ice cores do not represent overlapping time intervals for some unknown reason.

Another explanation for the difference between the Greenland and Byrd core data is that, although hard to understand, there has been a larger difference between the CO_2 concentrations in the two hemispheres during the Wisconsin glaciation, due to a different distribution of CO_2 sources and sinks and different atmospheric circulation dynamics.

If the rapid CO_2 changes really did occur, they may have been caused by the release of huge amounts of CO_2 by decomposition of biological material that had been slowly accumulated in the tundra regions during the preceding cold periods. Such a process could also explain changes in the atmospheric CH_4 concentration. Another explanation would be that the shifts between mild and cold periods were caused by changes in the North Atlantic ocean currents, which then induced a different functioning of the "biological pump" in the ocean which controls the total CO_2 content and, consequently, the CO_2 partial pressure in the surface water.

Cosmogenic Radioisotopes

Due to the interaction of cosmic rays and their secondaries with atomic nuclei of the elements N, O, Ar, etc. in the atmosphere, radioactive isotopes are produced continuously. Of major importance for studies of their production variations and of Earth system processes are the isotopes ^{3}H, ^{10}Be, ^{14}C, ^{26}Al, ^{36}Cl, ^{39}Ar and ^{129}I. In Table 2 information on properties of these radioisotopes, their terrestrial behavior, and applications are given. One can distinguish essentially three different types of behavior after production (Oeschger 1982).

^{39}Ar is mixed with the stable argon in the atmosphere. As a noble gas isotope it is not involved in any chemical reaction. Until now it has been measured by counting the decay rate of an Ar sample of a given size. It is occluded with the air in ice during the firn-ice transition. In fact, Ar extracted from old polar ice, in which most of the ^{39}Ar had already decayed, served as the background gas of the counting system with which ^{39}Ar was detected for the first time, and samples extracted from boreholes and dated independently by annual $\delta^{18}O$ layer counting helped to establish the ^{39}Ar dating method.

Table 2 Radioisotopes produced by cosmic radiation in the atmosphere as studied in ice samples; measuring techniques and applications

Isotopes	$T_{1/2}$	Analytics	Application
^{3}H	12.4 yr	low level counting	water cycle studies, ice dating
^{14}C	5730 yr	low level counting, accelerator mass spectroscopy	carbon cycle studies, ice dating
^{39}Ar	270 yr	low level counting	water and ice dating
^{10}Be	1.5×10^{6} yr	accelerator mass	cosmic radiation variations,
^{26}Al	7×10^{5} yr	spectroscopy	pathways of aerosols,
^{36}Cl	3×10^{5} yr		ice dating
^{81}Kr	2×10^{5} yr	low level counting, resonance ionization spectroscopy	water and ice dating

^3H is oxidized in the atmosphere to ^3HHO and mixes with the water vapour. ^{14}C is transformed into $^{14}CO_2$ and mixes with the inactive CO_2. ^3H and ^{14}C thus are tracers of the water and the carbon cycles, respectively.

^{10}Be, ^{26}Al, ^{36}Cl and ^{129}I are attached to aerosols and removed from the atmosphere within months to a few years. Transport and deposition of ^{10}Be bound to aerosols depend on atmospheric mixing and circulation processes and on the distribution and the rate of precipitation. The concentration of these radioisotopes in precipitation therefore are determined by the production rate and terrestrial processes.

Polar ice cores enable one to study the concentration of the above mentioned group of radioisotopes in the precipitation of the past 100,000 years or so.

The concentrations of the cosmogenic radioisotopes are very small. Until about 10 years ago, they have been entirely measured by the counting of radioactive decays. Therefore, applications were essentially restricted to the relatively short-lived isotopes ^3H, ^{14}C, and ^{39}Ar. In 1977, particle accelerators, originally constructed for research in nuclear physics, were used for the first time to measure rare isotopes at a low abundance ratio (Gove et al. 1987). The long-lived radioisotopes ^{10}Be, ^{14}C, ^{26}Al, ^{36}Cl and ^{129}I can now be measured in small natural samples having isotopic abundances in the range 10^{-12} and 10^{-15} and as few as 10^5 and 10^6 atoms. At present, e.g., ^{10}Be and ^{36}Cl are routinely measured in ice samples of only 1 to 2 kg (Andrée et al. 1984). ^{14}C dating is now possible on the CO_2 extracted from about 10 kg of ice, instead of the several tons of ice needed for decay counting. In the following we will discuss ^{10}Be measurements performed on ice cores from Greenland and interpret them together with the ^{14}C tree-ring record (Beer et al. 1988).

The ^{10}Be record in ice cores. Samples from the ice core from Camp Century, Greenland enabled the study of long-term ^{10}Be concentration behavior in polar precipitation. The data obtained hitherto are shown in Fig. 14, together with the $\delta^{18}O$ value measured on the corresponding samples. The most striking feature is that below a flat part with minor fluctuations, both records show a rapid increase at 250 m above bedrock. As discussed earlier, at this height above bedrock the transition from Pleistocene to Holocene ice occurs. During the glacial period, with some exceptions, the ^{10}Be concentrations are considerably higher than during the Holocene. This observation can be explained by the assumption of a roughly constant ^{10}Be flux and a precipitation rate which was smaller during the cold glacial period than during the warm postglacial period. In fact, the variations of the ^{10}Be concentrations have been used to estimate accumulation rate changes during the past 160,000 yrs in the Vostok core (Raisbeck et al. 1987).

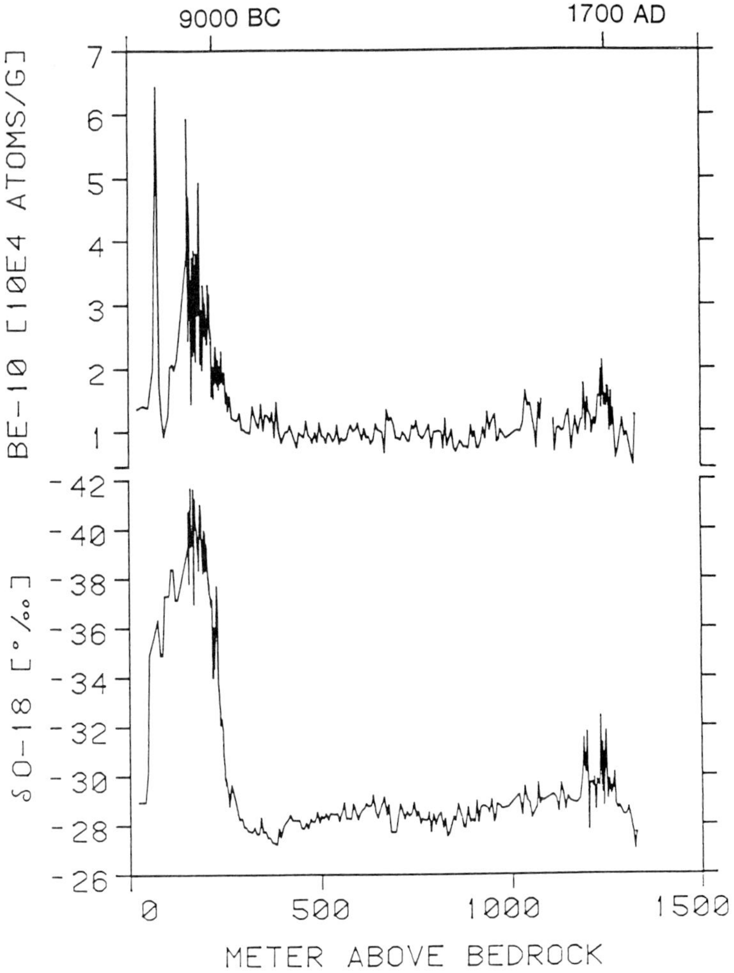

Fig. 14—^{10}Be concentration (10^4 atoms/g) and δ^{18}O (‰) of the Camp Century ice core. The strong increase of both curves at about 200 m represents the transition into the last glaciation (9,000 B.C.). The gap in the ^{10}Be curve around 1100 m is caused by four missing data points (from Beer et al. 1988).

There are also, however, some interesting differences in the ^{10}Be and δ^{18}O profiles. Below 100 m above bedrock ^{10}Be concentrations are low, comparable to the Holocene level (with the exception of a single point at 80 m), whereas δ^{18}O still indicates glacial conditions. The small variations in ^{10}Be and δ^{18}O during the Holocene are not correlated with each other, except for the range 1200 to 1250 m above bedrock which corresponds to the Maunder Minimum period (A.D. 1645–1715) known as a very cold period in Europe. We must therefore find another explanation for the Holocene ^{10}Be fluctuations.

Both ^{10}Be and ^{14}C are produced by cosmic radiation in the atmosphere. ^{14}C measured in tree-rings shows variations which are thought to reflect production variations. If the ^{14}C variations were indeed due to production variations, then the other cosmic ray-produced isotopes like ^{10}Be should exhibit similar fluctuations.

Figure 15 (Stuiver et al. 1986) shows the deviations of ^{14}C from a standard over the past 9000 yrs, measured on absolutely dated tree-rings and corrected for radiocative decay. Essentially, two types of variations were observed:

1. short-term fluctuations (wiggles) of ca. 20–30‰ (peak to peak) which, during certain time intervals, show a periodicity of ca. 200 yrs,
2. a long-term trend showing (on the reversed time axis) a slight decrease to ca. −15‰ around A.D. 500 followed by an increase to a maximum of +100‰ around 5000 B.C.

The short-term variations are generally attributed to variations in the production rate of ^{14}C due to solar modulation of cosmic radiation. Also,

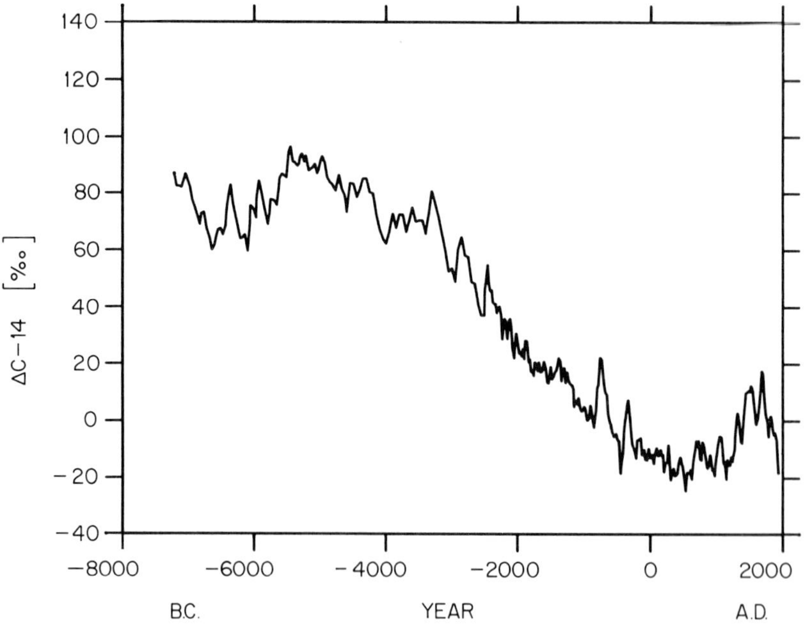

Fig. 15—Deviations of ^{14}C from a standard during the past 9000 yrs, measured on absolutely dated tree-rings and corrected from radioactive decay (after Stuiver et al. 1986).

the long-term trend is considered to reflect production rate changes due to a varying magnetic shielding of the Earth from cosmic radiation. However, as in the case of the ^{10}Be variations, terrestrial effects on the ^{14}C/C ratio might also exist, especially during periods of major climatic change. The comparison of the ^{14}C and ^{10}Be variations offers the possibility of unravelling production and terrestrial effects.

It can be assumed to a first order approximation that ^{10}Be and ^{14}C production variations are proportional but that the terrestrial behavior of

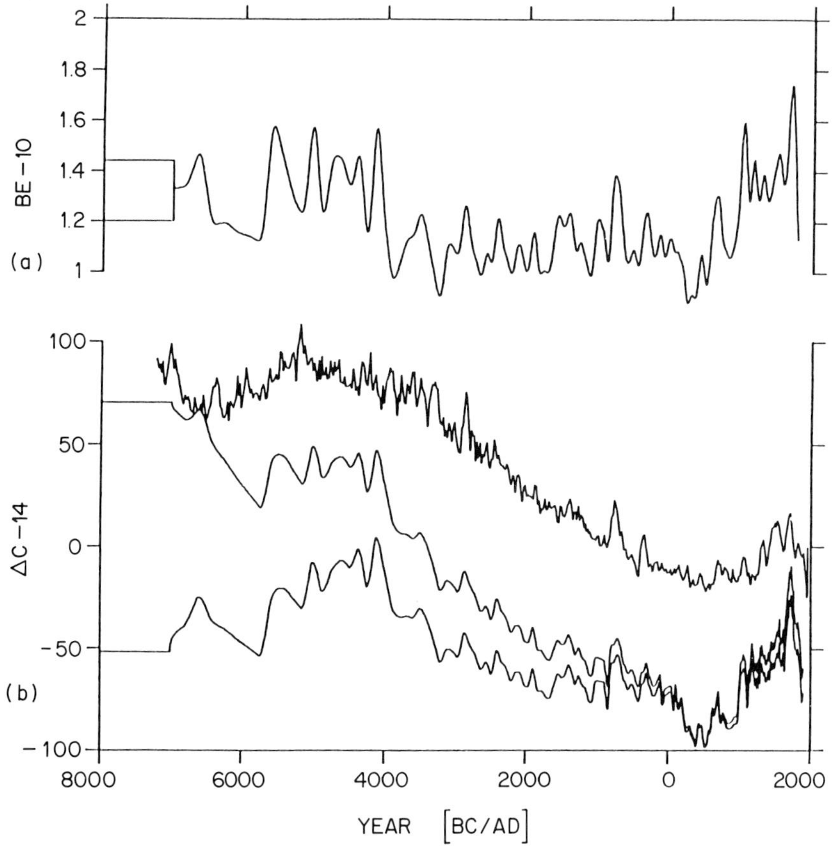

Fig. 16—*a*: ^{10}Be concentration from 7000 B.C. on, used as input function for the model calculations (units: 10^4 atoms/g). The data were slightly smoothed for the figure. The two straight lines before7000 B.C. represent the two initial conditions (1.20 and 1.44). *b*: Measured atmospheric Δ^{14}C data from tree-rings and calculated Δ^{14}C data using the ^{10}Be data of Fig. 16a as input to the box diffusion model. The calculations were carried out for the two different initial conditions as indicated in Fig. 4a. The resulting model curves were shifted by 70‰ for an easier comparison with the measured Δ^{14}C data (from Beer et al. 1988).

the two isotopes is different. Whereas ^{10}Be is deposited on the average about 1–2 years after production, newly produced ^{14}C is oxidized to $^{14}CO_2$, then mixed with atmospheric CO_2 and later with the carbon in the ocean and in the biomass on land. Carbon cycle models simulate this spreading into the exchanging carbon system. Thus, assuming proportionality of the ^{10}Be and ^{14}C production variations and neglecting climatic effects, the atmospheric ^{14}C variations corresponding to the ^{10}Be ice core fluctuations can be calculated and compared with the actually observed fluctuations.

For such a comparison the ^{10}Be data obtained from two Greenland ice cores have been compared with the ^{14}C tree-ring record (Beer et al. 1988). Below is a summary of the results obtained for the Camp Century core.

The carbon cycle model simulation of atmospheric ^{14}C variations was started 7000 B.C., when the ^{10}Be concentrations seemed to be no longer strongly influenced by the glacial-postglacial transition, which mainly took place 13,000 to 10,000 yrs B.P. In Fig. 16 the calculated ^{14}C concentrations

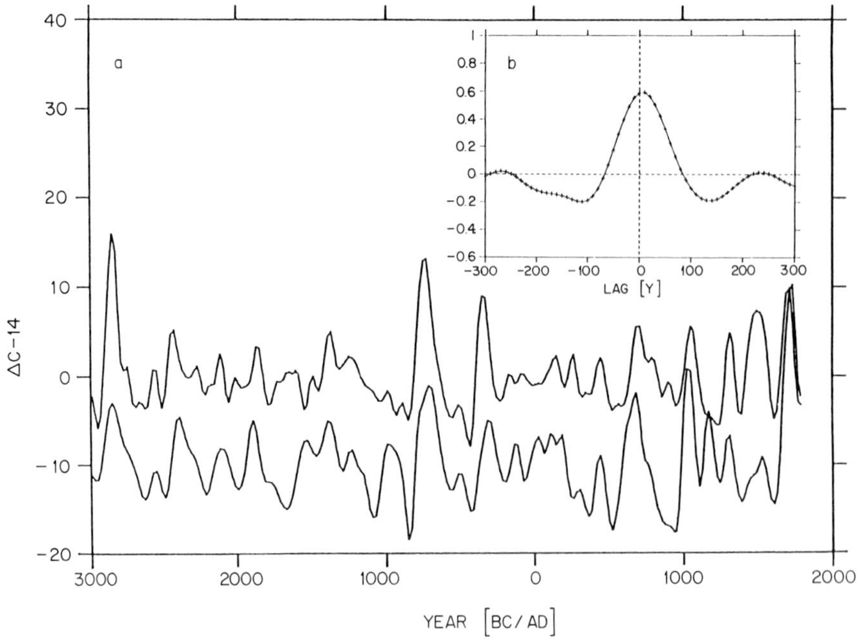

Fig. 17—*a*: Enlargement of the period 3000 B.C. to A.D. 2000 of Fig. 16b after removal of the long-term trends by a high-pass filter. The upper curve represents the measured $\Delta^{14}C$ variations; the lower shows calculated $\Delta^{14}C$ based on the ^{10}Be data using the higher initial conditions of Fig. 16a. The lower curve is shifted by 10‰ for easier comparison. Both curves are slightly smoothed. *b*: Cross-correlation coefficient between the measured and calculated $\Delta^{14}C$ curves (see *a*) as a function of lag time.

are plotted for two sets of initial conditions and are compared with the actually measured concentrations. We do not discuss the long-term trend because both the ^{10}Be and ^{14}C records are probably influenced to some degree by changes of the terrestrial system. However, we are able to compare the detrended short-term variations (Fig. 17) from the period 3000 B.C. to A.D. 1800. A high degree of similarity of the two curves is observed. As an example, the prominent maxima of the tree-ring ^{14}C record at ca. 2800 B.C., 1900 B.C., 700 B.C., 300 B.C., A.D. 800, A.D. 1100, and A.D. 1700 appear also in the ^{10}Be-based ^{14}C variation curve. This clearly suggests that these short-term variations of ^{14}C and ^{10}Be have a common cause: most probably, variations in the production rate due to solar modulation of cosmic radiation. Since the amplitudes of the variations in the two records agree well, as does the phase, we also conclude that the carbon cycle model for this type of variations leads to proper values of the attenuation and of the phase shift due to the spreading of the atmospheric ^{14}C disturbance into the carbon system.

REFERENCES

Andrée, M.; Beer, J.; Oeschger, H.; Broecker, W.; Mix, A.; Ragano, N.; O'Hara, P.; Bonani, G.; Hofman, H. J.; Morenzoni, E.; Nessi, M.; Suter, M.; and Wölfli, W. 1984. ^{14}C Dating of Polar Ice. *Nucl. Instr. Meth.* **B5**: 385–388.

Beer, J.; Siegenthaler, U.; Bonani, G.; Finkel, R. C.; Oeschger, H.; Suter, M.; and Wölfli, W. 1988. Information on past solar activity and geomagnetism from ^{10}Be in the Camp Century ice core. *Nature* **331**: 675–679.

Berner, W.; Oeschger, H.; and Stauffer, B. 1980. Information on the CO_2 cycle from ice core studies. *Radiocarbon* **22**: 227–235.

Broecker, W. S. 1987. Unpleasant surprises in the Greenhouse. *Nature* **328**: 123–126.

Budd, W. F., and Young, N. W. 1983. Application of modelling techniques to measured profiles of temperatures and isotopes. In: G. de Q. Robin, The Climatic Record in Polar Ice Sheets, ed. G. de Q.Robin, pp. 150–179. Cambridge: Cambridge Univ. Press.

Clausen, H. B.; Gundestrup, N.; Johnsen, S. J.; Bindschadler, R.; and Zwally, J. 1988. Glaciological investigations in the Crête-area, Central Greenland. A search for a new deep drilling site. *Ann. Glaciol.* **10**: 10–15.

CLIMAP Project Members. The surface of the ice-age earth. 1976. *Science* **191**: 1131–1137.

Craig, H. 1961. Isotope variations in meteoric waters. 1961. *Science* **133**: 1702–1703.

Dahl-Jensen, D. 1985. Determination of the flow properties at Dye 3, South Greenland, by bore hole tilting measurements and perturbation modeling. *J. Glaciol.* **31**: 92–98.

Dahl-Jensen, D., and Johnsen, S. J. 1986. Palaeotemperatures still exist in the Greenland ice sheet. *Nature* **320**: 250–252.

Dansgaard, W. 1964. Stable isotopes in precipitation. *Tellus* **16**: 436–468.

Dansgaard, W.; Clausen, H. B.; Gundestrup, N.; Hammer, C. U.; Johnsen, S. J.; Kristinsdottir, P. M.; and Reeh, N. 1982. A new Greenland deep ice core. *Science* **218**: 1273–1277.

Dansgaard, W.; Johnsen, S. J.; Clausen, H. B.; Dahl-Jensen, D.; Gundestrup, N.; Hammer, C. U.; and Oeschger, H. 1984. North Atlantic climatic oscillations revealed by deep Greenland ice cores. In: Climate Processes and Climate Sensitivity, M. Ewing

Symp., vol. 5, eds. J. E. Hansen and T. Takahashi. *Geophys. Monog.* **29**: 288–298. Washington, D.C.: Amer. Geophys. Union.

Dansgaard, W.; Johnsen, S. J.; Clausen, H. B.; and Gundestrup, N. 1973. Stable isotope glaciology. *Meddelelser om Grønland* **197**: 1–53.

Delmas, R. J.; Ascenio, J. M.; and Legrand, M. 1980. Polar ice evidence that atmospheric CO_2 20,000 y BP was 50% of present. *Nature* **284**: 155–157.

Fisher, D. A., and Koerner, R. M. 1986. On the special rheological properties of ancient microparticle-laden Northern Hemisphere ice as derived from bore hole and core measurements. *J. Glaciol.* **32**: 501–510.

Fisher, D.; Koerner, R.; Paterson, W.; Dansgaard, W.; Gundestrup, N.; and Reeh, N. 1983. Effect of wind scouring on climatic records from ice-core oxygen-isotope profiles. *Nature* **301**: 205–209.

Flint, R. F. 1971. Glacial and Quaternary Geology, p. 477. New York: Wiley.

Gove, H. E.; Litherland, A. E.; and Elmore, D. 1987. Accelerator mass spectrometry. Proc. of the Fourth International Symposium on Accelerator Mass Spectrometry. *Nucl. Instr. Meth. Phys. Res.* **29**: Nos. 1, 2. Amsterdam: North Holland Physics Publishing.

Gundestrup, N. S., and Hansen, B. L. 1984. Bore-hole survey at Dye 3, South Greenland. *J. Glaciol.* **30**: 282–288.

Hammer, C. U.; Clausen, H. B.; Dansgaard, W.; Neftel, A.; Kristinsdottir, P.; and Johnson, E. 1985. Continuous impurity analysis along the Dye 3 deep core. In: Greenland Ice Core: Geophysics, Geochemistry and Environment, eds. C. C. Langway, Jr., H. Oeschger and W. Dansgaard. *Geophys. Monog.* **33**: 90–94. Washington, D. C.: Amer. Geophys. Union.

Hammer, C. U.; Clausen, H. B.; and Tauber, H. 1986. Ice-core dating of the Pleistocene/Holocene boundary applied to a calibration of the ^{14}C time scale. *Radiocarbon* **28**: 284–291.

Herron, M. M., and Langway, C. C., Jr. 1985. Cloride, nitrate and sulfate in the Dye 3 and Camp Century, Greenland ice cores. In: Greenland Ice Core: Geophysics, Geochemistry and the Environment, eds. C. C. Langway, Jr., H. Oeschger and W. Dansgaard. *Geophys. Monog.* **33**: 77–84. Washington, D.C.: Amer. Geophys. Union.

Johnsen, S. J. 1977. Stable isotope homogenization of polar firn and ice. In: Isotopes and Impurities in Snow and Ice, Proc. IU66 Symp., Grenoble, 1975. IAHS-AISH Publ. **118**: 210–219.

Johnsen, S. J.; Dansgaard, W.; and White, J. 1988. The origin of Arctic precipitation under glacial and interglacial conditions. *Tellus*, in press.

Jouzel, J., and Merlivat, L. 1984. Deuterium and oxygen 18 in precipitation: modeling of the isotopic effect during snow formation. *J. Geophys. Res.* **89**: 11749–11757.

Jouzel, J.; Russell, G. L.; Suozzo, R. J.; Koster, R. O.; White, J. W. C.; and Broecker, W. S. 1987. Simulations of the HDD and $H_2{}^{18}O$ atmospheric cycles using the NASA/GISS general circulation model: the seasonal cycle for present day conditions. *J. Geophys. Res.* **92**: 14739–14760.

Majoube, M. 1971. Fractionnement en oxygen 18 et deuterium entre l'eau et sa vapeur. *J. Chim. Phys.* **10**: 1423–1436.

Merlivat, L., and Jouzel, J. 1979. Global climatic interpretation of the deuterium-oxygen-18 relationship for precipitation. *J. Geophys. Res.* **84**: 5029–5033.

Neftel, A.; Oeschger, H.; Schwander, J.; Stauffer, B.; and Zumbrunn, R. 1982. Ice core sample measurements give atmospheric CO_2 content during the past 40,000 y. *Nature* **295**: 220–223.

Neftel, A.; Oeschger, H.; Staffelbach, T.; and Stauffer, B. 1988. CO_2 record in the Byrd ice core 50,000–5,000 years BP. *Nature* **331**: 609–611.

Oeschger, H. 1982. The contribution of radioactive and chemical dating of the understanding of the environmental system. In: Nuclear and Chemical Dating Techniques: Interpreting the Environmental Record, ed. L. A. Currie, *ACS Sympos. Ser.* **176**: 5–42.

Oeschger, H.; Beer, J.; Siegenthaler, U.; Stauffer, B.; Dansgaard, W.; and Langway, C. C., Jr. 1984. Late glacial climate history from ice cores. In: Climate Precesses and Climate Sensitivity, M. Ewing Symp., vol. 5, eds. J. E. Hansen and T. Takahashi. *Geophys. Monog.* **29**: 299–306. Washington, D.C.: Amer. Geophys. Union.

Overgaard, S., and Gundestrup, N. 1985. Bedrock topography of the Greenland ice sheet in the Dye 3 area. In: Greenland Ice Cores: Geophysics, Geochemistry and Environment, eds. C. C. Langway, Jr., H. Oeschger and W.Dansgaard. *Geophys. Monog.* **33**: 49–56. Washington, D.C.: Amer. Geophys. Union.

Paterson, W. S. B. 1981. The Physics of Glaciers, 2nd ed. London: Pergamon.

Paterson, W. S. B.; Koerner, R. M.; Fisher, D.; Johnsen, S. J.; Clausen, H. B.; Dansgaard, W.; Bucher, P.; and Oeschger, H. 1977. An oxygen-isotope climatic record from the Devon Island ice cap, Arctic Canada. *Nature* **266**: 508–511.

Pisias, N. G.; Dauphin, J. P.; and Sancetta, C. 1973. Spectral analysis of late Pleistocene-Holocene sediments. *Quatern. Res.* **3**: 3–9.

Raisbeck, G. M.; Yiou, F.; Bourles, D.; Lorius, C.; Jouzel, J.; and Barkov, N. I. 1987. Evidence for two intervals of enhanced ^{10}Be deposition in Antarctic ice during the last glaciation period. *Nature* **326**: 273.

Raynaud, D., and Lorius, C. 1973. Climatic implications of total gas content in ice at Camp Century. *Nature* **243**: 283–284.

Reeh, N. 1984. Reconstruction of the glacial ice covers of Greenland and the Canadian Arctic Islands by three-dimensional, perfectly plastic ice-sheet modelling. *Ann. Glaciol.* **5**: 115–121.

Reeh, N.; Højmark Thomsen, H.; and Clausen, H. B. 1987. The Greenland ice-sheet margin — a mine of ice for Palaeo-environmental studies. *Paleogeog. Pal. Pal.* **58**: 229–234.

Ruddiman, W. F., and McIntyre, A. 1981. The North Atlantic ocean during the last deglaciation. *Paleogeog. Pal. Pal.* **35**: 145–214.

Staffelbach, T.; Stauffer, B.; and Oeschger, H. 1988. A detailed analysis of the fast changes in ice core parameters during the last ice age.*Ann. Glaciol.* **10**: 167–170.

Stauffer, B.; Fischer, G.; Neftel, A.; and Oeschger, H. 1985. *Science* **299**: 1386–1388.

Stauffer, B.; Hofer, H.; Oeschger, H.; Schwander, J.; and Siegenthaler, U. 1984. Atmospheric CO_2 concentration during the last glaciation. *Ann. Glaciol.* **5**: 160–164.

Stauffer, B.; Lochbronner, E.; Oeschger, H.; and Schwander, J. 1988. Methane concentration in the glacial atmosphere was only half that of the preindustrial Holocene. *Nature*, in press.

Stauffer, B.; Neftel, A.; Oeschger, H.; and Schwander, J. 1985. CO_2 concentration in air extracted from Greenland ice samples. In: Greenland Ice Core: Geophysics, Geochemistry and the Environment, *Geophys. Monog.* **33**. Washington, D.C.: Amer. Geophys. Union.

Stauffer, B., and Oeschger, H. 1985. Gaseous components in the atmosphere and the historic record revealed by ice cores. Ann. Glaciol. **7**: 54–59.

Stuiver, M.; Kromer, B.; Becker, B.; and Ferguson, C. W. 1986. Matching of the German Oak and U.S. Bristlecone Pine Chronologies. *Radiocarbon* **28**: 969–979.

Woillard, G. M., and Mode, W. D. 1982. Carbon-14 dates at Grande Pile: correlation of land and sea chronologies. *Science* **215**: 159–161.

The Environmental Record in Glaciers and Ice Sheets
eds. H. Oeschger and C.C. Langway, Jr., pp. 319–341
John Wiley & Sons Limited

Long-term Changes in the Concentrations of Major Chemical Compounds (Soluble and Insoluble) along Deep Ice Cores

R. J. Delmas and M. Legrand

Laboratoire de Glaciologie et Géophysique de l'Environnement
38402 St Martin d'Hères Cedex, France

Abstract. Changes in the Earth's environment over the last ~ 0.5 Myr are recorded in polar ice sheets. The chemical analysis of deep ice cores from both Greenland and Antarctica has already revealed valuable information on the composition of the past atmospheric aerosol and, consequently, on the evolution of global biogeochemical cycles. Despite a certain lack of knowledge regarding the fundamental mechanisms by which atmospheric impurities are included in snowflakes and subsequently buried in the snow and ice layers, the interpretation of chemical profiles obtained from five deep ice cores from Greenland and Antarctica is now possible, particularly when a balanced ion budget is available. For this, the measurement of six major cations (H^+, Na^+, K^+, NH_4^+, Mg^{2+} and Ca^{2+}) and three major anions (Cl^-, SO_4^{2-} and NO_3^-) is needed. Aluminium, as an indicator of the insoluble crustal-derived fraction, should also be measured. The data reveals that integlacial snow principally contains a mixture of small amounts of sea salt and two mineral acids, H_2SO_4 and HNO_3. The continental contribution is very weak. On the other hand, during the ice ages, the inputs of crustal material play a very significant role, while the sea salt contribution is also increased, but to a lesser extent. These variations are linked to meteorological changes at the surface of the continents and the oceans (winds, humidity, sea level). The sulfate measurements give insight into the marine biological productivity and paleovolcanic activity. Clearly, more work is still needed to understand fully the chemical records already published; however, the interpretation of the very recent Vostok ice cores has clearly demonstrated the great interest of such records for paleoenvironmental studies.

INTRODUCTION

Archives of numerous environmental parameters are recorded in polar ice sheets. Over the last decades, much research was devoted to water isotope records (^{18}O and ^{2}H), which provide global paleotemperature information. However, snow accumulating in polar areas also buries atmospheric impurities, which are then preserved for millennia in the ice layers. These impurities are incorporated in snowflakes when precipitation is formed in clouds. Significant amounts of atmospheric trace substances may also deposit directly on the snow surface. These two basic processes are referred to as wet and dry deposition, respectively. Both gaseous and solid atmospheric trace substances are stored in the ice crystals. It is generally recognized that the migration phenomena within the ice lattice may be considered as unimportant in paleoclimatic studies. That means that the precipitation layers contain trace components which were present in the atmosphere at the time of deposition. Contrary to what occurs for gases enclosed in bubbles, no significant postdepositional phenomenon has to be considered for the impurities found in the ice lattice (except H_2O_2, see below).

In the first glaciochemical studies, one or at best a few chemical elements were investigated in sections selected along the ice cores. The recent use of rapid and sensitive analytical techniques like ion chromatography has made it possible to determine several major ions together in small amounts ($<$10 ml) of meltwater. It is now relatively easy to obtain the ion balance of an ice sample. This is of great help in determining the origin of the various measured chemical elements and the overall composition of polar precipitation. It is, for instance, of primary importance to know the counter ion of a given sulfate, whether this ion was deposited as H_2SO_4 or a salt. In the past, the interpretation of sulfate concentrations along deep ice cores was nearly impossible owing to the lack of this very essential information.

No valuable conclusions can be drawn from long-term chemical profiles unless a very comprehensive study of the present-day polar precipitation is first conducted in the geographical area of the deep ice core site. We will therefore first examine the status of our present knowledge of polar precipitation chemistry.

The deep ice cores considered for this review are from relatively coastal sites in Greenland and at more central locations in Antarctica (Table 1). Elevations range from 1530 m (Byrd station) to 3480 m (Vostok station). Present-day snow accumulation rates are markedly higher in Greenland (32–50 $g.cm^{-2}.a^{-1}$) than in Antarctica (2.2–16 $g.cm^{-2}.a^{-1}$).

Figure 1 reports on the chemical composition of recent snow at two North and South Pole sites. Some of the chemical profiles obtained for these different ice cores have been put together in Figs. 2 (Greenland) and 3 (Antarctica). Mean values of the major chemical constituents for both Holocene and glacial ages are recapitulated in Table 2.

Table 1 Deep ice core locations

	Elevation (m a.s.l.)	Distance from coast (km)	Present-day accumulation rate	Mean annual temperature °C
Greenland				
Camp Century (CC)				
77°10′N, 61°08′W	1885	200	32	−24
Dye 3 (D3)				
65°11′N, 43°50′W	2480	400 (West coast) 150 (East coast)	50	−19.6
Antarctica				
Vostok (VK)				
78°28′S, 106°48′E	3480	1300	2.2	−57
Dome C (DC)				
74°40′S, 124°10′E	3240	1070	3.7	−53
Byrd Station (BY)				
79°59′S, 120°01′W	1530	670	16	−28

PRECIPITATION CHEMISTRY IN CENTRAL GREENLAND AND ANTARCTIC AREAS

The Atmosphere-Snow Relationship

It is known that in polar atmospheres, scavenging within the clouds (rainout) is dominant in comparison with below-cloud processes (washout); however, the physical processes leading to the inclusion of the impurities in snowflakes or ice crystals in polar conditions is poorly documented (Junge 1977). One has to recognize that this very fundamental aspect has been generally overlooked and that part of related glaciochemical work is still empirical. Discussion arises particularly when the results of chemical analyses of ice cores are interpreted straight off in terms of past atmospheric composition, without considering all possible physical phenomena at the atmosphere-snow interface. However, considering the lack of knowledge on cloud chemistry in polar conditions, we have to assume that concentrations in the snow and air vary in a parallel manner. This is probably true when a given compound or ion is considered, but we know that significant differences between the overall chemical compositions of the aerosol and the precipitation are caused by the deposition of trace gases such as HNO_3 and HCl. Despite all these difficulties, limitations, and uncertainties, the chemical records obtained from ice core studies are very relevant to cloud and atmospheric chemistry and are useful in investigating the impact of climatic changes on the Earth's environment.

Table 2 Mean values of major chemical parameters obtained for five ice cores: Camp Century (CC), Dye 3 (D3), Dome C (DC), Vostok (VK), and Byrd Station (BY). LGM is Last Glacial Maximum and GA is Glacial Age. Na_m is sea salt sodium and SO_4* is non-sea salt (or excess) sulfate. Data is taken from the papers of Cragin et al. (1977), Herron and Langway (1985), Hammer et al. (1985), Legrand (1985), Palais and Legrand (1985), Legrand (1987), Legrand et al. (1988), and Legrand and Delmas (1988).

		CC	D3	DC	VK	BY
	Al	10	3.3	2.7	2.4	1.5
	Na_m	1.05**	0.6**	0.7	1.2	1.1
Holocene (H)	Cl	0.9	0.5	0.9	0.5	1.2
	SO_4	0.7	0.5	1.6	2.8	0.8
	SO_4^*	0.6	0.4	1.3	2.6	0.6
	NO_3	~1.3	0.8	0.3	0.25	0.6
	Al	6	—	6.4	—	—
	Na_m	3.4**	4.6**	3.4	—	1.5
T/H	Cl	3.4	4.6	1.8	—	1.8
	SO_4	~8	~6.5	1.8	—	1.2
	SO_4^*	9.2	6.8	1.2	—	1.3
	NO_3	~1.2	1.5	0.9	—	1.4
	Al	62	—	17.4	—	—
	Na_m	3.6**	2.7**	2.3	—	1.6
Transition (T)	Cl	3.1	2.3	—	—	2.1
	SO_4	~6	3	2.85	—	1.0
	SO_4^*	5.7	2.8	1.6	—	0.8
	NO_3	~1.5	1.2	0.3	—	0.9
	Al	11	—	27	37	8
	Na_m	1.4**	3.2**	5.6	3.7	2.1
LGM/H	Cl	~1.5	3.2	6.3	10.2	2.4
	SO_4	7.5	6.4	2.5	1.9	1.6
	SO_4^*	3	3.2	2.4	1.8	1.7
	NO_3	~0.3	0.75	2	5.6	1.3
	Al	110	—	74	88	11.5
	Na_m	1.5**	1.9**	3.8	4.5	2.3
LGM	Cl	~1.3	1.6	4.6	5.1	2.85
	SO_4	~5.3	~3	3.9	5.3	1.3
	SO_4^*	1.9	1.3	3.2	4.8	1.0
	NO_3	~0.4	0.6	0.6	1.4	0.8
	Al	4.7	—	8	~10	—
	Na_m	—	—	4.9	3.1	1.3
GA/H	Cl	—	—	—	4.2	1.5
	SO_4	—	—	2.0	1.6	1.3
	SO_4^*	—	—	1.0	1.5	0.95
	NO_3	—	—	0.8	1.6	0.15
	Al	47	—	21.5	~20	—
	Na_m	—	—	3.3	3.7	1.4
GA	Cl	—	—	—	3.1	1.8
	SO_4	—	—	3.1	4.4	1.0
	SO_4^*	—	—	1.4	3.9	0.85
	NO_3	—	—	0.3	0.4	0.7

(**) calculated from Cl

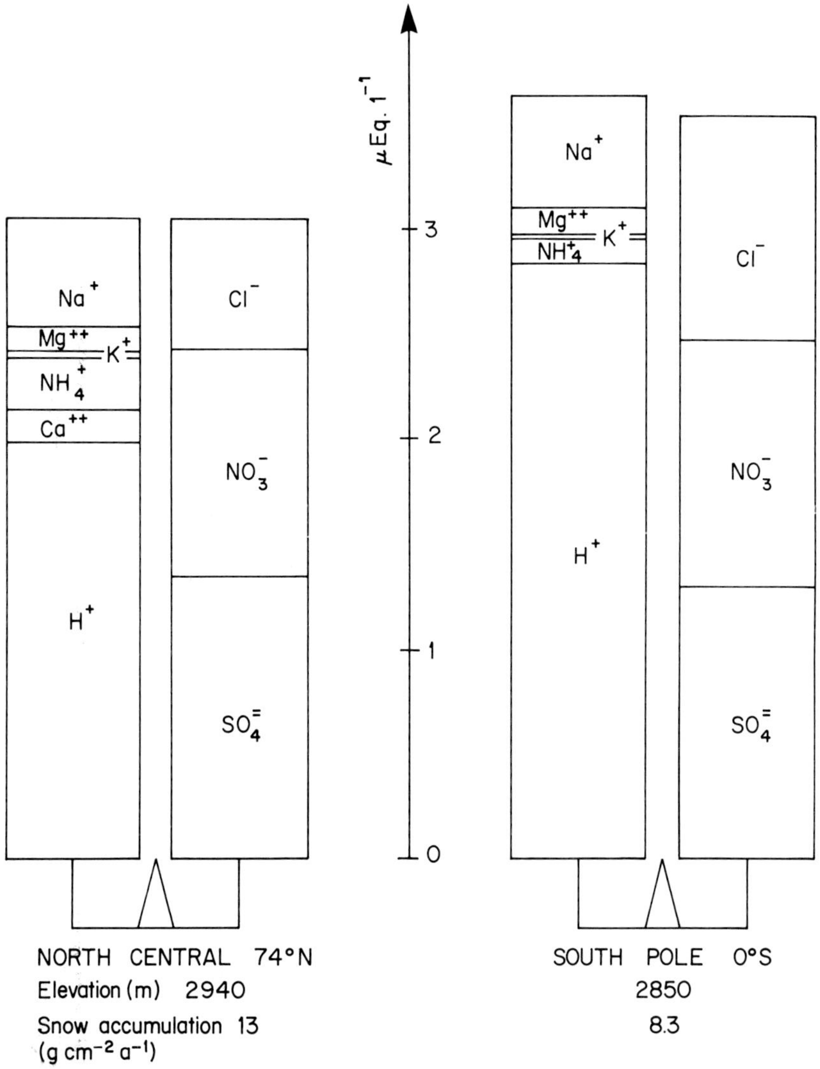

Fig. 1—Ion balance of Greenland and Antarctic snows. As no H^+ data are available for North Central, the H^+ calculated here corresponds to an exact anion-cation balance. Na^+, Mg^{2+} and K^+ have been calculated from Cl values.

The above reserves could be removed if the following fundamental processes were investigated both for gases and particles:

1. inclusion in snowflakes at cloud level around and over polar regions,
2. direct deposition on the snow surface,
3. reworking in the shallowest snow layers.

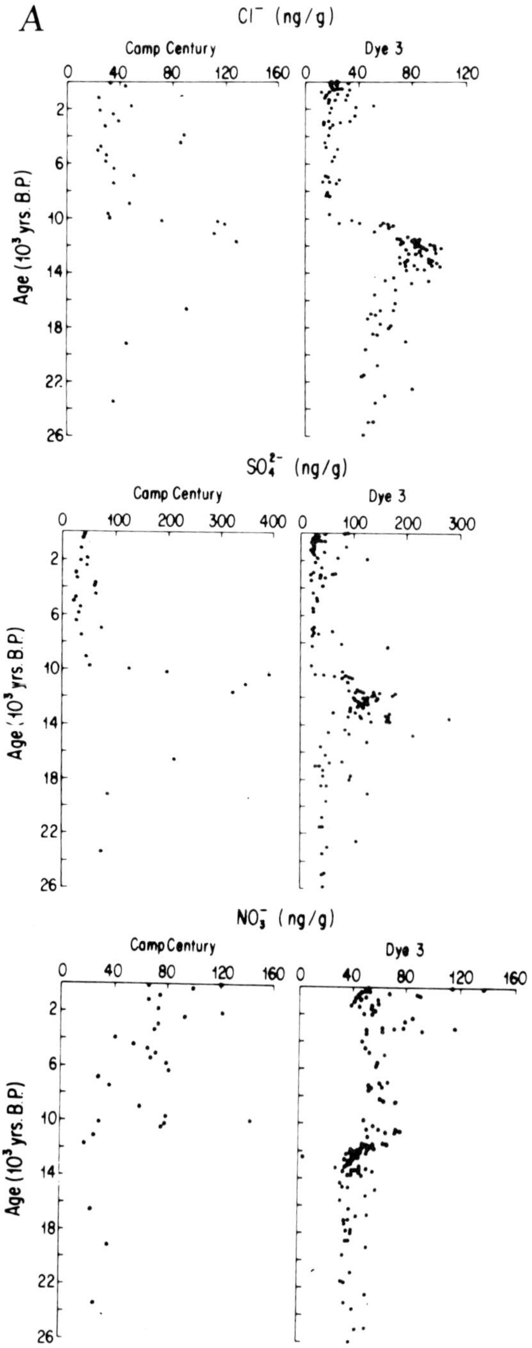
A
Cl⁻ (ng/g)
Camp Century
Dye 3
Age (10³ yrs. B.P.)
SO₄²⁻ (ng/g)
Camp Century
Dye 3
Age (10³ yrs. B.P.)
NO₃⁻ (ng/g)
Camp Century
Dye 3
Age (10³ yrs. B.P.)

B

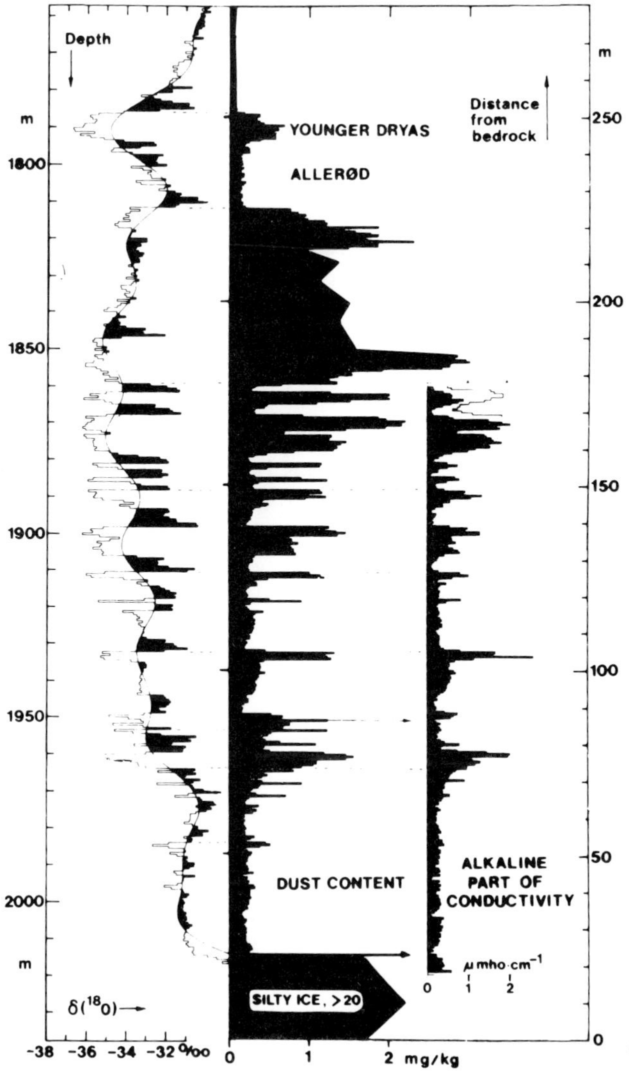

Fig. 2—Chemical profiles from deep Greenland ice cores. 2A: Cl^-, SO_4^{2-} and NO_3^- at Camp Century and Dye 3 (Herron and Langway 1985). 2B: Dust content at Dye 3 for the lower part of the ice core. Dust is low for relatively warm periods, high for cold periods. Lower than 1786 m (the depth of the Holocene-Wisconsin transition) the ice is alkaline. The $\delta^{18}O$ profile indicates the climatic periods during the glacial age (Hammer et al. 1985).

Chemistry of Polar Precipitation: Basic Equations

Polar snow is extremely pure compared to precipitation collected at midlatitudes: its ion budget is generally lower than 10 μEq.l^{-1}. It is slightly acidic due to the weakness of primary aerosol contributions (alkaline sea salt and crustal dust), in comparison to secondary (gas-derived) acid aerosol. At Dye 3 (Table 2), dust represents about 30% of the deposits (Hammer et al. 1985). At the South Pole, this contribution, calculated for a 30 yr time period, amounts to only 4% (Delmas et al. 1982). This figure is only true for the present time since in glacial-age weather conditions, the insoluble fraction was much higher (see below).

The ion balance of present Antarctic snow has been obtained only recently (Legrand and Delmas, 1984) (concentrations in μEq.l^{-1}):

$$[Na^+]+[K^+]+[NH_4^+]+[Ca^{2+}]+[Mg^{2+}]+[H^+] = [SO_4^{2-}]+[NO_3^-]+[Cl^-]. \quad (1)$$

The imbalance between cations and anions has been accurately determined on several thousands of samples. It has been found to be generally lower than 0.7 μEq.l^{-1}, i.e. less than 10% of the ion budget (Legrand 1987). This strongly indicates that all major ionic species have been taken into account. New measurements indicate that organic acids (formic and methane sulfonic particularly) are negligible when one calculates the ion budget. Recent Antarctic snow contains, therefore, essentially mineral acids (H_2SO_4, HNO_3, and sometimes HCl) and sea salt. For Greenland snow, no actual ion

A

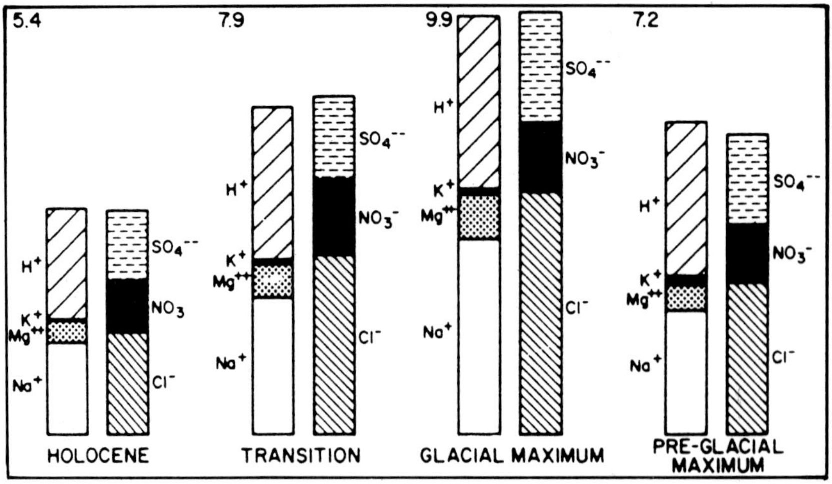

Fig. 3—(see legend on p. 328).

B

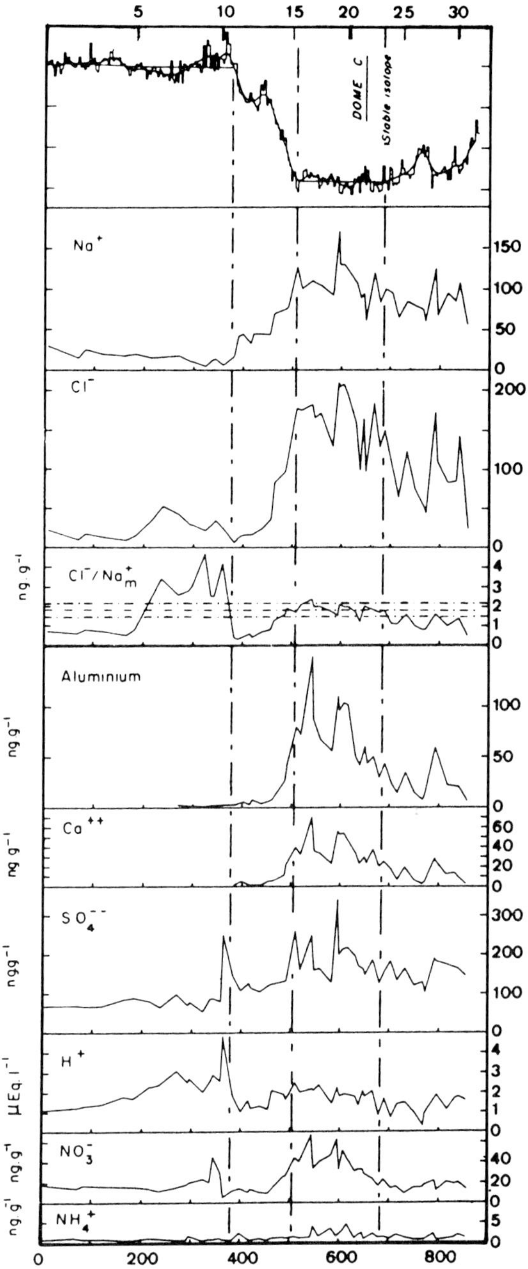

Fig. 3—(see legend on p. 328).

C

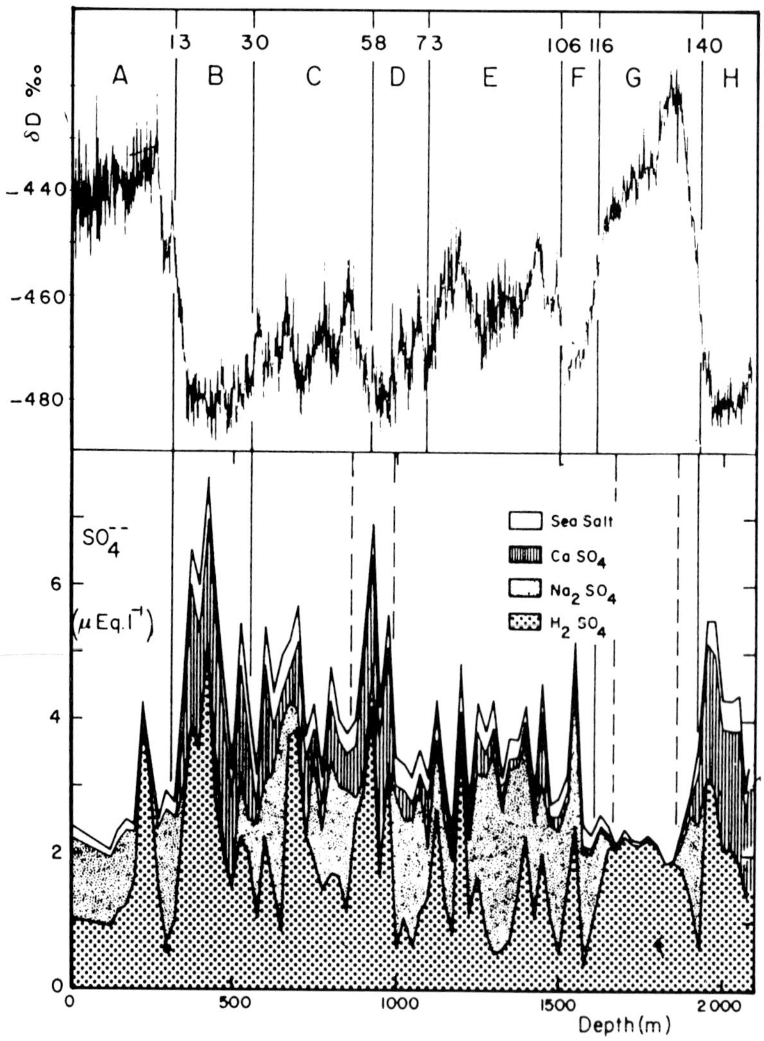

Fig. 3—Chemical data from Antarctic ice cores. A: Ion balance along the Byrd Station Ice Core (Palais and Legrand 1985). B: Concentrations of major ions, Al and the Cl/Na ratio along the Dome C core (Legrand and Delmas 1988). The isotope profile (upper curve) indicates the climate. C: Sulfate contributions along the Vostok ice core. Climatic variations are indicated on the upper curve (deuterium content in δD%). Ages B.P. in Kyr are given at the top (Legrand et al. 1988).

balance has been published but it is possible to construct an approximate diagram, on the basis of a few available data, in order to compare preindustrial ionic compositions in both polar regions (Fig. 1). Ice-age ice contains much greater quantities of chemical impurities than interglacial ice (by a factor of 3 for the soluble fraction at Vostok, where the increase is particularly high). The first measurements by Cragin et al. (1977) clearly show that continental and sea salt-derived ions or particulates are particularly abundant in all ice core levels corresponding to cold periods. The chemical system is therefore intricate, each ion originating from multiple sources. In this case, a balanced ionic budget is of primary importance in the understanding of the possible associations of the various ions in the aerosol. Figure 3B shows the ion balances at Byrd Station for four time periods along the ice core. To illustrate the use of the ion balance in interpreting the chemical composition of ice-age ice, we propose the example of the Vostok ice core (Legrand et al. 1988).

The ion balance of the soluble impurities may be written as

$$[Na_m^+]+[Ca^{2+}]+[Mg^{2+}]+[H^+] = [Cl^-]+[NO_3^-]+[SO_4^{2-}] \qquad (2)$$

with the following assumptions:

1. Na_m, sea salt sodium, has to be calculated from measured total sodium by subtracting the crustal fraction Na_t (Na_t = 0.29 Al, concentrations in $ng.g^{-1}$). Na_t is not water soluble, but is nevertheless obtained by ion chromatography (Legrand 1987).
2. $[K^+]$ and $[NH_4^+]$ may be neglected since their sum represents less than 2% of the ion budget.

Moreover, calcium and magnesium are contributed not only by sea salt, as is the case during the Holocene, but also by a powerful continental source. Legrand (1987) has demonstrated that the amount of excess Ca^{2+} and excess Mg^{2+} (the term "excess" is used to designate the non-sea salt fraction) is clearly correlated with aluminium, a strong indication of their terrestrial origin. In the absence of carbonates in Antarctic ice (Legrand 1987)—this observation is, however, not valid for Arctic ice—these excess amounts of Ca^{2+} and Mg^{2+} have to be balanced in the aerosol phase to corresponding amounts of the three anions Cl^-, NO_3^- and SO_4^{2-}.

This finding is essential to the understanding of the chemical composition of terrestrial inputs during the ice age. In the absence of this atmospheric neutralization process, the ice may be found to be alkaline, which is the case for Greenland ice-age ice.

Present-day Precipitation

Primary aerosol. The low primary aerosol deposition may be explained by the relatively high elevations of the central polar areas where the deep ice cores are drilled. This parameter seems to determine the amount of sea salt transported inwards over the ice sheets. Herron (1982) proposed that the deposition of sea salt particles in polar snow may be expressed as:

$$C = Co\ e^{-Z/H}, \qquad (3)$$

where C is the concentration at an elevation Z (km), H is an empirically derived scale height equal to 1.52 km, and Co a constant, different for Greenland and the Antarctic, reflecting the activity of sea salt aerosol generation and the efficiency of transport to the polar ice sheets (Herron and Langway 1985).

Provided that the snow accumulation rate is not too small (> 8 g.cm^{-2} a^{-1}) NaCl concentration in snow would have to vary seasonally, with highest values when the sea is free of ice and the weather stormy, i.e., in autumn. A different pattern was obtained at Dye 3 by Langway et al. (1975): clear spring maxima were observed. As this maximum is also found for aluminium (see below), a meteorological phenomenon is likely responsible. The position of the northern polar front during the year could be an explanation for these seasonal patterns, as has been proven also for Arctic air pollution.

In Antarctica, the seasonal deposition patterns of sea salt are poorly documented for the central Antarctic sites investigated in this study, but measurements at the South Pole have shown that the transport of sea salt occurs essentially in the winter season (Legrand and Delmas 1984). Finally, the Cl/Na values, frequently different from the 1.8 reference value in sea salt, suggest strongly that a fractionation of sea salt may occur in the polar atmosphere (see next section).

The continental contribution may be assessed either by insoluble microparticle counting or by the content in a typical crust-derived element like Al, Si, Ca, etc. Alkalinity has been found to be an indirect indicator of this type of impurity (Hammer et al. 1985).

Additional information can be drawn from the particle-size distribution of dust. Microparticle content is higher at Camp Century than both at Byrd station and Dome C (Thompson and Mosley-Thompson 1981). The size distribution of microparticles in polar snow clearly indicates that they originate mainly from the surrounding continents and not from local sources.

The Al concentrations are higher in Greenland than in Antarctic precipitation, probably due to the larger continental areas in the Northern than in the Southern Hemisphere (Table 2). Al concentrations exhibit definite spring maxima at Dye 3 (Langway et al. 1975). Cyclical variations in particle concentrations were also found at Byrd station (Thompson 1977) and Dome C (Thompson and Mosley-Thompson 1981), with maximum values in winter or late winter.

Secondary aerosol (see figs. 2A, 3B, and 3C). The major part of chemical impurities present nowadays in polar precipitation is made up of gas-derived atmospheric acids (mainly H_2SO_4 and HNO_3). Therefore, sulfate and nitrate are the key ions to be measured in snow and ice samples. Present-day values (before anthropogenic influence on Greenland) are reported in Table 2.

Sulfate is generally the major ion and H_2SO_4 the major acid measured in polar precipitation in Antarctica (but not in Greenland). The sulfate contributed by sea salt (in the form of salt) is minor (generally $< 10\%$) due to the low sea salt content of snow. It may represent, however, a significant fraction of the sulfate deposited in autumn and winter, when the sea salt input is at its highest and the excess sulfate at its lowest. It is generally assessed using Na as the marine reference element ($SO_4/Na = 0.25$ in bulk seawater, weight ratio). Total sulfate minus sea salt sulfate is referred to as excess sulfate (SO_4^*).

Excess sulfate is ubiquitous throughout the global troposphere as the end product of the oxidation series of various gaseous sulfur-bearing compounds produced by biogenic activity, in particular dimethylsulfide emitted by marine phytoplankton. In the Antarctic, the biogenic activity of the sub-Antarctic ocean is an obvious candidate as a source of the sulfate measured in the snow. In the North, ice-free marine areas at high latitudes are much more limited, so the continents (in particular continental shores) may also contribute to the natural sulfur budget in Greenland. The higher sulfate concentrations found in Antarctica, in comparison to Greenland, may be partly explained by the larger areas covered by oceans in the Southern Hemisphere. However, the effect of the dry deposition of sulfate may well also play a role in explaining this difference (Legrand 1985; Delmas et al. 1985).

Herron (1982) proposed a different interpretation for sulfate: he suggested a relatively constant deposition flux of excess sulfate on both polar regions, the Greenland flux being about a factor of 3 higher than the Antarctic flux. It may be argued that Herron's calculation of excess sulfate was made using chloride as the marine reference element, which may lead to errors for cases where sea salt fractionation is important (central Antarctic sites). Moreover, theoretical considerations (Junge 1977) favor atmospheric processes where the concentrations in the precipitation are nevertheless relatively proportional to the air concentrations. Fundamental research on the inclusion mechanisms of sulfate in the snow is clearly needed.

Seasonal variations have been found at sites where accumulations are sufficiently high. For Dye 3, Finkel et al. (1986) reported a weak summer or spring maximum whereas it was demonstrated for South Pole snow that sulfate is maximum in summer, most presumably in relation to the photochemical formation of excess sulfate from its organic gaseous precursors (Legrand and Delmas 1984).

The biogenic sulfate background is sporadically disturbed by large volcanic eruptions which represent transient but very significant sources of gaseous sulfur compounds for the background troposphere. In fact, large volcanic events generally inject SO_2 into the stratosphere from where H_2SO_4 fallout occurs during the 1–2 years following the eruption. This phenomenon is clearly recorded in polar snow. For Greenland this was first pointed out by Hammer (1977), with the aid of electroconductometric measurements, and then demonstrated in Antarctica (Delmas et al. 1985) using various techniques, in particular the determination of exceptional sulfate levels (Legrand and Delmas 1987a).

However, there is no absolute method for identifying unambiguous volcanic and biogenic sulfate. Volcanic H_2SO_4 signals are discovered only on the basis of their suddenness and their short duration, which is a rather subjective method, particularly for minor eruptions. Isotopic methods (on S and O in SO_4) could possibly be used to solve this problem.

Generally, large historical events are easily recognized. At the central Antarctic site of Vostok, extremely large eruptions like that of Tambora (1815) can raise the sulfate concentrations in snow from 2 $\mu Eq.l^{-1}$ up to 18 $\mu Eq.l^{-1}$! For Greenland locations, there is a strong influence of the latitude eruptions, in particular of the Icelandic volcanoes (Hammer 1984). This volcanic sulfate, clearly deposited in snow as H_2SO_4, is the second important contributor to the sulfate budget of polar snow. For the Holocene, this contribution is probably less than 20% of total sulfate, but no accurate estimate is available at present.

Nitrate has been recognized as a major ion both in Greenland and Antarctic precipitation. On a global scale, HNO_3 is the second most important mineral acid implicated in the acid rain problem. The data reported in Table 2 indicates, at least for the stations considered, that the NO_3^- concentrations are higher in Greenland than in Antarctica. This conclusion must not be considered as general, since the value obtained at the South Pole is very similar to the Camp Century figure. As a first approximation, Herron (1982) found that the nitrate concentrations vary as a -0.5 power of the snow accumulation rate in both polar regions. He assumed that this dependence may reflect a combination of incorporation mechanisms. However, Legrand and Delmas (1986), on the basis of their own data set, disagree with Herron's conclusion and have demonstrated that no clear relationship exists for the Antarctic.

The origin of the nitrate found in polar regions is, at present, not fully understood and a matter of serious discussions. The controversies focus on the following questions:

1. What is the stratospheric contribution to the nitrate deposited in polar regions?
2. Do the nitrate profiles record solar activity changes, nuclear explosions, meteor falls, supernovae, etc.?

On the basis of the marked summer maximum observed for the deposition of NO_3^- in Greenland, in phase with "the opening of a tropospheric-stratospheric exchange every spring," Risbo et al. (1981) concluded that nitrate probably has a stratospheric origin. Parker et al. (1982) also favor a stratospheric source, modulated by solar activity (Zeller and Parker 1981). However, Risbo et al. (1981), Herron (1982), and Rasmussen et al. (1984) reported data which disagree not only with the solar influence, but also with various assumptions such as the existence of NO_3^- spikes in Antarctic ice caused by supernovae, aurorae, cosmic ray bursts, and the 1908 Tunguska meteor fall. Legrand and Delmas (1986) have reviewed the different NO_3^- production mechanisms generally invoked for explaining the nitrate levels in polar snow and conclude that a tropospheric source due to lightning, followed by a long-range transport poleward, is more likely. They have also demonstrated that Antarctic nitrate is deposited as HNO_3, this acid being present in the air in the gaseous state. However, in Greenland nitrate may also exist as particulates.

A third mineral acid (HCl) has sometimes been detected in central Antarctic ice at the South Pole (Legrand and Delmas 1984). Its presence has been explained by the interaction of excess sulfate (H_2SO_4) on sea salt particles according to the reaction:

$$H_2SO_4 + 2\,NaCl \rightarrow Na_2SO_4 + 2\,HCl. \tag{4}$$

The reaction occurs in the atmosphere in cloud droplets. Besides HCl, Na_2SO_4 is also formed. According to their different nature (HCl is a gas, Na_2SO_4 an aerosol), these two compounds have different residence times and are deposited separately in Antarctic snow. As indicated above, the snow of certain areas may contain HCl while in other areas the presence of Na_2SO_4 has been proven. In other words, the Cl/Na ratio (R) in central Antarctic snow is frequently different from its reference value in bulk seawater (1.8). This fractionation is more significant for quiet weather conditions. It is generally negligible during winter when the air masses move rapidly over the Antarctic continent. At Dome C and Vostok, for the Holocene, an excess of sodium ($R < 1.8$) is generally observed. At Byrd station, R values are generally around 1.8. For Greenland, this parameter has not yet been studied, but combining the Na data of Cragin et al. (1975) and the Cl data of Herron and Langway (1985) at Camp Century, R is calculated to be equal to about 2.0. At Dye 3, using data from Herron and

Langway (1985) and Cragin et al. (1975), a value of 1.5 is obtained. The uncertainties are large for this calculation and additional measurements are needed on the Cl/Na ratio before concluding that sea salt is fractionated in Greenland.

TEMPORAL VARIATIONS OF MAJOR COMPOUNDS IN DEEP ICE CORES

Primary Aerosol (Continental Dust and Sea Salt)

Continental dust (see Figures 2B and 3B). The content of continental dust in polar ice has been found to be highly variable as a function of time. This contribution exhibits the largest differences when passing from the Last Glacial Maximum to the Holocene time period. This phenomenon was first pointed out by Cragin et al. (1977), using Al measurements on the Camp Century and Byrd station deep ice cores. Further papers have demonstrated that late Wisconsin ice contains much more crustal material than Holocene ice. The ratio of the particle concentrations between the last glacial and Holocene levels is 6, 3 and 12 for DC, BY, and CC, respectively. Thompson and Mosley-Thompson (1981) and Petit et al. (1981) have confirmed this figure for DC. During the ice age a strong contribution of alkaline, Ca-rich dust (3–70 times its Holocene level) was added to the precipitation, making the ice alkaline below 1786 m depth at Dye 3 (Hammer et al. 1985; see also Fig. 2B).

The aluminium measurements reported in Table 2 suggest that during the Last Glacial Maximum (LGM) the aluminium content was 8 to up to 37 times higher than during the Holocene, depending on the site considered. The change of concentration lasts several millennia, both in Antarctica (Petit et al. 1981; Legrand and Delmas 1987b) and Greenland (Hammer et al. 1985). However, the Younger Dryas (around 10,500 yr B.P.) is more significantly marked in Greenland than in Antarctic dust profiles with an extremely rapid (only a few tens of years) jump of concentration at the end of this time period. For the coldest part of the Wisconsin at Dye 3, Hammer et al. (1985) have determined dust contents to be between 1 and 2 mg kg^{-1}. This figure illustrates clearly the very intense atmospheric transport of dust which existed at this time. Petit et al. (1981), Briat et al. (1982), and De Angelis et al. (1987) discounted volcanism as a cause of the enhanced crustal deposition observed at Dome C during the LGM. This conclusion was confirmed at Vostok by elemental composition studies of microparticles as well as by Zn measurements (this metal in the atmosphere was assumed to have a significant volcanic source; De Angelis et al. 1984). Hammer et al. (1985) suggested that the alkaline dust source could be the continental shelves in combination with a lower sea level during the ice age (mainly the ocean bottom north of the USSR). The main causes of the high dust content of glacial age ice is a combination of three effects which characterize this time period: enhanced aridity (increased the desert areas), stronger winds (augmentation of the dust production), and stronger atmospheric circulation (facilitating the long-range transport of atmospheric particles toward polar regions). An additional factor, until now not taken into account, is the increase of the lifetime of aerosols as a function of the decreased precipitation rates. Moreover, snow accumulation changes may account for part of the concentration changes measured in ice, particularly for sites where accumulation rates are extremely low like DC and VK. For VK, De Angelis et al. (1984) assumed that half of the LGM/Holocene decrease could be due to this effect. Potential sources of dust during the ice age are located in Australia, Tasmania, and South America, which were arid, semi-arid, or deserts at this time. Dome C results (Petit et al. 1981) show a shift in the microparticle volume distribution when comparing the LGM and Holocene data: the presence of particles as large as 5 μm in radius only during the LGM suggests a more vigorous meridional circulation from the tropics towards Antarctica at this time.

Moreover, from the ion balance on the Vostok and Dome C data, Legrand (1985) calculated that the precipitation contained, in addition to insoluble continental dust, a large fraction (25% of the ion budget) of "terrestrial salts" formed from the neutralization of calcium and magnesium carbonates, compounds which are common in sedimentary soils.

From all these measurements, it can be deduced that the polar atmosphere during the ice ages was heavily loaded with crustal material, not only in the Northern Hemisphere where the continental surfaces are relatively close and extensive but even in the Southern Hemisphere where Antarctica is still separated by large marine areas from the South American, Australian and African coasts.

Sea salt (see Figures 2A and 3B). The strong winds of the ice ages also influenced atmospheric sea salt contents. This effect has been shown by both Na and Cl measurements (Table 2). The LGM/Holocene ratio may differ when calculated from Na or Cl due to the fractionation of sea salt observed at certain Antarctic locations for the Holocene. This ratio is particularly high at central Antarctic locations where the change in snow accumulation rate plays an important role (as for the deposition of dust, see above). However, all authors agree to globally higher sea salt generation rates and increased efficiency of aerosol transport in explaining the elevated concentrations of Na or Cl observed for the glacial ages. Petit et al. (1981) calculated that wind speeds were up to 5–8 m s^{-1} greater than now. From Vostok data, De Angelis et al. (1987) deduced a figure of 8 to 13 m s^{-1}, probably an upper limit which includes the changes in snow accumulation rates. Herron and Langway (1985) found that the transition at Dye 3 corresponds to a decline of wind speeds by a factor of 3.5 to 5.5 m s^{-1}.

The extension of sea ice around Antarctica by about 1000 km increased the distance between the source areas of sea salt particles and Antarctica, but this factor seems to have been secondary. Moreover, Herron and Langway suggested that elevation changes may be calculated from sea salt concentration variations using equation 3, provided that other factors have been evaluated. This method, applied to Byrd station, led to the conclusions that LGM ice was formed 1.2–1.6 km higher than the present elevation of 1520 m, a figure consistent with glaciological findings on the evolution of the West Antarctic ice sheet.

Finally the recent Vostok ice core has presented the opportunity to compare the sea salt inputs during the present and the past interglacials (De Angelis et al. 1987; Legrand et al. 1988). Surprisingly, it has been found that the Na_m concentrations are significantly higher over the Holocene (22 ± 4.5 ng g^{-1}) than during the last interglacial (9 ± 2 ng g^{-1}). These authors suggest that the meridonial circulation and/or marine source productivity was lower at that time than now (whereas the Al levels are identical for these periods).

The fractionation process of sea salt, which has been clearly shown for Holocene ice at Dome C and Vostok (the Cl/Na ratio is different from its reference value in bulk seawater), seems to be absent in ice-age ice (Legrand 1987; Legrand et al. 1988), as also observed nowadays for winter precipitation. This may be explained by the storminess of the weather at this time, preventing the reaction of H_2SO_4 with sea salt particles (equation 4).

Secondary Aerosol

Sulfate. In every ice core considered in this paper, the sulfate levels during the ice age were significantly higher than for the Holocene (Table 2). This effect, however, is due to the combination of various phenomena complicating the qualitative and quantitative evaluation of the environmental processes. Even if the composition of LGM sulfate is indeed very intricate, it can be nearly completely explained, provided that all major ions have been determined and that the ion budget is satisfactorily balanced. This has been fully developed by Legrand et al. (1988) for the Vostok core. The total sulfate has been decomposed into its various components which are sea salt sulfate, sodium sulfate, and calcium sulfate formed by the reaction of excess sulfate with sea salt particles and calcium carbonate particles, respectively, and finally H_2SO_4 directly linked with marine biogenic activity (Fig. 3C). The profile reported covers an entire ice age and two interglacials. Even if the high concentrations observed for the coldest time periods may be partly explained by accumulation changes (a factor 1.8 seems to be a reasonable estimate of the increase due to this reason), the excess sulfate concentrations (the non-sea salt fraction, sum of the H_2SO_4, Na_2SO_4, and $CaSO_4$ contributions) seem to be related to marine biogenic activity (which also influences the atmospheric CO_2 content, reported by Lorius et al., this volume).

The interpretation of these very recent sulfate results deserves more detailed discussions in order to demonstrate the link, which probably exists, between the C and S cycles through marine biogenic activity. Additional environmental information is probably attached to these various forms of sulfate, but more measurements are still needed for a full understanding.

In Greenland, the sulfate levels for the LGM are also relatively high, and the contrast between Holocene and LGM levels is even more marked than in Antarctica if we take into account that the accumulation effect is probably higher in Antarctica than in Greenland. This could certainly be explained if the ion budget measurements, using the method developed by Legrand for Dome C and Vostok ice, were available. The exceptionally high sulfate levels observed at D3 and CC for the transition are intriguing since they do not correspond to a similar trend in the Southern Hemisphere. This indicates that the environment does not evolve in an entirely parallel manner in both hemispheres, even if the large climate change 15,000–11,000 yr B.P. was global.

Finally, the interpretation of the sulphate results obtained at Vostok minimizes the role of explosive volcanism as a source of excess sulfate in the global atmosphere, as already proposed in the past. We cannot entirely rule out, however, the possibility that volcanic activity may have been the cause of exceptional high sulfate levels for periods of centuries or even millennia along the ice cores analyzed. In this case, it would be of primary importance to investigate whether this increased volcanic activity was really global or restricted only to a limited geographical area.

Nitrate. Nitrate behaves differently in Northern and Southern Hemisphere ice cores. As pointed out by Herron and Langway (1985), nitrate is the only compound to exhibit lower concentrations in Greenland ice during the ice age than during the Holocene. On the other hand, in the Vostok core, as at Dome C and Byrd Station, this ion is definitely low for warm periods and high for ice ages. As the origin of nitrate in Holocene ice has still not been clearly explained, it appears difficult at first glance to interpret the variations of this ion along the deep ice cores. However, a first step has been made for Antarctic ice cores with available ion budgets. The problem may indeed be simpler than for sulfate, since nitrate is most likely introduced in the Antarctic atmosphere in only two forms: HNO_3 (as a gas) and terrestrial nitrate (as particles).

The high values of nitrate found in central Antarctic ice for the ice age are clearly related to the deposition of nitrate salts and not an increase in the HNO_3 input (Legrand 1985; Legrand et al. 1988). This was demonstrated by correlating nitrate and aluminium concentrations

$$(C_{NO_3} = 9 + 0.4\ C_{Al}) \tag{5}$$

for the LGM. We can assume that this phenomenon is directly linked to the reaction of HNO_3 with terrestrial $CaCO_3$ particles (as was also the case for H_2SO_4), resulting in the formation of $Ca(NO_3)_2$. The LGM/H ratio is particularly high at Vostok (see Table 2), with a value of 5.6. Even if this value is partly due to the change in accumulation rates, it may be calculated that nitrate deposition was about 3 times higher during the Holocene than during the LGM. This figure seems to be directly linked to the transport of crustal material.

At Byrd station, the LGM/H ratio is only 1.3, corresponding to a low influence of crustal deposits at this site. Changes in HNO_3 deposition could be an indication of changes in the productivity of HNO_3 by lightning. The Vostok results, however, show no such changes along the profile. For Greenland ice, there is apparently no noticeable link between dust particles and NO_3. Are the low values of nitrate observed during the LGM at D3 and CC related to the significant decrease of the ice-free continental areas at high latitudes in this hemisphere (lightning occurs essentially over land areas)?

THE CASE OF HYDROGEN PEROXIDE

The case of H_2O_2 has to be presented separately for several reasons:

1. It is a species generally considered as unstable and reactive.
2. Its deposition mechanism seems to be quite different from the other atmospheric gases or aerosol particles.
3. It is soluble but not dissociated in ions in meltwater.
4. It is not long-range transported but formed in the local atmosphere.
5. It has been detected very recently (Neftel et al. 1984) so that the information available about its past variations is scarce.

The concentration of hydrogen peroxide in polar ice is high (in the 50–200 ppb range, i.e., several micromoles per liter). There is, however, still uncertainty as to the true concentration levels of H_2O_2, because of analytical problems. Most recent measurements (Sigg and Neftel 1988) indicate similar mean values in both polar regions, whereas the first results led to the conclusion that the concentrations are lower in the Antarctic than in Greenland (Neftel et al. 1986). Despite these difficulties, the long-term records show markedly lower levels for the ice age in the Camp Century, Dye 3, and Byrd ice cores. One part of this effect may certainly be attributed to a possible slow disintegration of H_2O_2 in the ice; even if this phenomenon is taken into account, the initial concentrations of H_2O_2 in the snow were probably significantly lower during the ice age than during the Holocene. Moreover, the H_2O_2 profile obtained in the Byrd core exhibits very elevated values during the time period of the last climatic transition (about 3 times the Holocene mean value).

As H_2O_2 is a strong oxidizer in cloud droplets and also very active in gaseous phase photochemical processes, H_2O_2 results obtained from polar precipitation analyses are of interest for atmospheric chemists. Two points, however, have to be investigated seriously before considering ice core data as entirely reliable for atmospheric studies:

1. the stability of H_2O_2 records as a function of time,
2. the transfer mechanism of H_2O_2 from the atmosphere to the snow (Sigg and Neftel (1988) propose a co-condensation of H_2O_2 and H_2O molecules during the snow formation from the vapor phase, without a fractionation).

CONCLUSIONS

The chemical analysis of deep ice cores began about 15 years ago. Due to recent progress in analytical methods and also in the knowledge of the chemical composition of rain in remote areas, it is now possible to propose interpretations in terms of past environmental changes for the data sets obtained from deep ice cores. The information archived in ice sheets is not only relevant to pure climatological investigations but also to atmospheric chemistry processes. This type of work is still in its infancy. It is now time to investigate seriously how atmospheric information is recorded in the snow layers. Even without accurate knowledge of these fundamental processes, polar ice core analysis can be considered a very powerful tool for reconstructing the history of the vegetation on the continents, the meteorological conditions that prevailed not only on the polar areas themselves but also at midlatitudes, past marine biological activity, ... The list goes on.

REFERENCES

Briat, M.; Royer, A.; Petit, J. R.; and Lorius, C. 1982. Late glacial of eolian continental dust in the Dome C ice core: additional evidence from individual microparticle analysis. *Ann. Glaciol.* **3**: 27–31.

Cragin, J. H.; Herron, M. M.; and Langway, C. C., Jr. 1975. The chemistry of 700 years of precipitation at Dye 3, Greenland, CRREL Research Rep. 341.

Cragin, J. H.; Herron, M.M.; Langway, C. C., Jr.; and Klouda, G. 1977. Interhemispheric comparison of changes in the composition of atmospheric precipitation during the late cenozoic area. In: Polar Oceans, ed. M. J. Dunbar, pp. 617–631. Calgary: Arctic Institute of North America.

Dansgaard, W.; Johnsen, S. J.; Clausen, H. B.; Dahl-Jensen, D.; Gundestrup, N.; Hammer, C. U.; and Oeschger, H. 1984. North Atlantic climatic oscillations revealed by deep Greenland ice cores. In: Climate Processes and Climate Sensitivity, M. Ewing Symp., vol. 5, eds. J. E. Hansen and T. Takahashi. *Geophys. Monog.* **29**: 288–298. Washington, D.C.: Amer. Geophys. Union.

De Angelis, M.; Barkov, N. I.; and Petrov. V. N. 1987. Aerosol concentrations over the last climatic cycle (1660 kyr) from an Antarctic ice core. *Nature* **325**: 318–321.

De Angelis, M.; Legrand, M.; Petit, J. R.; Barkov, N. I.; Korotkevich, Ye. S.; and Kotlyakov, V. M. 1984. Soluble and insoluble impurities along the 950 m deep Vostok ice core (Antarctica) — climatic implications. *J. Atmos. Chem.* **1**: 215–239.

Delmas, R. J.; Briat, M.; and Legrand, M. 1982. Chemistry of South Polar snow. *J. Geophys. Res.* **87**: 4314–4318.

Delmas, R. J.; Legrand, M.; Aristarain, A. J.; and Zanolini, F. 1985. Volcanic deposits in Antarctic snow and ice. *J. Geophys. Res.* **90**: 901–920.

Finkel, R. C., and Langway, C. C., Jr. 1985. Global and local influences on the chemical composition of snowfall at Dye 3, Greenland: the record between 10 Ka BP and 40 Ka BP. *Earth Plan. Sci. Lett.* **73**: 196–206.

Finkel, R. C.; Langway, C. C., Jr.; and Clausen, H. B. 1986. Changes in precipitation chemistry at Dye 3, Greenland. *J. Geophys. Res.* **91**: 9849–9855.

Hammer, C. U. 1977. Past volcanism revealed by Greenland ice sheet impurities. *Nature* **270**: 482–486.

Hammer, C. U. 1984. Traces of Icelandic eruptions in the Greenland ice sheet. *Jokull* **34**: 51–65.

Hammer, C. U.; Clausen, H. B.; Dansgaard, W.; Neftel, A.; Kristinsdottir, P.; and Johnsen, E. 1985. Continuous impurity analysis along the Dye 3 deep core. In: Greenland Ice Cores: Geophysics, Geochemistry and the Environment, eds C. C. Langway, Jr., H. Oeschger and W. Dansgaard. *Geophys. Monog.* **33**: 90–94. Washington, D.C.: Amer. Geophys. Union.

Herron, M. M. 1982. Impurity sources of F^-, Cl^-, NO_3^- and SO_4^{2-} in Greenland and Antarctic precipitation. *J. Geophys. Res.* **87**: 3052–3060.

Herron, M. M., and Langway, C. C., Jr. 1985. Chloride, nitrate and sulfate in the Dye 3 and Camp Century, Greenland ice cores. In: Greenland Ice Cores: Geophysics, Geochemistry and the Environment, eds. C. C. Langway, Jr., H. Oeschger and W. Dansgaard. *Geophys. Monog.* **33**: 77–84. Washington, D.C.: Amer. Geophys. Union.

Junge, C. E. 1977. Processes responsible for the trace concent in precipitation. In: Isotopes and Impurities in Snow and Ice, Proc. IUGG Symp., Grenoble, 1975, *IAHS-AISH Publ.* **118**: 63–77.

Langway, C. C., Jr.; Cragin, J. H.; Klouda, G. A.; and Herron, M. M. 1975. Seasonal variations of chemical constituents in annual layers of Greenland deep ice deposits. CRREL Research Rep. 347.

Legrand, M. 1985. Chimie des neiges et glaces polaires: reflet de l'environnement. Thèse de Doctorat d'Etat, Univ. Scientifique et Médicale de Grenoble.

Legrand, M. 1987. Chemistry of Antarctic snow and ice. *J. Physique* **48**: 77–86.

Legrand, M., and Delmas, R. J. 1984. The ionic balance of Antarctic snow: a 10 yr detailed record. *Atmos. Envir.* **18**: 1867–1874.

Legrand, M., and Delmas, R. J. 1986. Relative contributions of tropospheric and stratospheric sources to nitrate in Antarctic snow. *Tellus* **38B**: 236–249.

Legrand, M., and Delmas R. J. 1987a. A 220 yr continuous record of volcanic H_2SO_4 in the antarctic ice sheet. *Nature* **327**: 671–676.

Legrand, M., and Delmas, R. J. 1987b. Environmental changes during last deglaciation inferred from chemical analysis of the Dome C ice core. In: Biviers NATO Workshop "Abrupt Climatic Change," eds. W. H. Berger and L. G. Labeyrie, pp. 247–259. Dordrecht: Reidel.

Legrand, M., and Delmas, R. J. 1988. Soluble impurities in four Antarctic ice cores over the last 30,000 yr. *Ann. Glaciol.* **10**: 116–120.

Legrand, M. R.; Lorius, C.; Barkov, N. I.; and Petrov, V. N. 1988. Atmospheric chemistry changes over the last climatic cycle (160,000 yr) from Antarctic ice. *Atmos. Envir.* **22**: 317–331.

Neftel, A.; Jacob, P.; and Klockow, D. 1984. Measurements of hydrogen peroxide in polar ice samples. *Nature* **311**: 43–45.

Neftel, A.; Jacob, P.; and Klockow, D. 1986. Long term record of H_2O_2 in polar ice cores. *Tellus* **38B**: 262–270.

Palais, J. M., and Legrand, M. 1985. Soluble impurities in the Byrd Station ice core. Antarctica; their origin and sources. *J. Geophys. Res.* **90**: 1143–1154.

Parker, B. C.; Zeller, E. J.; and Gow, A. J. 1982. Nitrate fluctuations in Antarctic snow and firn: potential sources and mechanisms of formation. *Ann. Glaciol.* **3**: 243–248.

Petit, J. R.; Briat, M.; and Royer, A. 1981. Ice age aerosol content from East Antarctic ice core samples and past wind strength. *Nature* **293**: 391–394.

Rasmussen, K. L.; Clausen, H. B.; and Risbo, T. 1984. Nitrate in the Greenland ice sheet in the years following the 1908 Tunguska event. *Icarus* **58**: 101–108.

Risbo, T.; Clausen, H. B.; and Rasmussen, K. L. 1981. Supernovae and nitrate in the Greenland ice sheet. *Nature* **294**: 637–639.

Sigg, A., and Neftel, A. 1988. Seasonal variations of hydrogen peroxide in polar ice cores. *Ann. Glaciol.* **10**: 157–162.

Thompson, L. G., 1977. Variations in microparticle concentration, size distribution and elemental composition found in Camp Century, Greenland, and Byrd Station, Antarctica, deep cores. In: Isotopes and Impurities in Snow and Ice, Proc. IUGG Symp., Grenoble, 1975. *IAHS-AISH Publ.* **118**: 351–364.

Thompson, L. G., and Mosley-Thompson, E. 1981. Microparticle concentration variations linked with climatic change: evidence from polar ice cores. *Science* **212**: 812–815.

Zeller, E. J., and Parker, B. C. 1981. Nitrate ion in Antarctic firn as a marker for solar activity. *Geophys. Res. Lett.* **8**: 895–898.

The Environmental Record in Glaciers and Ice Sheets
eds. H. Oeschger and C.C. Langway, Jr., pp. 343–361
John Wiley & Sons Limited

Long-term Environmental Records from Antarctic Ice Cores

C. Lorius[1], G. Raisbeck[2], J. Jouzel[3], and D. Raynaud[1]

[1] *Laboratoire de Glaciologie et Géophysique de l'Environnement 38402 St. Martin d'Hères Cedex, France*
[2] *Laboratoire René Bernas, 91406 Orsay, France*
[3] *Laboratoire de Géochimie Isotopique, IRDI-DESICP-CEA 91191 Gif sur Yvette Cedex, France*

Abstract. Paleoclimate and environmental conditions in Antarctica have been reconstructed from three deep ice cores extending back at least into the last glaciation, one core covering the entire last climatic cycle (160,000 yrs). In comparison with current Holocene conditions, ice core studies suggest that the Last Glacial Maximum (LGM), around 20 kyr B.P., was characterized by a colder temperature, lower snow accummulation (possibly leading to a thinner ice sheet), higher continental and marine aerosol contents, and lower atmospheric CO_2 concentration.

Over the last climatic cycle, large periodic temperature changes were observed. During the coldest stages, atmospheric conditions were similar to those observed during the LGM. In particular, a very close association between CO_2 and climate is observed throughout the record.

The use of cosmogenic isotopes (especially ^{10}Be) is discussed with regard to their potential for determining the time variability of primary cosmic ray intensity, solar activity and geomagnetic field intensity, as well as for estimating paleoprecipitation rates. The possibility of using ice cores to obtain information on temporal variations of extraterrestrial (in particular cometary) matter input to the Earth is noted. Total gas content indicates that the Antarctic ice sheet may have been somewhat thinner in central parts and thicker in coastal areas during the LGM. The study of air bubbles may also bring information on past oxygen cycle.

Most of the recorded changes are, at least qualitatively, of global significance.

INTRODUCTION

Deep ice cores from polar ice sheets provide a unique record of past atmospheric conditions. In these archives, the isotopic composition of the snow reflects the temperature of the atmosphere while air bubbles trapped in the ice preserve a record of past CO_2 levels. The impurity of ice is affected by rates of precipitation (accumulation), the contribution of various natural and man made sources, and the intensity of airborne transport.

There are of course limitations in extracting atmospheric time series of relevant parameters; some of them will be addressed in the sections addressing how glaciers record and preserve information and the uncertainties in dating ice cores. This paper will focus on records relevant to climate (temperature, accumulation, ice thickness) and the environment (aerosol loading, atmospheric composition) obtained from three Antarctic deep ice cores; Byrd, Dome C, and Vostok.

DEEP ANTARCTIC ICE CORES

Although several ice cores drilled to bedrock in coastal areas contain very ancient ice, the recovery of continuous detailed long-term environmental records has been principally derived from drillings performed on the high plateau. Three such series reach back at least into the last glaciation (Table 1). The first one has been obtained at Byrd Station in West Antarctica. The drilling reached bedrock in 1968 at a depth of 2163 m. Both the Camp Century ice core drilled in 1966 and the Byrd core provide data which demonstrated the potentialities and complementarities of ice data from both poles. In both cases, the ice flow conditions led to a rather uncertain chronology beyond about 20 kyr B.P.

Twelve years later, at Dome C in East Antarctica, a deep ice core was drilled during the summer season. The low annual snow accumulation rate made it possible to reach back into the last ice age, avoiding possible ice dynamics disturbances which occur near bedrock.

At Vostok Station, East Antarctica, a first series of drillings reached a depth of 950 m in 1974. At the same site a new deep hole was drilled in 1980 and reached 2083 m in 1982. The depth of penetration was later

Table 1

Station	Elevation (m)	Temperature mean annual	Depth (m)	Snow Accumulation Rate ($g.cm^{-2}.a^{-1}$)	Time Scale (kyr)
Byrd Station (W. Antarctica)	1530	−28°C	2163	16	ca. 70
Dome C (E. Antarctica)	3240	−53°C	905	3.4	ca. 40
Vostok (E. Antarctica)	3490	−55.5°C	2200	2.3	ca. 160

extended to 2200 m. An ice core chronology has been obtained from a two-dimensional steady state glaciological model using the known bedrock and surface profiles and accounting for changes in the rate of snow accumulation (see discussion below). The results indicate a rather linear increase of age with depth, reflecting in particular favorable conditions linked to low accumulation ($2.3\,g\,cm^{-2}\,a^{-1}$) and large ice thickness (about 3500 m). For the first time an ice core extended as far back as about 160,000 yrs, completely through the last glacial-interglacial cycle.

CLIMATIC DATA

Isotope Temperature Records

We discuss in this section how the long-term change in surface temperature (Ts) may, at least for sites on the East Antarctic Plateau, be inferred from the isotope record (deuterium/hydrogen and oxygen 18/oxygen 16 ratios denoted δD and $\delta^{18}O$ and expressed in per mil with respect to SMOW, the Standard Mean Ocean Water). A transfer function based on modern snow δ-Ts gradients is first briefly presented. This is followed by a discussion of the Vostok temperature record (~ 160 kyr) and a comparison of the climatic records for the three deep Antarctic ice cores reaching back to the LGM over their common interval (the last ~ 40 kyr). Special emphasis is placed on examining the validity of such a temperature interpretation of ice core isotope data.

The transfer function. Isotope fractionation occurs during the atmospheric water cycle owing essentially to differences in the vapor pressure (equilibrium effect) and additionally to differences in molecular diffusivity (kinetic effect) of the three molecules involved (H_2O, HDO, and $H_2{}^{18}O$). These processes lead to a general decrease of the δs with temperature, which for modern polar precipitation is a linear relationship between the annual averages of the δs and Ts (Dansgaard et al. 1973; Lorius and Merlivat 1977). We propose to use this relationship, at least for interpreting Dome C and Vostok isotope data, because the experimental isotope-temperature relationship is particularly well defined over this part of East Antarctica where climatological conditions are relatively simple (Lorius and Merlivat 1977). In addition, the observed slopes (6‰/°C for δD; 0.75‰/°C for $\delta^{18}O$) agree with the values derived from a one-dimensional atmospheric model provided, i.e., that the temperature prevailing just above the inversion layer is used as the formation temperature and that the kinetic effect during snow formation is considered (Jouzel and Merlivat 1984). Note that the relative influence of kinetic effects which are not directly temperature-related is weaker for δD than for $\delta^{18}O$.

For this reason, we prefer to use the δD rather than the $\delta^{18}O$ record for temperature reconstruction.

The 160 kyr temperature record. Preliminary estimates of atmospheric temperature changes in the Vostok area were first obtained from $\delta^{18}O$ data (Lorius et al. 1985). A relatively continuous temperature record (Fig. 1b) was later deduced from the deuterium profile shown in Fig. 1a; this was done using the above 6‰/°C gradient after correcting for the change in δD of the oceanic water (Jouzel et al. 1987).

The record shows the existence of two interglacials (A, the current Holocene, and G) and extends back to the previous ice age (H). The peak of the previous interglacial is significantly warmer (about 2°C) than the Holocene period.

Conditions equivalent to those prevailing during the LGM (B) were encountered only at the end of the penultimate glacial, around 150 kyr B.P. The last deglaciation is clearly a two step process, with two warming periods interrupted by a 2°C temperature reversal lasting about 1 Ka.

The last glacial period is characterized by three well marked temperature minima (with the one occurring around 110 kyr B.P. about 2°C warmer than full glacial conditions), separated by two interstadials (4 and 6°C, respectively) above the LGM.

These data offer a unique opportunity to examine, using a continental record, the link existing between an atmospheric time series and astronomical forcing. Spectral analysis (Jouzel et al. 1987) shows that in addition to the ~ 100 kyr oscillation, the isotope record is dominated by a strong ~ 40 kyr component (clearly seen by visual inspection of the record). The analysis also shows a weak 20 kyr component. The two frequency bands can be associated with the obliquity and precession cycles respectively.

The last 40 kyr period. Applying the same transfer function, the Dome C deuterium profile (Lorius et al. 1984) has been transformed (J. Jouzel, personal communication) into a temperature record given in Fig. 2 with a revised Dome C time scale which puts the bottom of the core at ~ 40 kyr (instead of 32 kyr). This Dome C redating has been obtained by estimating the accumulation changes as for the Vostok core and remarkably confirmed by a common ^{10}Be peak (Raisbeck et al. 1987) around 35 kyr.

The Dome C and Vostok temperature records (Fig. 2) compare quite well in terms of the magnitude of glacial-interglacial changes (~ 9°C), the rate and magnitude of the last deglaciation, and last glacial period. Note in particular the close temperature correlation around 35 kyr B.P. a period over which the ^{10}Be peak provides an independent and isochronous stratigraphic marker. Furthermore, a quantitative interpretation of the Byrd isotope profile (Johnsen et al. 1972; Fig. 2) suggests a comparable glacial-interglacial

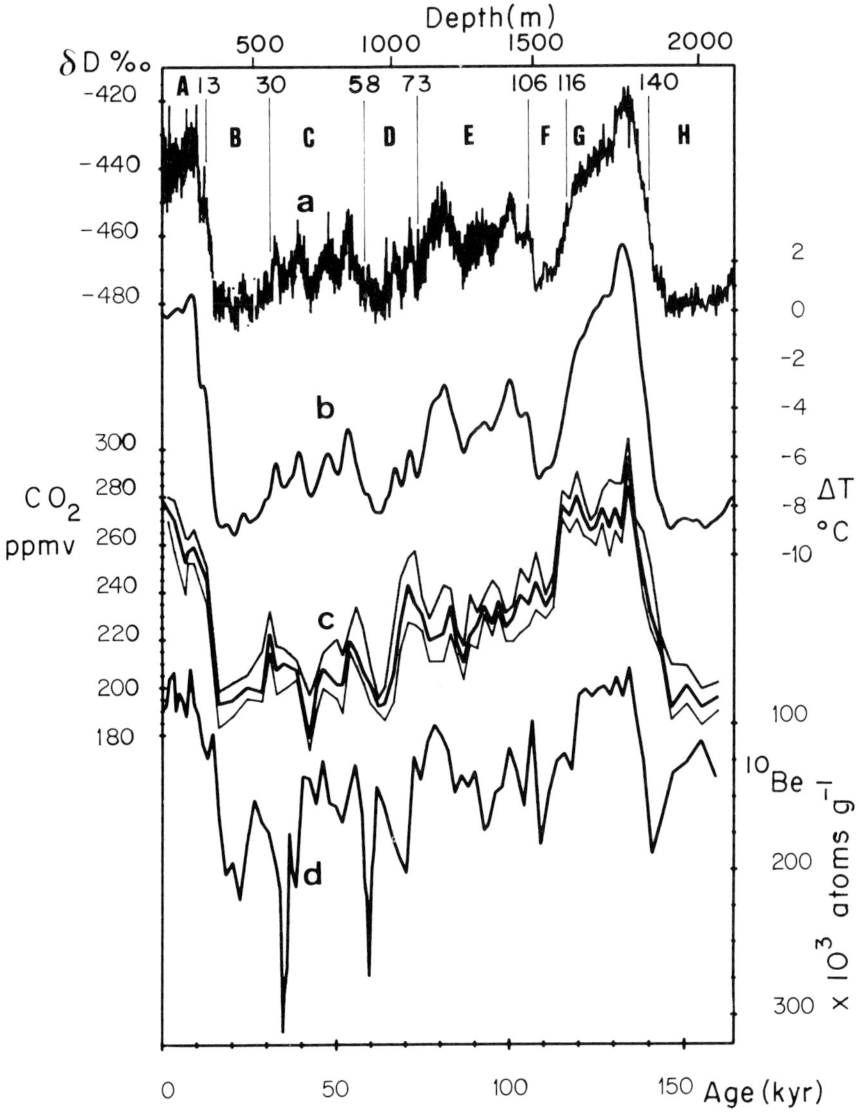

Fig. 1—Vostok ice core records. The upper and lower scales give the depth and age, respectively. *a*: Deuterium content with successive climatic stages (A to H) (from Jouzel et al. 1987). The relation between Vostok stages A to H and marine stages 1 to 6 is discussed in Lorius et al. (1985). *b*: Smoothed temperature in °C expressed as a difference with respect to current surface temperature value (from Jouzel et al. 1987). *c*: CO_2 concentrations (ppmv = parts per million per volume). The best estimates are shown by the thick line with the uncertainty bands on either side (from Barnola et al. 1987). *d*: ^{10}Be concentration. Note inverted scale to facilitate comparison with other records (from Raisbeck et al. 1987).

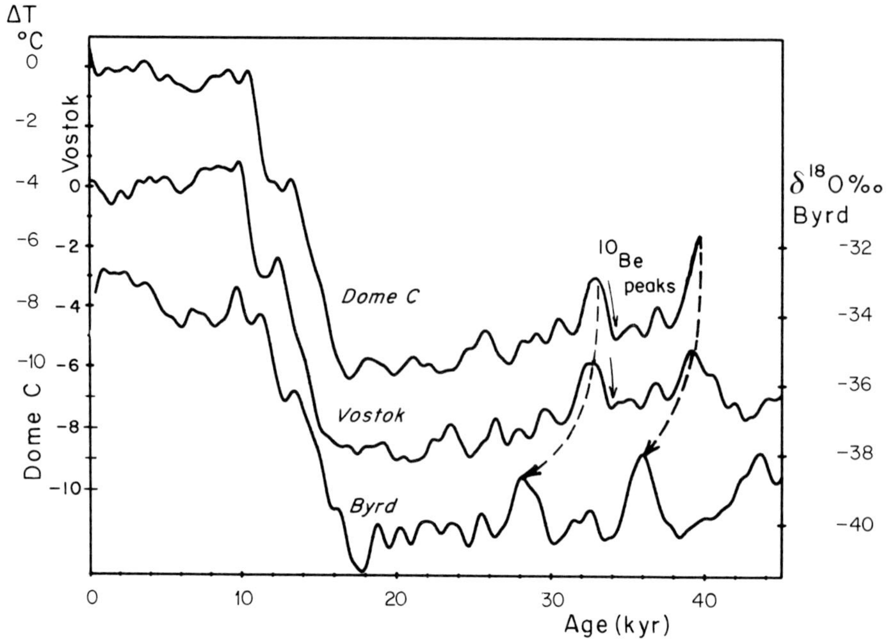

Fig. 2—Antarctic ice core climatic records over the last 40,000 years. Dome C and Vostok: left scale in °C expressed as a difference with respect to current surface temperature value. Byrd: right scale in $\delta^{18}O$ ‰ (from Johnsen et al. 1972). The solid arrows indicate observed ^{10}Be peaks. The dotted arrows suggest a possible correspondence of climatic events in the three cores.

change as shown by the similar $\delta^{18}O$ shift in the three cores (the dotted lines in Fig. 2 tentatively link similar events in Vostok, Dome C, and Byrd records).

Validity of the temperature interpretation. The above interpretation deals with surface temperature change at the deposition site. This signal includes the climatically induced temperature change as well as the change due to elevation variation either as a consequence of ice flow or of general changes in the ice sheet thickness. These changes in elevation may be evaluated from the total gas content of the ice (see below). Preliminary results (Raynaud, personal communication) suggest that at Vostok, elevation and deuterium changes are roughly in phase with lower elevation during glacial periods. Correcting for this effect would increase the amplitude of the temperature signal but leave its shape essentially unchanged.

Beyond this necessary correction, the validity of applying the present temperature gradient relevant to a spatial scale to an interpretation of the isotope content of past precipitation (that is, on a temporal basis) must be examined. This interpretation requires (*a*) a negligible or statistically nullified influence of factors other than the temperature of formation of precipitation (i.e., conditions of temperature and relative humidity in the moisture source regions, dynamical, and microphysical history of the air masses), (*b*) an unchanged relation between inversion and surface temperatures, and (*c*) no additional isotopic noise (e.g., through variations in the summer to winter precipitation ratio). The complexity of the various processes involved may be taken into account by incorporating the water isotopes into General Circulation Models of the atmosphere (Joussaume et al. 1988). Hopefully, this approach may answer some of the above questions in the near future.

There are, however, some results supporting this temperature interpretation:

1. Recently, Petit et al. (1987) proposed that changes of the size of crystals in cores are related to surface temperature change. Applied to Dome C and Vostok ice cores, the model shows good agreement between temperature changes derived independently from the crystal size and the isotope profiles with glacial interglacial shifts of ~ 10°C (Dome C) and 11°C (Vostok), comparing quite well with our value of ~ 9°C.
2. The general agreement of two independent estimates of the accumulation rate based on the temperature record and the ^{10}Be profile (see discussion below).
3. The similarities between the Vostok and sea surface temperature records, as obtained from the subpolar Indian Ocean core RC 11–120 (Jouzel et al. 1987 and references therein) and more recently from a core within the Antarctic waters (J.J. Pichon et al., personal communication).

Still, the interpretation of isotope records in terms of temperature changes remains open to discussion and more work is required in (*a*) documenting the distribution of isotopes in Antarctic surface snow and its relation to meteorological conditions prevailing around and over Antarctica, and (*b*) developing isotopic models.

Major improvements in estimating paleotemperatures from ice cores could come from total gas content data, making it possible to separate climatically induced temperature changes from those due to elevation changes, and independent temperature indicators.

Precipitation

Determination of paleoprecipitation rates is important because (*a*) precipitation is a fundamental climatic parameter and (*b*) past precipitation (or more accurately accumulation) rates are a key input parameter to ice flow modeling; these in turn are essential for making environmental interpretations of ice core records. Over large regions of the Antarctic plateau, the annual precipitation rate is sufficiently low so that annual signals of stable isotopes are not likely to be preserved. Also, the few attempts of direct observation of other annual signals have often been contradictory.

The present precipitation pattern over large regions of Antarctica can be correlated with the saturation vapor pressure of water above the inversion temperature layer, which is itself correlated with the stable isotope record of the precipitation (Robin 1977). If one makes the assumption that this same correlation has remained valid in the past, then the stable isotope ratio $\delta^{18}O$ or δD can be used to calculate the paleoprecipitation rate. Note that there is really very little justification for the above assumption and that, in particular, it implies that either atmospheric circulation patterns have remained stable over time or that they do not significantly affect the temperature-precipitation correlation. Perhaps for these reasons the procedure for estimating polar precipitation was not taken seriously until recent observations of ^{10}Be variations.

In 1981 Raisbeck et al. reported the first measurements of cosmogenic ^{10}Be in the 906 m Dome C ice core. Surprisingly, they found the ^{10}Be concentration to be larger in glacial ice (by a 2–3 factor) than in Holocene ice. Among the possible explanations offered was that the precipitation rate in Antarctica during the glacial period was approximately 50% of that during the Holocene. It was noted that such an interpretation was consistent with the precipitation change calculated from the saturated water vapor pressure argument mentioned above.

When later ^{10}Be measurements (Fig. 1d) were made in the Vostok core, the good correlation of ^{10}Be and $\delta^{18}O$ was observed to hold throughout the last glacial-interglacial cycle. Also, a direct comparison of the paleoprecipitation rates estimated from the ^{10}Be and water vapor pressure procedures were in good agreement (Yiou et al. 1985). In Fig. 3 we show the calculated precipitation rates based on the most recent ^{10}Be and paleotemperature profiles (Raisbeck et al. 1987; Jouzel et al. 1987). Although there are differences (amounting to as much as 50% at some depths), the overall agreement is quite remarkable.

There are several possible reasons why ^{10}Be concentrations might be approximately inversely correlated with precipitation rate (Raisbeck and Yiou 1985). Nevertheless, it is obvious that precipitation rates based on ^{10}Be cannot be accurate in detail since, as we will see below, there are other causes of ^{10}Be concentration variations. In addition, changes in atmospheric circulation could also significantly affect the ^{10}Be deposition at a given location (Raisbeck et al. 1981).

In summary, we have a situation in which two relatively independent procedures for estimating polar precipitations, neither on a particularly solid

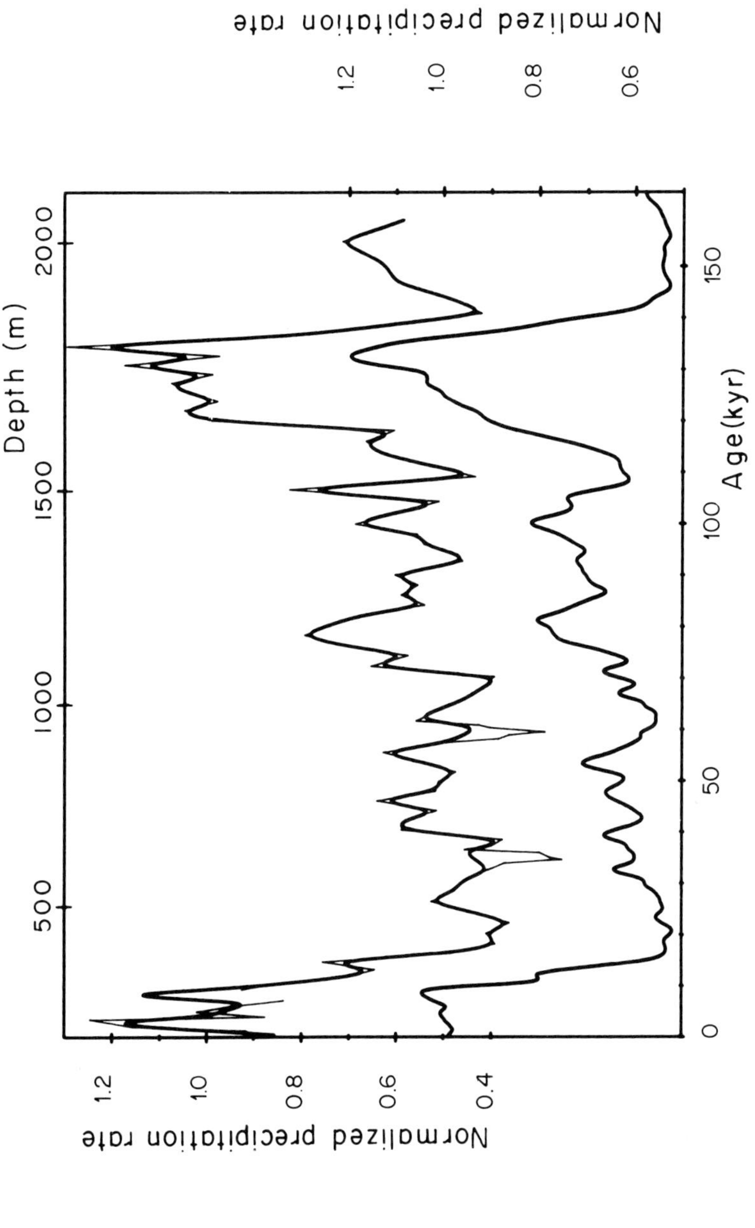

Fig. 3—Vostok ice core: normalized precipitation rate with respect to Holocene mean value: upper curve from the —^{10}Be concentration record (Raisbeck et al. 1987); lower curve from the isotope temperature record (Jouzel et al. 1987).

theoretical footing, give similar and apparently reasonable results. The degree to which one is convinced by such concordance is obviously a question of personal taste (not necessarily the same for all the authors of this report!).

THE AEROSOL RECORD

Major Aerosols

In this section we focus on the climatic information which can be extracted from the analysis of impurities contained in ice, since marine and terrestrial contributions are possibly dependent upon various parameters such as remote source strength, atmospheric transport efficiency, sea ice extent, and rate of snow accumulation. Chemical analyses of these materials are discussed elsewhere (Delmas and Legrand, this volume).

A common characteristic to the Byrd, Dome C, and Vostok cores (see reference in Delmas and Legrand, this volume) is that they show much larger impurity concentrations during the LGM than during the current Holocene interglacial. These changes which affect both the continental and, to a lesser extent, the marine components (possibly due to a larger sea ice area) are too large to be explained by accumulation variations. They have been interpreted (Petit et al. 1981) as reflecting more extensive arid areas over the surrounding continents, giving a greater exposure of the continental shelves due to a lower sea level and more efficient meridional transport linked to higher wind speeds (likely induced by higher temperature gradients with latitude). Anion and cation profiles are now documented over the full glacial-interglacial cycle (De Angelis et al. 1987; Legrand et al. 1988).

As an illustration, the aluminium (an indicator of continental source), marine sodium, and acidity (which is potentially linked to volcanic activity) profiles for Vostok are given (Fig. 4 b, c, and d) together with the temperature record (Fig. 4a). The main features of the Al and especially Na records are rather closely related to climate, with the main high concentrations associated with the coldest stages, However, this overall agreement is not marked for the relatively cold period around 110 kyr B.P.; the absence of high Al values could possibly indicate that the amplitude of stage F is a regional climatic feature. As opposed to marine and terrestrial inputs, there is no apparent relationship between acidity and the temperature record; this suggests no long-term correlation between volcanism and climate, as already deduced from mineralogical studies of insoluble particles. Further data showing a shift in the microparticle size distribution with larger particles during the LGM (Petit et al. 1981) and a Cl/Na ratio close to the bulk seawater value (Legrand et al. 1988) support the idea of more intense meridional circulation and transport efficiency during glacial periods. The

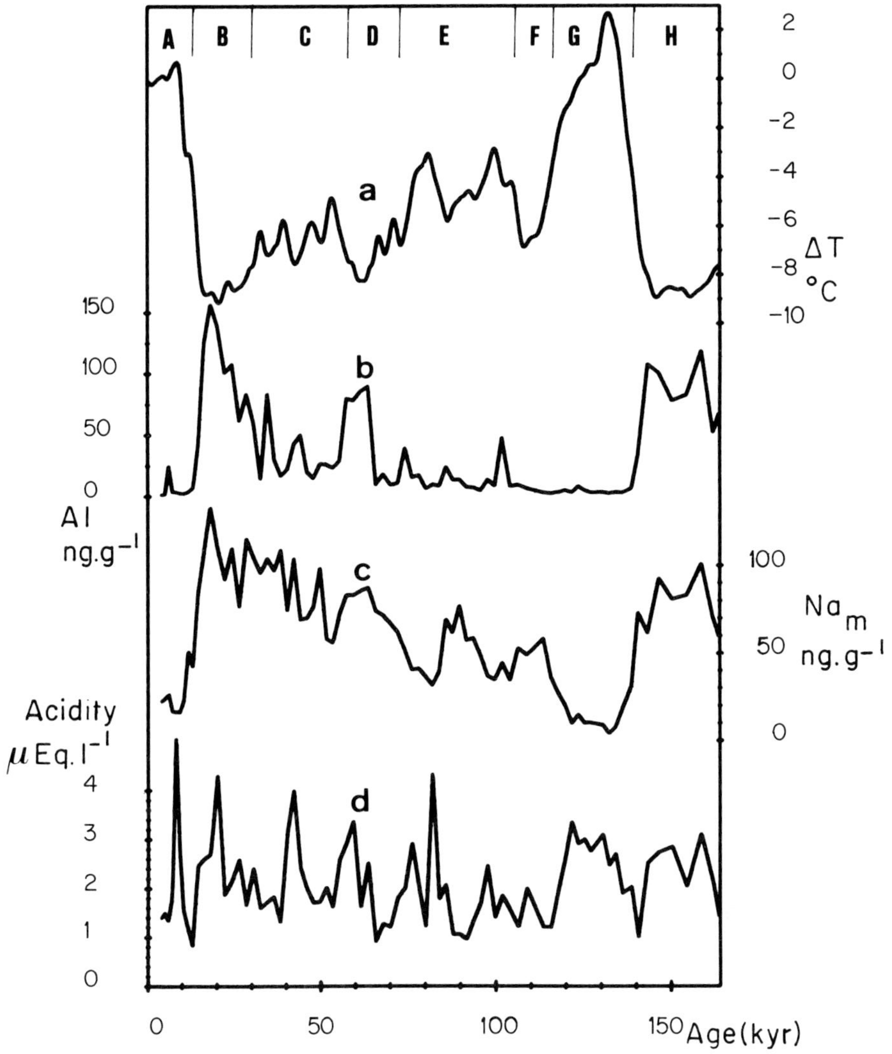

Fig. 4—Vostok ice core records. *a*: Smoothed temperature in °C (as in Fig. 1b). *b*: Aluminium content (from Legrand et al. 1988). *c*: Marine sodium content (from Legrand et al. 1988). *d*: Acidity (from Legrand et al. 1988).

wind speed increase over the source areas has been tentatively estimated at a factor of 1.3–1.8 (Petit et al. 1981), but these values are highly uncertain due to unknown factors such as the origin of the continental dust.

To further test this hypothesis, simulations have been performed by using a General Circulation Model of the Atmosphere (Joussaume et al. 1988) for modern day and 18 kyr B.P. boundary conditions (February and August). During the last Ice Age, a small increase (15%) of the extent of source areas and an increase of dust deposits is obtained over East Antarctica but is smaller than observed; this discrepancy might be due to an underestimation of the Australian desert source or to the relative shortness of the simulations (Joussaume et al. 1988).

Cosmogenic Isotopes

A number of long-lived radioactive nuclides are formed by cosmic ray induced nuclear reactions in the atmosphere (Raisbeck and Yiou 1984). These "cosmogenic" isotopes are then either transported to ice surfaces in precipitation or dry fallout (for nonvolatile species) or occluded, along with trapped air in ice bubbles (volatile species). The technique of accelerator mass spectrometry (AMS) now makes it feasible to measure several of these isotopes in reasonably sized ($\sim$ 1 kg) samples of polar ice. These isotopes have two potential applications in ice core studies: (*a*) they might possibly be used to "date' (either in an absolute or relative way) ice cores, and (*b*) they might give a time record of the cosmogenic production rate. In fact, these two potential applications are strongly coupled, and exploitation of one necessitates some independent information about the other. Since dating methods are covered elsewhere in this volume, we restrict our comments here to the question of production rate changes. Also, because ^{10}Be (half-life 1.5 My) is the most abundant long-lived cosmogenic nuclide in ice, and the one for which the most extensive measurements have been made, we will concentrate on this isotope while noting that others (^{26}Al, ^{36}Cl) might give similar or complementary information.

There are three factors which can influence cosmogenic production rates (Raisbeck and Yiou 1984):

1. Primary cosmic ray intensity. The origin and acceleration mechanism of galactic cosmic rays is still uncertain. However, certain theories allow, or even predict, changes in the flux of high energy particles and/or gamma rays on a variety of time scales. Such changes might show up as variable ^{10}Be concentrations in polar ice cores.
2. Solar modulation. It is believed that interplanetary magnetic fields associated with solar activity presently exclude from the solar system a significant fraction of the interstellar cosmic ray flux, particularly particles having energies $<$ 1 GeV. Variations in the strength of this "modulation" with the 11 (or 22) year solar cycle are known to change cosmogenic production rates by 30–40%. The same type of effects are believed to have caused the fluctuations in $^{14}C/^{12}C$ ratios observed in tree-rings on time scales of 100–200 yrs. Detailed profiles of ^{10}Be concentration in

the Dome C ice core show similar fluctuations, at least some of which appear to correlate reasonably well with the ^{14}C peaks during the Holocene (Raisbeck and Yiou, 1988). ^{10}Be profiles in deep ice cores thus offer the potential of extending such studies well beyond the range accessible with tree-rings. It is possible, for example, that the two periods of enhanced ^{10}Be concentration occurring at ~ 35 and 60 kyr B.P. (Raisbeck et al. 1987)] may be due to extended periods of reduced solar modulation.

3. Geomagnetic field intensity. The shielding effect of the geomagnetic field on the primary cosmic ray particles presently reduces cosmogenic production rates by a factor of 2 to 3, compared to what they would be in its absence. Since it is essentially the horizontal component of this field which deflects the cosmic ray particles, the effects are small above geomagnetic latitudes of ~ 60°. In part this may be the reason why the long-term trend of the ^{14}C tree-ring record (often attributed to geomagnetic modulation) is not seen in the 4000 yr ^{10}Be profile at Dome C (Raisbeck and Yiou 1988). However, since about 70% of cosmogenic production takes place in the stratosphere, where meridional mixing can be important, some of the geomagnetic modulation may show up in ice core cosmogenic records. Thus a "calibration" of this dependence will be necessary before quantitative conclusions regarding the geomagnetic field can be inferred from ice core ^{10}Be records (Raisbeck and Yiou 1988).

In addition to these anticipated applications of cosmogenic isotopes, initial results of ^{10}Be in the Dome C and Vostok ice cores unexpectedly demonstrated that this isotope may also give potentially interesting climatic information (see section on *Precipitation*).

Flux of extraterrestrial matter. Since Antarctica is far removed from most sources of terrestrial erosion products, it is obviously an attractive location to look for evidence of extraterrestrial matter. The specific attraction of studies in ice cores (as opposed to sediments or surface ice) is that they can give high resolution information on the time variability of the input of such matter. This is important not only in looking for possible brief "events" (such as passage through a dense comet stream), but also because several studies suggest that at least portions of the interplanetary dust complex is not in dynamical equilibrium on a time scale of ~ 10^5 yrs.

Estimates of extraterrestrial input can either be based on the flux of certain isotopes (such as ^{53}Mn) or elements (such as iridium), which are highly depleted in terrestrial matter compared to extraterrestrial, or on the detailed composition of individual particles. For example, Yiou and Raisbeck (1988) recently found five "cosmic' spherules in ice deposited at Dome C over the past ~ 4000 yrs. While the flux of these objects is very low, they are of great interest because measurements of $^{26}Al/^{10}Be$ in such spherules collected elsewhere have shown that they very likely represent cometary debris. They also represent the size of objects which appear to be in disequilibrium in the interplanetary dust spectrum. It would be very interesting to have a profile of these objects, or the other extraterrestrial indicators mentioned above, over the last ~ 10^5 yrs.

AIR BUBBLES IN ICE; ICE SHEET BEHAVIOR AND ATMOSPHERIC COMPOSITION

When snow is transformed into ice by sintering near the surface of ice sheets, a portion of the surrounding atmosphere is trapped in small pockets (several hundred per g of ice) which appear as bubbles in the ice. Study of the air thus trapped may provide information on a variety of parameters, such as the elevation at which the ice was formed, the past atmospheric pressure pattern over the ice sheet, and the paleocomposition of the atmosphere. The knowledge of these parameters should help in understanding the relationship between climate and biogeochemical cycles as well in evaluating climatic forcings (in particular from radiatively active gases) which may contribute to changes in global surface temperature.

Among the existing ice sheets, Antarctica is unique in terms of gas records because of the wide area where the snow is transformed into ice by pure sintering, i.e., in the absence of seasonal melting at the surface. We will next review the dominant results from the existing Antarctic records and stress some of the potential implications for future studies.

Total Gas Content and Past Changes of the Antarctic Ice Sheet

The amount of air trapped in the ice formed in the absence of melting depends on the porosity of ice when the air bubbles become isolated from the surrounding atmosphere and on the atmospheric pressure and temperature prevailing at the ice formation site. Based on present-day observations, empirical transfer functions have been developed linking the measured total gas content (V) with atmospheric pressure and temperature, and showing the high sensitivy of V to atmospheric pressure changes (Raynaud and Lebel 1979). Because of the V-atmospheric pressure relationship, measurements of V along the ice cores have been used as indicators of the elevation at which the ice was originally formed. The deduced elevation changes may be due to the horizontal advection of the ice (upslope origin due to ice flow) or to changes in ice thickness. By coupling the results of V records with ice flow modeling in steady state conditions, it is possible to infer past changes in ice thickness. The few V data available (see a review in Lorius et al. 1984) support the picture of a present-day Antarctic ice sheet not dramatically different from the one during the LGM, but somewhat thicker in central parts and thinner in more coastal areas. Such changes in the shape of the ice sheet are in agreement with sensitivity studies (C. Ritz, personal communication), stressing the quick response of the central part to changes in the rate of snow accumulation (which was smaller during the ice age) compared to coastal areas, which react in priority to sea level and temperature changes.

Interpretations of V variations in terms of elevation changes have assumed that the atmospheric pressure field over Antarctica remains unchanged. In fact, part of the variance observed could be due to changes in the atmospheric pressure pattern, and a more definite interpretation of V data requires further efforts regarding a better definition of the transfer function and independent information (from General Circulation Models) on possible changes in the pressure field between an ice age and an interglacial.

The CO_2 Records

Initial studies of the air trapped in Greenland and Antarctic ice cores revealed that the atmospheric CO_2 content could have been around 200 ppmv during the LGM compared with an average close to 270 during the Holocene period (Delmas et al. 1980; Neftel et al. 1982). These studies were later confirmed by Antarctic Siple data showing a marked increase in CO_2 due to anthropogenic disturbances, corresponding closely to the atmospheric record obtained since 1958 (Neftel et al. 1985). More recently the record of CO_2 in ice cores has been extended as far back as the last glacial-interglacial cycle, with a very detailed set of data for the LGM.

The CO_2 record has been extended over the last 160 kyr from the Vostok core (Barnola et al. 1987). Results are shown in Fig. 1c, where best estimates of the CO_2 concentration are plotted together with the associated uncertainty bands and where the ages of the samples have been adjusted to account for the gradual enclosure of atmospheric air in ice.

The main features of the profile include (*a*) a high correlation with the isotope temperature signal (Fig. 1b) measured on the same core, and (*b*) besides the large glacial-interglacial signal, a dominant ~ 20 kyr periodic component shown by spectral analysis and close to the orbital precession cycle. These features strongly support the coupling of atmospheric CO_2 and temperature changes over the last climatic cycle. The potentiality of a significant effect of CO_2 on temperature change is emphasized by the statistical fitting of the Vostok temperature record by linear combinations of the CO_2 and Northern and Southern Hemispheres insolation forcings (Genthon et al. 1987).

Some differences are nevertheless observed between the Vostok temperature and CO_2 records (Fig. 1b and 1c). For instance, there is no very low CO_2 value associated with the rather cold stage depicted around 110 kyr B.P.; also, both parameters vary simultaneously during transitions from glacial to interglacial while CO_2 clearly lags behind the climatic record when proceeding from warm to cold stages. This suggests that different mechanisms are involved in these different modes of climate and carbon cycle interaction.

A recent detailed reconstruction of atmospheric CO_2 for the 15 to 50 kyr B.P. period has been obtained from the Byrd ice core (Neftel et al. 1988). This record agrees well with the mean trends of the longer but scarcer Vostok data, although it casts some doubt on the significance of some individual measurements such as the low CO_2 value found around 42 kyr in the Vostok ice. Eliminating this value does not affect the significance of the variance density peak around ~ 20 kyr.

An important aim of determining the CO_2 content with as much detail as possible during the 5,000–50,000 yrs B.P. period is to get more information on the possible atmospheric significance of "abrupt" CO_2 changes recorded in the Greenland Dye 3 ice core around 30–40 kyr B.P. (Stauffer et al. 1984) and in the Antarctic Dome C ice core around the end of the LGM-Holocene transition (Raynaud and Barnola 1985). So far measurements performed on the Byrd ice core do not show the large and abrupt CO_2 variations observed in the Dye 3 core. Much more can be learned about the carbon cycle-climate system which is of global significance, for instance by investigating in more detail the transition periods and in determining the carbon isotope composition. In view of evaluating the possible impact of atmospheric forcings of the climate, we should in priority obtain records of other radiatively and chemically active trace gases such as CH_4, N_2O, CO....

Oxygen Cycle

Recent measurements of the isotope composition of the oxygen ($\delta^{18}O$) contained in the air bubbles of the Dome C core revealed that valuable information related to the global primary productivity can also be obtained (Bender et al. 1985). It was found that the $\delta^{18}O$ change of the atmospheric air associated with the Ice Age-Holocene transition was the same as the isotope change in the seawater due to the melting of large ice sheets at the end of the ice age. This finding indicates that the isotope composition of atmospheric O_2 is strongly controlled by the oceanic isotope composition. The oceanic change is transferred to the atmosphere via photosynthesis. A detailed record of the $\delta^{18}O$ change between glacial and interglacial periods could furthermore be used to calculate the rate of global primary productivity by looking at how quickly the oceanic variation is transferred to the atmosphere. The $\delta^{18}O$ of the atmosphere depends also on biological isotope fractionation associated with photosynthesis and respiration, and the ^{18}O record may also provide information on the respective participation of continental and marine productivity.

CONCLUSION

Despite the very small number of deep Antarctic ice cores available, many significant results dealing with environmental matters have been obtained. Technological progress allows the determination of an increasing number of parameters required for a better understanding of climatic changes, either from natural or human influences. The polar location, climatological conditions, ice sheet characteristics, and relative isolation of Antarctica offer some advantages in the recovery of environmental history: large amplitude of effects, good preservation of long-term records, and relative simplicity of interpretation. The good agreement between ice records from different areas and with other paleorecords (although not discussed in this paper) support the wide significance of ice core records.

There are, however, some weaknesses in these records related, for instance, to transfer functions and chronology problems. Further important information (e.g., changes in various greenhouse gases) need to be extracted and more detailed records should be obtained during climatic transitions. It would also be rewarding to obtain longer records covering at least two climatic cycles.

Beyond their specific interest, quantitative data from ice cores are important both for possible use as inputs in General Circulation Models and, as these may be subsequently applied to long-term climatic forecasts, for checking their capability of simulating the Earth's climate for boundary conditions very different from current ones.

REFERENCES

Barnola, J.M.; Raynaud, D.; Korotkevich, Y.S.; and Lorius, C. 1987. Vostok ice core provides 160000 year record of atmospheric CO_2. *Nature* **329**: 408–414.

Bender, M.; Labeyrie, L.D.; Raynaud, D.; and Lorius, C. 1985. Isotopic composition of atmospheric O_2 in ice linked with deglaciation and global primary productivity. *Nature* **318**: 349–352.

Dansgaard, W.; Johnsen, S.J.; Clausen, H.B.; and Gunderstrup, N. 1973. Stable isotope glaciology. *Meddeleser om Gronland* **197, 2**: 1–53.

De Angelis, M.; Barkov, N.I.; and Petrov, V.N. 1987. Aerosol concentrations over the last climatic cycle (160 kyr) from an Antarctic ice core. *Nature* **325**: 318–321.

Delmas, R.J.; Ascencio, J.M.; and Legrand, M. 1980. Polar ice evidence that atmospheric CO_2 20,000 yr BP was 50% of present. *Nature* **284**: 155–157.

Genthon, C.; Barnola, J.M.; Raynaud, D.; Lorius, C.; Jouzel, J.; Barkov, N.I.; Korotkevich, Y.S.; and Kotlyakov, V.M. 1987. Vostok ice core: climatic response to CO_2 and orbital forcing changes over the last climatic cycle. *Nature* **329**: 414–418.

Johnsen, S.J.; Dansgaard, W.; Clausen, H.B.; and Langway, C.C., Jr. 1972. Oxygen isotope profiles through the Antarctic and Greenland ice sheets. *Nature* **235**: 429–434.

Joussaume, S.; Jouzel, J.; and Sadourny, R. 1988. Simulations of the Last Glacial Maximum with an atmospheric General Circulation Model including paleoclimatic tracers. In: Contribution of Geophysical Sciences to Climate Change Studies, Proc. IUGG Symp. Washington, D.C.: Amer. Geophys. Union, in press.

Jouzel, J.; Lorius, C.; Petit, J.R.; Genthon, C.; Barkov, N.I.; Kotlyakov, V.M.; and Petrov, V.N. 1987. Vostok ice core: a continuous isotope temperature record over the last climatic cycle (160,000 years). *Nature* **329**: 403–409.

Jouzel, J., and Merlivat, L. 1984. Deuterium and oxygen 18 in precipitation: modelling of the isotopic effect during snow formation. *J. Geophys. Res.* **89**: 11749–11757.

Legrand, M.; Lorius, C.; Barkov, N.I.; and Petrov, V.N. 1988. Vostok (Antarctica ice core): atmospheric chemistry changes over the last climatic cycle (160,000 yr). *Atmos. Envir.* **22**: 317–331.

Lorius, C.; Jouzel, J.; Ritz, C.; Merlivat, L.; Barkov, N.I.; Korotkevich, Y.S.; and Kotlyakov, V.M. 1985. A 150,000 year climatic record frrom Antarctic ice. *Nature* **316**: 591–596.

Lorius, C., and Merlivat, L. 1977. Distribution of mean surface stable isotope values in East Antarctica: observed changes with depth in a coastal area. In: Isotopes and Impurities in Snow and Ice, Proc. IUGG Symp., Grenoble, 1975. *IAHS-AISH Publ.* **118** 127–137.

Lorius, C.; Raynaud, D.; Petit, J.R.; Jouzel, J.; and Merlivat, L. 1984. Late Glacial Maximum-Holocene atmospheric and ice thickness changes from Antarctic ice core studies. *Ann. Glaciol.* **5**: 88–94.

Neftel, A.; Moor, E.; Oeschger, H.; and Stauffer, B. 1985. Evidence from polar ice cores for the increase in atmospheric CO_2 in the past two centuries. *Nature* **315**: 45–47.

Neftel, A.; Oeschger, H.; Schwander, J.; Stauffer, B.; and Zumbrunn, R. 1982. Ice core sample measurements give atmospheric CO_2 content during the past 40,000 yr. *Nature* **295**: 220–223.

Neftel, A.; Oeschger, H.; Staffelbach, T.; and Stauffer, B. 1988. CO_2 record in the Byrd ice core 50,000–5,000 years B.P. *Nature* **311**: 609–611.

Petit, J.R.; Briat, M.; and Royer, A. 1981. Ice age aerosol content from East Antarctic ice core samples and past wind strength. *Nature* **293**: 391–394.

Petit, J.R.; Duval, P.; and Lorius, C. 1987. Long-term climatic changes indicated by crystal growth in polar ice. *Nature* **326**: 62–64.

Raisbeck, G.M., and Yiou, F. 1984. Production of long-lived cosmogenic nuclei and their applications, *Nucl. Instr. Meth. Phys. Res.* **B5**: 91–99.

Raisbeck, G.M., and Yiou, F. 1985. ^{10}Be in polar ice and atmospheres. *Ann. Glaciol.* **7**: 138–140.

Raisbeck, G.M., and Yiou, F. 1988. ^{10}Be as a proxy indicator of variations in solar activity and geomagnetic field intensity during the last 10,000 years. In: Secular Solar and Geomagnetic Variations in the Last 10,000 Years, eds. F.R. Stephenson and A.W. Wolfendale. Dordrecht: Reidel, in press.

Raisbeck, G.M.; Yiou, F.; Bourles, D.; Lorius, C.; Jouzel, J.; and Barkov, N.I. 1987. Evidence for two intervals of enhanced ^{10}Be deposition in Antarctic ice during the last glacial period. *Nature* **326**: 273–277.

Raisbeck, G.M.; Yiou, F.; Fruneau, M.; Loiseaux, J.M.; Lieuvin, M.; Ravel, J.C.; and Lorius, C. 1981. Cosmogenic ^{10}Be concentrations in Antarctic ice during the past 30 000 years. *Nature* **292**: 825–26.

Raynaud, D., and Barnola, J.M. 1985. CO_2 and climate: information from Antarctic ice core studies. In: Current Issues in Climate Research, eds. A. Ghazi and R. Fantechi, pp. 240–246. Dordrecht: Reidel.

Raynaud, D., and Lebel, B. 1979. Total gas content and surface elevation of polar ice sheets. *Nature* **281**: 289–291.

Robin, G. de Q. 1977. Ice cores and climatic changes. *Phil. Trans. R. Soc. Ln. B.* **280**: 143–168.

Stauffer, B.; Hofer, H.; Oeschger, H.; Schwander, J.; and Siegenthaler, U. 1984. Atmospheric CO_2 concentration during the last glaciation. *Ann. Glaciol.* **5**: 160–164.

Yiou, F., and Raisbeck, G.M. 1988. Cosmic spherules from an Antarctic ice core. *Meteoritics* **22**: 539–540.

Yiou, F.; Raisbeck, G.M.; Bourles, D.; Lorius, C.; and Barkov, N.I. 1985. ^{10}Be at Vostok Antarctica during the last climatic cycle. *Nature* **316**: 616–617.

The Environmental Record in Glaciers and Ice Sheets
eds. H. Oeschger and C.C. Langway, Jr., pp. 363–377
John Wiley & Sons Limited

Studies of Polar Ice: Insights for Atmospheric Chemistry

M. B. McElroy

Department of Earth and Planetary Sciences and
Division of Applied Sciences
Harvard University
Cambridge, MA, U.S.A.

Abstract. Results are presented for a model of the troposphere characterized by different values of the concentration of CH_4. The concentration of OH is expected to increase with decreasing levels of CH_4, while HO_2 and H_2O_2 are predicted to decrease. The lifetime of CH_4 may be as short as 4 years near the glacial maximum when the concentration of CH_4 was about 0.3 ppm. The corresponding source of CH_4 would be about 210 Tg yr^{-1}, as compared with about 290 Tg yr^{-1} for the preindustrial Holocene period when the concentration of CH_4 was about 0.6 ppm. It is suggested that Arctic tundra or a similar yet unidentified source may make a significant contribution to the budget of atmospheric CH_4. If the associated emissions were isotopically heavy, $\delta^{13}C$ of about $-40^0/_{00}$, then they could account for what appears to be a yet unidentified source of heavy CH_4. The rise in CH_4 over the past 100 years may reflect, in part at least, enhanced emissions stimulated by climatic warming at high latitudes. It is suggested also that precipitation of HNO_3 from the springtime Antarctic stratosphere could make an important contribution to the abundance of NO_3^- in the Antarctic ice: the contemporary flux is estimated at 10 kg^{-2} would be expected to be deposited primarily between October and December.

INTRODUCTION

Studies of gases and chemicals trapped in polar ice have provided a remarkable perspective on the changing status of the global environment. It is clear that the concentration of CO_2 has risen from a preindustrial value of about 280 ppm to a contemporary level near 350 ppm (Siegenthaler and Oeschger 1987). Methane has increased over the same period from about 600 ppb to near 1700 ppb (Khalil and Rasmussen 1982; Craig and Chow 1982; Rasmussen and Khalil 1984), while N_2O has risen from about 285 ppb to about 307 ppb (Khalil and Rasmussen 1988). The long-term record of CO_2 from the Vostok core, extending back 160,000 years, indicates a strong correlation between CO_2 and global average temperature (Barnola et al. 1987; Lorius et al., this volume). There are indications that CO_2 varying from about 180 ppm to 290 ppm, may provide a trigger, or at least a leading indicator, for major shifts in climate (Genthon et al. 1987). A similar association appears to link climate and CH_4: Stauffer et al. (1988), in an analysis of data from Dye 3 and Byrd Station, showed that the concentration of CH_4 was about 500 ppb 100,000 years ago, and that it fell to near 300 ppb near the peak of the last ice age 18,000 years ago, before rising to a typical Holocene value of about 650 ppb. Measurements by Raynaud et al. (1988) on gas trapped in the Vostok ice core have extended the perspective on CH_4 back to 160,000 years before present, indicating that the concentration of CH_4 was similarly low at the end of the preceding ice age. These data provide an invaluable record of changes in atmospheric chemistry. They provide important clues to the complex links that bind the atmosphere to the biosphere, soils and ocean, and they explode once and for all the myth that the biogeochemical cycles that regulate life on this planet should be expected to conform to the simple construct of a steady state.

Methane is removed from the atmosphere initially by reaction with OH (Levy 1972; McConnell et al. 1971; Wofsy et al. 1972). The relevant chemistry is reviewed in the next section, which explores some of the consequences of the low concentrations of CH_4 observed near the end of the last ice age. The abundance of OH in the contemporary environment is controlled largely by a balance of production by

$$O(^1D) + H_2O \rightarrow OH + OH \tag{1}$$

and loss by

$$OH + CO \rightarrow H + CO_2, \tag{2}$$

with $O(^1D)$ formed by photolysis of O_3,

$$h\nu + O_3 \rightarrow O(^1D) + O_2. \tag{3}$$

Combustion of fossil fuel and biomass burning accounts for about 30% of the contemporary source of CO (Logan et al. 1981). Enhanced release of CO may be expected to cause a net reduction in the concentration of OH (Sze 1977). As much as 50% of the rise in CH_4 since the industrial revolution may be attributed to the CO-related drop in OH. The balance must be due to an increase in production. As we shall see, there are reasons to expect that the concentration of OH at the end of the last ice age may have been significantly larger than today. It is likely, nonetheless, that the source of CH_4 was less, perhaps by as much as a factor of 2.

As ice cores can reveal much about the environment of the past, so also can they aid in interpretation of results from the present. In the last section I summarize briefly the phenomenon that has come to be known as the Antarctic ozone hole. Precipitation of nitrates from the stratosphere plays an important role in several of the current theories (Crutzen and Arnold 1986; McElroy et al. 1986; Toon et al. 1986; Molina and Molina 1987; Molina et al. 1987; McElroy et al. 1988; Wofsy et al. 1988) for the rapid loss of O_3 observed over Antarctica in spring over the past decade (Farman et al. 1985). I shall argue that a significant fraction of the nitrate removed from the stratosphere may end up in surface ice. Measurements of nitrate in polar ice cores could answer a critical question: Is precipitation of nitrate from the stratosphere a recent phenomenon? Information on this would add valuable perspective to current interpretations of the ozone hole.

CHEMISTRY OF ATMOSPHERIC METHANE

As noted earlier, oxidation of atmospheric CH_4 is initiated by reaction with OH,

$$OH + CH_4 \rightarrow H_2O + CH_3. \tag{4}$$

If the concentration of NO_x ($NO + NO_2$) is low, it proceeds through the following sequence of reactions leading to production of CH_2O:

$$\begin{aligned} CH_3 + O_2 + M &\rightarrow CH_3O_2 + M \\ CH_3O_2 + HO_2 &\rightarrow CH_3OOH + O_2 \\ CH_3OOH + OH &\rightarrow CH_2OOH + H_2O \\ CH_2OOH + M &\rightarrow CH_2O + OH + M \end{aligned} \tag{5}$$

If the concentration of OH is high, CH_2O is removed by

$$\begin{aligned} OH + CH_2O &\rightarrow H_2O + CHO \\ CHO + O_2 &\rightarrow CO + HO_2. \end{aligned} \tag{6}$$

Otherwise oxidation of CH_2O may proceed by

$$\begin{aligned} CH_2O + h\upsilon &\rightarrow HCO + H \\ H + O_2 + M &\rightarrow HO_2 + M \\ HCO + O_2 &\rightarrow CO + HO_2. \end{aligned} \tag{7}$$

In either case, oxidation of CH_4 leads to production of CO which is removed by (2), with H atoms converted subsequently to HO_2 by

$$H+O_2+M \rightarrow HO_2+M \tag{8}$$

At high concentrations of NO_x, initial oxidation of CH_4 is by (4) followed by

$$\begin{aligned} CH_3+O_2+M &\rightarrow CH_3O_2+M \\ CH_3O_2+NO &\rightarrow CH_3O+NO_2 \\ CH_3O+O_2 &\rightarrow CH_2O+HO_2. \end{aligned} \tag{9}$$

The oxidation path (4)+(5)+(6)+(2)+(8) is equivalent to

$$3OH+2O_2+CH_4 \rightarrow 3H_2O+HO_2+CO_2. \tag{10}$$

while (4)+(5)+(7)+(2)+(8) is equivalent to

$$2OH+3O_2+CH_4 \rightarrow 2H_2O+2HO_2+CO_2. \tag{11}$$

Oxidation at high NO_x proceeds by (4)+(9) followed either by (6)+(2)+(8) or by (7)+(2)+(8). The net reactions in these cases are

$$3OH+4O_2+NO+CH_4 \rightarrow 2H_2O+NO_2+3HO_2+CO_2 \tag{12}$$

and

$$2OH+5O_2+NO+CH_4 \rightarrow H_2O+NO_2+4HO_2+CO_2. \tag{13}$$

We note in all instances that oxidation of CH_4 serves to convert OH to HO_2. The reaction path (10) provides a net sink for odd hydrogen ($OH+HO_2$), while reaction path (13) results in a net source. Reaction paths (12) and (13) are responsible for net production of O_3 since NO_2 may be removed by

$$\begin{aligned} h\nu+NO_2 &\rightarrow NO+O \\ O+O_2+M &\rightarrow O_3+M \end{aligned} \tag{14}$$

Additional production of O_3 is associated with

$$NO+HO_2 \rightarrow NO_2+OH \tag{15}$$

followed by (14).

The sensitivity of OH to CH_4 is illustrated for altitudes of 0 and 6 km in figures 1 and 2. Results presented here were obtained with a one-dimensional model incorporating a relatively complete description of tropospheric chemistry. It was assumed that oxidation of CH_4 was the only significant source of CO. Inclusion of an additional source of CO associated, for example, with combustion or with oxidation of naturally formed hydrocarbons such as isoprene, would tend to lower the computed values for the concentration of OH. The correction would be significant if the additional source of CO were larger than that from CH_4, as is believed to be the case in the contemporary environment (Logan et al. 1981).

Concentrations of O_3 were obtained by solving the appropriate continuity equation with transport treated as a diffusive process. Diffusion coefficients and the flux of O_3 at the tropopause were assigned values applicable to the contemporary atmosphere. Mixing ratios of NO_x were specified for the lower troposphere at values between 3 and 6 ppt.

Note that the increase expected for OH at low concentrations of CH_4 is associated with a corresponding drop in the concentrations of HO_2 and H_2O_2. This may be readily understood if we recognize that oxidation of CH_4 and NO_x proceeds mainly by (10). Odd hydrogen formed by (1) is removed by (10), with additional loss due to rain out of H_2O_2 and the reactions

$$OH + HO_2 \rightarrow H_2O + O_2 \qquad (16)$$

and

$$OH + H_2O_2 \rightarrow H_2O + HO_2. \qquad (17)$$

The importance of rain out diminishes at low concentrations of CH_4. A summary of various sources and sinks for odd hydrogen is presented in figures 3–6.

The flux required to maintain a specified concentration of CH_4 is illustrated in figure 7. Note the relatively flat shape of the curve for mixing ratios of CH_4 between 0.3 and 1.5 ppm. The flux changes by about a factor of 2 corresponding to a factor of 5 change in the concentration of CH_4. It is tempting, as Stauffer et al. (1988) did, to attribute the reduction in CH_4 at the peak of the last glaciation to a reduction in the area occupied by wetlands. It is difficult, however, to interpret the isotopic composition of CH_4 in the present atmosphere.

The contemporary value for $\delta^{13}C$ is $-47.0‰$ (Stevens and Rust 1982). The increase in CH_4 over the past 100 years or so was accompanied by a rise in $\delta^{13}C$ of about 2‰ according to Craig (1988). Biogenic sources surveyed to date show typical values for $\delta^{13}C$ of about $-60‰$. To account for the preindustrial value of $\delta^{13}C$ would require a combination of sources

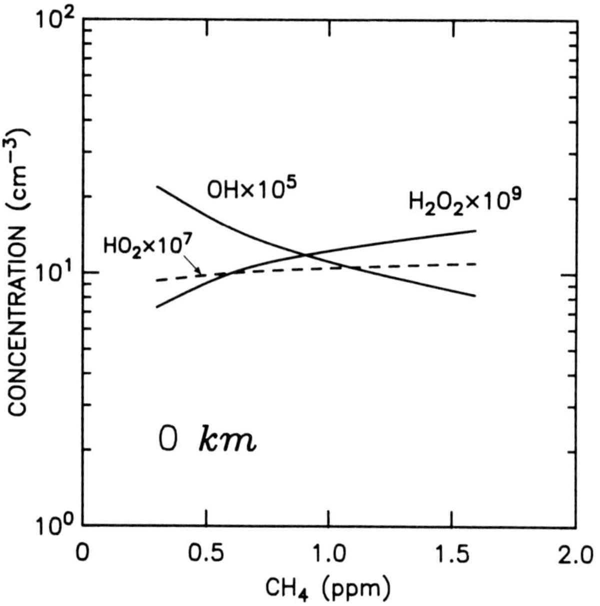

Fig. 1—Mean values for the concentrations of OH, HO_2, and H_2O_2 near the surface as a function of assumed values for the concentration of CH_4. Assumptions of the model are described in the text.

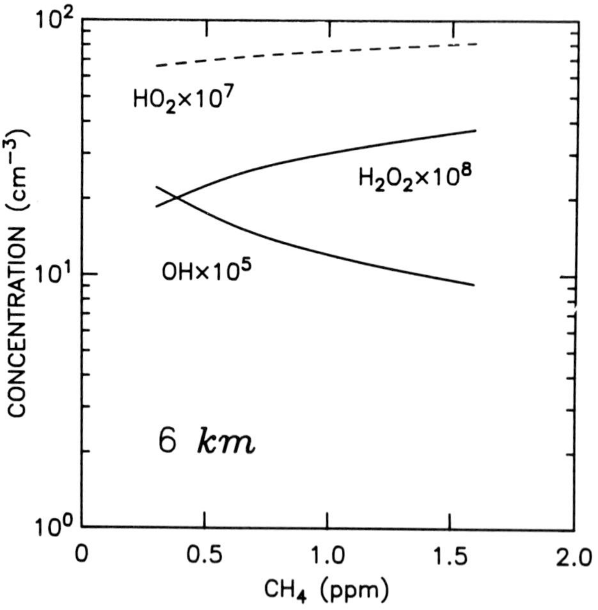

Fig. 2—Similar to Fig. 1 for an altitude of 6 km.

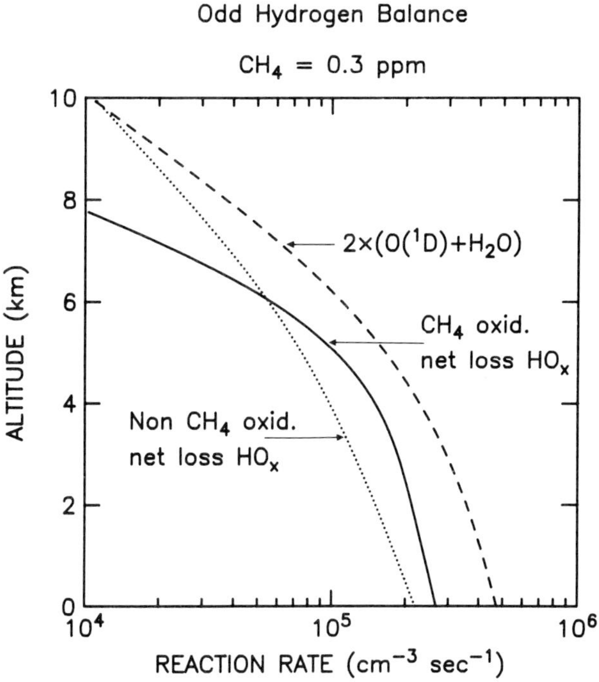

Fig. 3—The source of odd hydrogen, $2 \times (O(^1D) + H_2O)$, compared with sink due to oxidation of CH_4 and other sinks involving mainly HO_2 and H_2O_2 as described in the text. Results are shown for a concentration of CH_4 of 0.3 ppm.

with an effective $\delta^{13}C$ of about −52‰ if we adopt the value obtained by Rust and Stevens (1980) for the kinetic isotope effect associated with reaction (4). This could be explained if 40% of the source had a $\delta^{13}C$ of about −40‰ with the balance of about −60‰. The heavy component could be associated with seasonal release of CH_4 from Arctic tundra. This could provide a simple explanation for the drop in CH_4 concentration observed at the end of the last ice age: we would expect the area occupied by tundra to be much less during peak glaciation than today. If this suggestion were valid it would imply that the tundra source should be significant, both now and in the past, approaching perhaps as much as 150 Tg/yr today. On the other hand, if the isotope effect for (4) were as large as 10‰ as reported by Davidson et al. (1987) (they quote a value of 10±7‰ for the channel favoring removal of $^{12}CH_4$) the need to identify a source of heavy CH_4 would be effectively eliminated. Measurements reported by Sebacher et al. (1986) indicate that tundra sources may indeed make an important contribution to the contemporary budget of CH_4. Additional measurements of the flux of CH_4 from tundra including analyses of isotopic composition

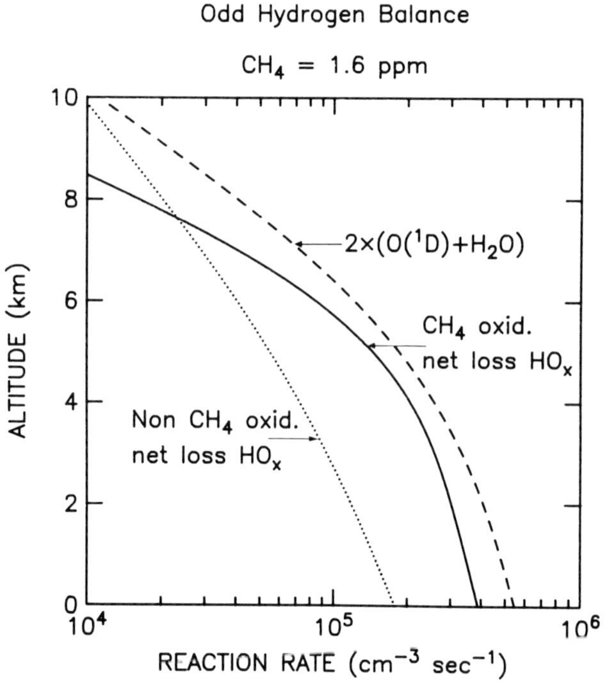

Fig. 4—Similar to Fig. 3 but with CH_4 equal to 1.6 ppm.

would clearly be of considerable interest. It would be useful to investigate also the possible importance of the sink for CH_4 associated with forest soils as observed by Keller et al. (1983). Selective removal of $^{12}CH_4$ in these environments would affect the isotopic composition of atmospheric CH_4 even though the associated gas loss might be trivial in terms of the overall atmospheric budget.

The recent rise in CH_4 and $\delta^{13}C$ may be attributed to a combination of biomass burning (with a value of $\delta^{13}C$ of −25‰), natural gas (−44‰), coal (−35‰), rice cultivation (−67‰), and cattle (−59‰) (Stevens and Engelkemeir 1986). There could be a contribution also due to enhanced release from tundra associated with recent climatic warming.

ANTARCTIC OZONE

There is convincing evidence that the springtime decline in O_3 over Antarctica is the result of catalytic reactions involving radicals formed by decomposition of industrial halocarbons. Two specific schemes have been suggested, one involving reactions of ClO with BrO (McElroy et al. 1986).

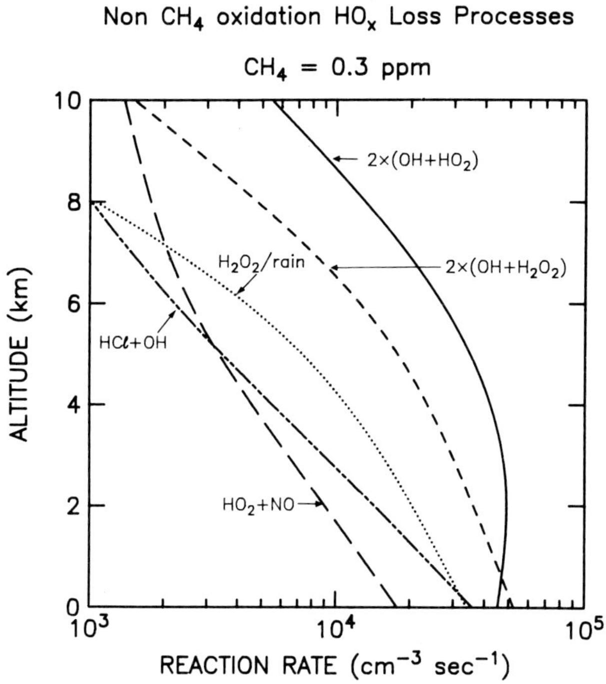

Fig. 5—Components of the nonmethane sink for odd hydrogen shown in Fig. 3 with CH_4 equal to 0.3 ppm.

$$\begin{aligned} Cl+O_3 &\rightarrow ClO+O_2 \\ Br+O_3 &\rightarrow BrO+O_2 \\ ClO+BrO &\rightarrow Cl+Br+O_2, \end{aligned} \tag{18}$$

the second a reaction of ClO with itself forming the ClO dimer (Molina and Molina 1987),

$$\begin{aligned} ClO+ClO+M &\rightarrow Cl_2O_2+M \\ h\nu+Cl_2O_2 &\rightarrow Cl+ClO_2 \\ ClO_2+M &\rightarrow Cl+O_2+M \\ 2\left[Cl+O_3\right. &\rightarrow \left.ClO+O_2\right]. \end{aligned} \tag{19}$$

These schemes are equivalent to

$$O_3+O_3 \rightarrow 3O_2 \tag{20}$$

and could be effective at relatively low altitude.

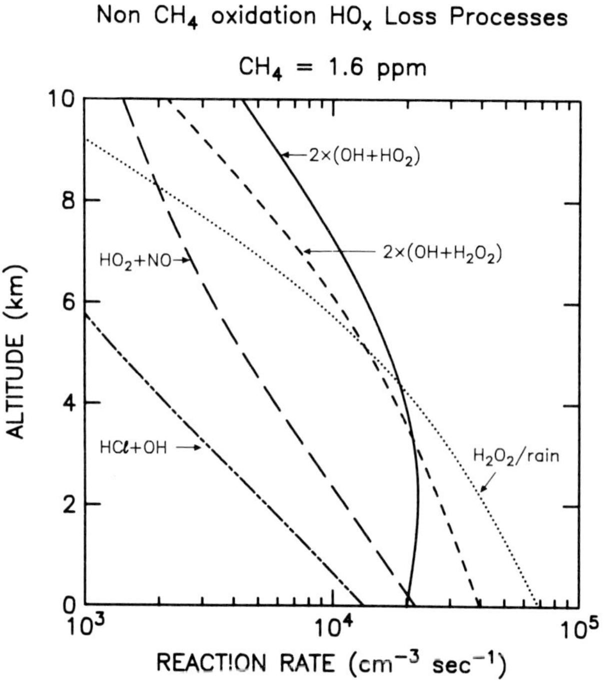

Fig. 6—Similar to Fig. 5 with CH_4 equal to 1.6 ppm.

Results from the 1987 expedition to McMurdo Station have provided the first direct measurements of atmospheric BrO (Carroll et al., submitted). The concentration of BrO is about 10 ppt, similar to values inferred by Salawitch et al. (1988a) based on earlier measurements of OClO (Solomon et al. 1987). *In situ* measurements of ClO by Anderson and Brune (personal communication) from the ER-2 indicate that the concentration of this species approached levels of 1 ppb within the polar vortex in September of 1987. Concentrations of ClO and BrO in this range are sufficient to account for the observed decline in O_3, as indicated in figure 8.

Concentrations of ClO and BrO at the levels measured by Anderson and Brune (personal communication) and Carroll et al. (submitted) required that a significant fraction of the total inorganic chlorine and bromine in the Antarctic stratosphere be converted to the form of radicals. It follows that the concentration of NO_2 must be very low; otherwise, ClO and BrO would be converted, essentially irreversibly, to $ClNO_3$ and $BrNO_3$. It appears, as suggested by Solomon et al. (1986), that heterogeneous reaction of $ClNO_3$ with HCl must play an important role in the production of chlorine radicals, through the reaction

$$ClNO_3 + HCl(s) \rightarrow Cl_2 + HNO_3(s) \tag{21}$$

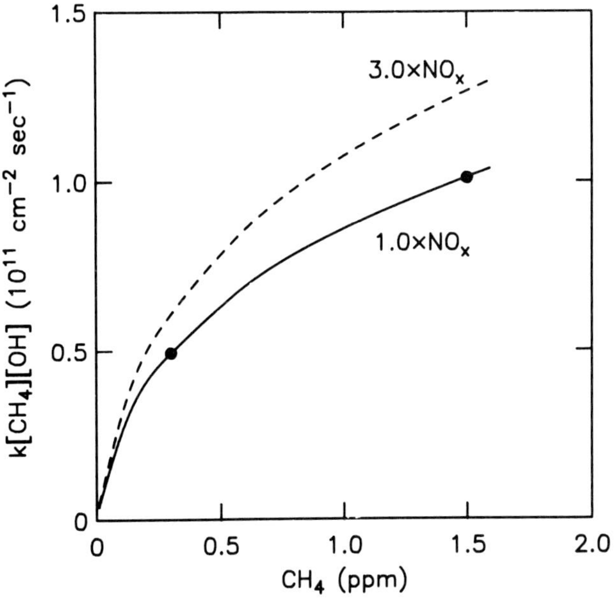

Fig. 7—The source of CH_4 required to maintain given levels of CH_4, expressed as the altitude integral of the rate for the reaction of OH with CH_4. The curve labeled $3.0 \times NO_x$ employs a level of NO_x three times that adopted for the standard model described in the text. If the concentration of NO_x were larger in the past, as might be the case if levels of N_2O were higher or if the troposphere were more stagnant, the curve describing flux vs. concentration would be even shallower than indicated here. A similar result could arise if levels of tropospheric O_3 were higher in the past.

followed by

$$h\nu + Cl_2 \rightarrow Cl + Cl. \tag{22}$$

Here (s) denotes a species present in the solid phase. Molina et al. (1987) have shown that (21) involving reaction of gas phase $ClNO_3$ with HCl contained in ice is exceptionally fast, with the product Cl_2 released to the gas phase while HNO_3 is trapped by the ice. Additional production of chlorine radicals (McElroy et al. 1986b; Solomon et al. 1986) occurs through

$$ClNO_3 + H_2O(s) \rightarrow HNO_3(s) + HOCl \tag{23}$$

followed by

$$h\nu + HOCl \rightarrow OH + Cl. \tag{24}$$

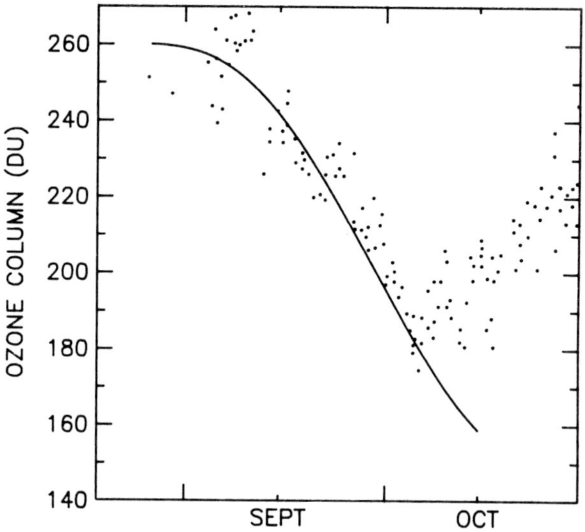

Fig. 8—Depletion of the O_3 column density as a function of time during the months of August–October at a latitude of about 78°S. The observational data are indicated by dots. The solid line summarizes results of a computation by Wofsy et al. (1988), in which loss of O_3 occurs mainly by a combination of the CLO–ClO (75%) and the ClO-BrO (25%) catalytic cycles. Taken from Wofsy et al. (1988).

Rates for (21) and (23) have been measured in the laboratory by Molina et al. (1987) and by Leu (1988).

McElroy et al. (1988), from an analysis of infrared data of Farmer et al. (1987), concluded that there was a net loss of HNO_3 from the Antarctic stratosphere during the spring of 1987. The loss is equivalent to 10 kg km^{-2} and would be more than sufficient to account for the mean annual deposition of NO_3^- in ice observed by Legrand and Delmas (1986) at D 55 (2.3 kg $km^{-2}yr^{-1}$), D 80 (13.1 kg $km^{-2}yr^{-1}$), Dome C (0.36 kg $km^{-2}yr^{-1}$), and South Pole (7.5 kg $km^{-2}yr^{-1}$) over the period 1959 to 1969. It is presumed that HNO_3 is incorporated in polar stratospheric clouds (PSCs), as discussed by Toon et al. (1986), McElroy et al. (1986a), and Crutzen and Arnold (1986). Under appropriate conditions a fraction of the particles in PSCs grows large enough to fall, carrying HNO_3 to the troposphere and ultimately to the surface. The depth and temporal extent of the ozone hole is a critical function of the efficiency with which HNO_3 is removed from the stratosphere, as discussed by Salawitch et al. (1988b). It appears from preliminary reports of results from the ER-2 missions in 1987 (NASA 1987) that removal was essentially complete during the spring of 1987.

Indirect support for the hypothesis that nitrate precipitates from the stratosphere and that it reaches the ground can be adduced from the measurements of particulate nitrate by Wagenbach et al. (1988) at the

German Antarctic research station (70°S, 8°W) near Atka Bay. These data cover the period 1983 to 1985. They observed strong peaks in NO^-_3 in spring. Concentrations were particularly large in October and November of 1983, though evidence for the spring maximum was inescapable also in 1984 and 1985. It is unclear how this behavior relates to the conclusions reached by Legrand and Delmas (1986), Zellar et al. (1986),and Herron (1980) who studied the seasonal variability of NO_3^- in ice at South Pole and at the Ross Ice Shelf. These authors found a maximum concentration of NO_3^- in ice cores in summer. The seasonal variation of NO_3^- may reflect the seasonal pattern of precipitation as pointed out by Laird (1986). It may be necessary to correct also for effects of sublimation (Gow 1965). It is important that these matters be resolved.

Measurements of NO_3^- in ice could serve to provide a long-term record of polar stratospheric temperatures and as such could make an invaluable contribution to current research on stratospheric O_3. It is clear though that further work must be done to elucidate the factors influencing the seasonal variation of the concentrations in Antarctic ice. The data for 1986 and 1987 should be particularly illuminating: one would expect the stratospheric source to be especially apparent in the record for 1987.

REFERENCES

Barnola, J. M.; Raynaud, D.; Korotkevich, Y. S.; and Lorius, C. 1987. Vostok ice core provides 160,000 year record of atmospheric CO_2.*Nature* **329**: 408–414.

Carroll, M. A.; Sanders, R. W.; Solomon, S.; and Schmeltekopf, A. L. 1988. Visible and near-ultraviolet spectroscopy at McMurdo Station, Antarctica. 5. Observations of BrO. *J. Geophys. Res.*, in press.

Craig, H.; 1988. Methane in polar ice cores. Seminar, Harvard Univ.

Craig, H., and Chow, C. C. 1982. Methane: the record in polar ice cores. *Geophys. Res. Lett.* **9**: 1221–1224.

Crutzen, P. J., and Arnold. F. 1986. Nitric acid cloud formation in the cold Antarctic stratosphere: a major cause for the springtime "ozone hole". *Nature* **324**: 651–655.

Davidson, J. A.; Cantrell, C. A.; Tyler, S. C.; Shetler, R. E.; Cicerone, R. J.; and Calvert, J. G. 1987. Carbon kinetic isotope effect in the reaction of CH_4 with HO. *J. Geophys. Res.* **92**: 2195–2199.

Farman, J. C.; Gardiner, B. G.; and Shanklin, J. D. 1985. Large losses of total ozone in Antarctica reveal seasonal ClOx/NOx interaction. *Nature* **315**: 207–210.

Farmer, C. B.; Toon, G. C.; Schaper, P. W.; Blavier, J. F.; and Lowes, L. L. 1987. Stratospheric trace gases in the spring 1986 Antarctic atmosphere. *Nature* **329**: 126–130.

Genthon, C.; Barnola, J. M.; Raynaud, D.; Lorius, C.; Jouzel, J.; Barkov, N. I.; Korotkevich, Y. S.; and Kotlayakov, V. M. 1987. Vostok ice core: climatic response to CO_2 and orbital forcing over the last climatic cycle. *Nature* **329**: 414–418.

Gow, A. J. 1965. On the accumulation and seasonal stratification of snow at the South Pole. *J. Glaciol.* **5**: 467.

Herron, M. M. 1980. Glaciochemical dating techniques. *ACS Sympos. Ser.* **176**: 303–318.

Keller, M.; Goreau, T. J.; Wofsy, S. C.; Kaplan, W. A.; and McElroy, M. B. 1983. Production of nitrous oxide and consumption of methane by forest soils. *Geophys. Res. Lett.* **10**: 1156–1159.

Khalil, M. A. K., and Rasmussen, R. A. 1982. Secular trends of atmospheric methane. *Chemosphere* **11**: 877–883.

Khalil, M. A. K., and Rasmussen, R. A. 1988. Nitrous oxide: trends and global mass balance over the last 3,000 years. *Ann. Glaciol.* **10**: 73–79.

Laird, C. M. 1986. Nitrate deposition in Antarctica: temporal and spatial variations. Ph.D. diss. Univ. of Kansas, Lawrence.

Legrand, M. R., and Delmas, R. J. 1986. Relative contributions of tropospheric and stratospheric sources to nitrate in Antarctic snow. *Tellus* **38B**: 236–249.

Leu, M. -T. 1988. Laboratory studies of sticking coefficients and heterogeneous reactions important in the Antarctic stratosphere. *Geophys. Res. Lett.*, in press.

Levy, H., II. 1972. Photochemistry of the lower troposphere. *Planet. Space Sci.* **20**: 919–935.

Logan, J. A.; Prather, M. J.; Wofsy, S. C.; and McElroy, M. B. 1981. Tropospheric chemistry: a global perspective. *J. Geophys. Res.* **86**: 7210–7254.

McConnell, J. C.; McElroy, M. B.; and Wofsy, S. C. 1971. Natural sources of atmospheric CO. *Nature* **233**: 187–188.

McElroy, M. B.; Salawitch, R. J.; and Wofsy, S. C. 1986a. Antarctic O_3: chemical mechanisms for the spring decrease. *Geophys. Res. Lett.* **13**: 1296–1299.

McElroy, M. B.; Salawitch, R. J.; Wofsy, S. C.; and Logan J. A. 1986b. Reductions of Antarctic ozone due to synergistic interactions of chlorine and bromine. *Nature* **321**: 759–762.

McElroy, M. B.; Salawitch, R. J.; and Wofsy, S. C. 1988. Chemistry of the Antarctic stratosphere. *Planet. Space Sci.* **36**: 73–87.

Molina, L. T., and Molina, M. J. 1987. Production of Cl_2O_2 from the self-reaction of the ClO radical. *J. Phys. Chem.* **91**: 433–436.

Molina, M. J.; Tso, T. -L.; Molina, L. T.; and Wang, F. C. -Y. 1987. Antarctic stratospheric chemistry of chlorine nitrate, hydrogen chloride and ice: release of active chlorine. *Science* **238**: 1253–1257.

NASA. 1987. Proceedings of press conference presenting results of 1987 ER-2 flight over Antarctica, Sept. 30, 1987. Goddard Space Flight Center, Greenbelt, MD.

Rasmussen, R. A., and Khalil, M. A. K. 1984. Atmospheric methane in the recent and ancient atmospheres. *J. Geophys. Res.* **89**: 11599–11605.

Rust, F., and Stevens, C. M. 1980. Carbon kinetic isotope effect in the oxidation of methane by hydroxyl. *Int. J. Chem. Kinet.* **12**: 371–377.

Salawitch, R. J.; Wofsy, S. C.; and McElroy, M. B. 1988a. Chemistry of OClO in the Antarctic stratosphere: implications for bromine. *Planet. Space Sci.* **36**: 213–224.

Salawitch, R. J.; Wofsy, S. C.; and McElroy, M. B. 1988b. The chemistry of Antarctic ozone 1960–1970. Abstract of paper to be presented at the special ozone conference, Aspen, CO, May 1988.

Sebacher, D. I.; Harriss, R. C.; Bartlett, K. B.; Sebacher, S. M.; and Grice S. S. 1986. Atmospheric methane sources: Alaskan tundra bogs, an alpine fen, and a subarctic boreal marsh. *Tellus* **38B**: 1–10.

Siegenthaler, U., and Oeschger, H. 1987. Biospheric CO_2 emissions during the past 200 years reconstructed by deconvolution of ice core data. *Tellus* **39B**: 140–154.

Solomon, S.; Garcia, R. R.; Rowland, F. S.; and Wuebbles, D. J. 1986. On the depletion of Antarctic ozone. *Nature* **321**: 755–758.

Solomon, S.; Mount, G. H.; Sanders, R. W.; and Schmeltekopf, A. L. 1987. Visible spectroscopy at McMurdo Station, Antarctica, 2, Observations of OClO. *J. Geophys. Res.* **92**: 8329–8338.

Stauffer, B.; Lochbronner, E.; Oeschger, H.; and Schwander, J. 1988. Methane concentration in the glacial atmosphere was only half that of the preindustrial Holocene. *Nature*, in press.

Stevens, C. M., and Engelkemeir, A. 1986. Results reported in Global ozone research and monitoring project, Report No. 16, Atmospheric Ozone 1985. Geneva: World Meteorological Organization.

Stevens, C. M., and Rust, F. W. 1982. The carbon isotopic composition of atmospheric methane. *J. Geophys. Res.* **87**: 4879–4882.

Sze, N. D. 1977. Anthropogenic CO emissions: implications for atmospheric CO-OH-CH_4 cycle. *Science* **195**: 673–675.

Toon, O. B.; Hamill, P.; Turco, R. P.; and Pinto, J. 1986. Condensation of HNO_3 and HCl in the winter polar stratospheres. *Geophys. Res. Lett.* **13**: 1284–1287.

Wagenbach, D.; Gorlach, U.; Moser, K.; and Munnich, K. 1988. Coastal Antarctic aerosol: the seasonal pattern of its chemical composition and radionuclide content. *Tellus*, in press.

Wofsy, S. C.; McConnel, J. C.; and McElroy, M. B. 1972. Atmospheric CH_4, CO, and CO_2, *J. Geophys. Res.* **77**: 4477–4493.

Wofsy, S. C.; Molina, M. J.; Salawitch, R. J.; Fox, L. E.; and McElroy, M. B. 1988. Interactions between HCl, NO_x, and H_2O ice in the Antarctic stratosphere: implications for ozone. *J. Geophys. Res.* **93**: 2442–2450.

Zeller, E. J.; Dreschhoff, G. A. M.; and Laird, C. M. 1986. Nitrate flux on the Ross ice shelf, Antarctic and its relation to solar cosmic rays. *Geophys. Res. Lett.* **13**: 1264–1267.

Standing, left to right:
Michael McElroy, Grant Raisbeck, Sigfus Johnsen, Chester Langway, Alan Hecht, Manfred Lange, John Eddy

Seated, left to right:
Hans Oeschger, Willi Dansgaard, Claude Lorius, Peter Schlosser

The Environmental Record in Glaciers and Ice Sheets
eds. H. Oeschger and C.C. Langway, Jr., pp. 379–388
John Wiley & Sons Limited

Group Report
Long-term Ice Core Records and Global Environmental Changes

A.D. Hecht, Rapporteur
W. Dansgaard
J.A. Eddy
S.J. Johnsen
M.A. Lange
C.C. Langway, Jr.
C. Lorius
M.B. McElroy
H. Oeschger
G. Raisbeck
P. Schlosser

Abstract. Physical and chemical information in ice cores from both Greenland and Antarctica span the last full interglacial-glacial-interglacial cycle. The availability of these data allow new questions to be addressed that specifically deal with the dynamics of climate change, the interaction between climate and chemical changes in the atmosphere, the timing and phase relationship of climate and chemical changes in both hemispheres, and possible effects of solar variability on climate. The wealth of information that has been extracted from a relatively few ice cores suggests that the next ten years will be a period of intense study of biogeochemical cycles and climate dynamics.

In consideration of the increasingly urgent need to understand the complex interactions of the physical-biological-chemical system of the Earth, the information retained in ice cores constitutes a unique and essential "Rosetta Stone" with which to decipher the long-term record of climate variations.

HISTORICAL BACKGROUND

The first modern-day multiple parameter ice core study was performed on a continuous 411 m deep ice core from Site 2, Greenland (Langway 1967). The first continuous ice core to be recovered that completely penetrated a polar ice sheet (1480 m) and into the bedrock material was obtained at Camp Century, Greenland in 1966 (Hansen and Langway 1966); the first comprehensively integrated and internationally coordinated polar deep ice core drilling program was completed at Dye 3, Greenland (2037 m) in 1981

(Langway et al. 1985). The Dye 3 core was the most completely recovered (99.9%), thoroughly recorded, and extensively investigated deep ice core yet to reach bedrock; its field core study program established a new standard of efficiency for such work. It is now 18 years since the first papers on the Camp Century ice core opened the door to reconstructing the climate history of polar regions (Dansgaard et al. 1969, 1971, 1973). The first published results showed:

1. the general pattern of climate fluctuations of the north polar region from the last interglacial to the present,
2. that the $^{18}O/^{16}O$ ratio in ice sheets was lighter during the glacial period,
3. a detailed record of Holocene climate events, including the Little Ice Age and Younger Dryas cold interval,
4. continuous dated chronology, at least for times of recent climate change,
5. several possible periodicities in the climate record.

Much of the research between 1971 and 1981 focused on two basic problems, namely, (*a*) obtaining a reliable chronology and (*b*) deriving accurate estimates of climate variables. While significant progress has been made, both remain central to the problem of interpreting the environmental change in ice cores. The chronology problem is now being addressed through a combination of detailed stratigraphy, radioactive dating, and correlation with well-dated reference horizons (see Budd et al., this volume). Accurate estimates of climatic variables from ice cores also require detailed knowledge of atmospheric and chemical processes. Much progress has also been made in understanding the relationships between source functions and their signature in ice cores (see White et al., this volume).

Over the last ten years, new findings in ice cores have greatly expanded their use in the study of Earth history. A remarkable series of achievements has come from the ice cores studies. These studies have:

1. identified Milankovitch cycles and other higher frequency climatic fluctuations,
2. documented changing levels of CO_2 and methane in the atmosphere,
3. provided measurements of H_2O_2, an important product of tropospheric photochemistry,
4. provided measurments of aerosols and major ions (e.g., Na^+, NH_4^+, Ca^{+2}, Mg^{+2}, H^+, Cl^-, NO_3^-, SO_4^{-2},
5. shown fluctuations of cosmogenic isotopes in ice cores.

In less than 20 years a wealth of environmental information has been extracted from only five long ice cores. Many of the major environmental questions facing society today, such as climate change due to anthropogenic greenhouse gases, can be directly studied from ice cores. Only from this source can past fluctuations of atmospheric CO_2 be documented and only from this source can the combined climatic efects of solar variability and chemical change be studied in a unified way.

Drawing on this past history and current opportunities, discussion in our group centered around the following questions:

1. How well documented are short-term climate changes in Greenland and Antarctica?
2. What can we learn about long-term climate changes and climate dynamics (timing, forcing functions, response, and feedbacks) from ice cores, including changes from glacial to interglacial intervals?
3. How can we reconstruct and explain past chemical conditions of the atmosphere and how does the changing chemistry affect global climate?
4. Can records of cosmogenic isotopes in ice cores be used to infer paleoprecipitation and production variations due to primary cosmic ray intensity, solar activity, and geomagnetic variation?
5. What can we learn about past changes in ice sheets from ice cores?

SUMMARY OF DISCUSSION ON SHORT-TERM CLIMATIC CHANGES

There are clear indications of rapid climate changes at the end of the Pleistocene, about 13,000 to 10,000 yrs B.P., in both Arctic ice cores, deep-sea cores, and lake sediment records (Dansgaard and Oeschger, this volume; Ruddiman and McIntyre 1981). These rapid changes are clearly evident in $\delta^{18}O$ records from the ice and are inferred from CO_2 measured in air bubbles trapped in the ice. Unfortunately, the amount of existing ice cores with adequate air bubbles is insufficient to document in detail CO_2 changes over this important interval.

The time of this strong oscillation (Bølling-Allerød-Younger Dryas) is well defined in Greenland cores and illustrates a remarkable period of rapid climate change at least in northern high latitude regions. Reasonably reliable dating suggests that 10,000 yrs ago, within a period of about 40 years, the temperature in the vicinity of Greenland changed by 5–6°C. The record of rapid climate change in the Greenland isotope record matches in detail many climate records from Europe. In North America, however, this event has been recorded at only a few locations.

The type of rapid climate event illustrated by the Bølling-Allerød-Younger Dryas oscillation is not unique. It appears to be the last of a long series of similar events in Greenland and Canada throughout a major part of the glaciation. Between 20,000 and 40,000 yrs B.P., $\delta^{18}O$ varied between about

−35.5‰ to −32.0‰ in Dye 3 and −42‰ to 38‰ in Camp Century. These differences correspond to a 5–6°C temperature change.

Over this interval the changes in $\delta^{18}O$ are in phase with changes in CO_2 levels and are antiphase to dust concentration. The glacial world was characterized by significantly lower levels of CO_2 and higher dust levels indicative of greater atmospheric turbidity. It is not possible with current resolution, however, to determine whether CO_2 lags or leads the changes in $\delta^{18}O$.

Although reasons for such rapid climate oscillations are unknown, the most recent one (13,000–10,000 yrs B.P.) occurred at a time of major change in the position of the polar front and appears related to the formation of deep water, as deduced from deep-sea sediments. Reconstruction of climate changes from deep-sea cores in the North Atlantic show a picture of movement of the polar front over an 11,000 yr period from 20,000 to 9,000 yrs ago. From 13,000 to 10,000 yrs ago the polar front moved from a position near Iceland to Northern Spain and, as a possible consequence of this, the climate of the Northern Hemisphere changed significantly. In southwest Norway, for example, the Scandinavian ice sheet advanced (about 11,000–10,000 yrs B.P.) to coastal positions abandoned 2,000 yrs earlier (Ruddiman and McIntyre 1981).

While there is strong evidence for rapid climate change in northern latitudes, comparable climatic events in Antarctica have not been similarly identified. While Vostok, Dome C, and Byrd cores give some evidence of a possible related event (Jouzel et al. 1987), it remains problematical as to whether the rapid climate oscillation that took place 13,000–11,000 yrs B.P. (or others identified in Greenland at earlier times) was global in extent.

In summary, there is evidence in Greenland ice cores of rapid climate change during the last glaciation. Intensive examination of Younger Dryas and earlier times of rapid climate change using all available ice cores and deep-sea data are needed to understand the significance of these important climate events.

SUMMARY OF DISCUSSION ON LONG-TERM CHANGES AND CLIMATE DYNAMICS

Periodicities of changes in the orbit of the Earth are clearly evident from oxygen isotope variations stored in deep-sea sediments. These isotopic variations predominantly reflect changes in the volume of ice on the continents, which are largely related to changes in solar radiation. The strong effect of this external forcing function on the Earth's climate system is also evident from isotope fluctuations in the Vostok core where frequencies of 40,000 and, to a lesser degree, 20,000 yr periods are evident. It is therefore apparent that long-term variations in the surface temperature of

Antarctica are, to a first approximation, closely related to orbital and solar radiation changes.

Qualitatively, there is agreement between the Vostok isotopic record and the climate record established from deep-sea sediments at least for the last 110,000 yrs. Before this period, the Vostok core is characterized by a longer interglacial period than is evident in deep-sea cores, assuming the current chronology for the Vostok core. (The Vostok time scale is independent of orbital tuning, as is used in deep-sea sediments.) It is possible that the deuterium data are indicating a phase shift in the response of surface temperatures in Antarctica to changes in solar radiation due to orbital changes. Another possible explanation is that the oceans remained warm long after the continents cooled off, which would also help to explain the fast build-up of continental ice.

Direct evidence of CO_2 changes is also available from the Vostok ice core for the past 160,000 yrs (Barnola et al. 1987). The CO_2 record is strongly correlated linearly with the isotopic temperature record suggesting a first order response to orbital forcing. However, spectral analysis of the isotopic and CO_2 data indicate that these records are responsive to different parts of the orbital forcing. The isotopic record is strongly dominated by a 40,000 yr frequency, whereas the CO_2 record is primarily dominated by a 20,000 yr frequency.

It is apparent from comparing the CO_2 trends with variations in surface radiation, as deduced from orbital changes, that CO_2 trends are related to seasonal variations (Budd, personal communication). One possible hypothesis for relating these factors is that seasonal variations of the radiation regime cause similar variations in the seasonality of the Antarctic sea ice, which can be expected to affect the CO_2 ocean exchange in the south polar region.

Comparison of the CO_2 and isotopic temperature curves also show differences which reflect the dynamics of the ocean-atmosphere chemistry system. There is a difference in the relative timing between changes of atmospheric CO_2 concentration and Antarctic climate changes, depending on whether it is an interglacial to glacial change or vice versa. In Antarctica, temperature changes lead CO_2 change during the last interglacial to glacial transition by as much as 10 K/yr, whereas changes in CO_2 and deuterium concentration are almost simultaneous ($\pm$2,000 years) when going from glacial to interglacial conditions.

It is clear that the ocean plays a major role in climate change, although the relationship between ocean circulation and ice sheet changes is not well understood. Two general questions seem ripe for investigation:

1. Since the Younger Dryas climate signal is related to changes in the position of the N. Atlantic polar front, what are the expected climate signals in the Southern Hemisphere and how are the two hemispheres coupled?

2. Since changes in the extent of sea ice may affect CO_2 uptake, can such changes be inferred from ice cores? At present no clear proxy indicator of sea ice extent in cores has been exploited. There are, however, a range of diatoms that only bloom on or within sea ice that may serve this purpose.

For the first time, the availability of multiple proxy records such as dust content, CO_2, methane, δD, $\delta^{18}O$, chemical ions, etc., allows the development of models that link atmospheric circulation (turbidity, storminess), atmospheric chemistry (methane), biological productivity (CO_2, $\delta^{18}O$ in air bubbles), surface climatology (δD, $\delta^{18}O$), and solar variability (^{10}Be) into conceptual models. Many of these variables show characteristics that are different from one another in both time and space. Concerted efforts to synthesize these data and to promote model development should be encouraged. At the same time, it is clear that a foundation of adequate chronology is essential for progress in climate dynamics. For example, with existing time scales, there is evidence that CO_2 first increased in Greenland at about 13.5 K B.P. and in Vostok at about 15–16 K B.P. Since CO_2 is a rapidly mixed constituent of the global atmosphere, this issue is likely to remain as a chronological problem.

Correlations obtained by comparing ice core records from Greenland with all other available data, such as deep-sea cores, is an urgent need. Several approaches are available for dating ice cores in Antarctica and Greenland. These include ice modeling, identification of isochronous horizons such as changes in globally mixed chemicals, ^{10}Be, methane and CO_2, volcanic events, and radiometric dating, and $\delta^{18}O$ in air bubbles.

While the time scale for deep-sea cores is partly derived by tuning with the Milankovitch cycle, this approach is not used in Antarctica ice cores at this time (Jouzel et al. 1987). Independent estimates of Antarctic chronology should be established because there are clear differences between the Vostok ice core and deep-sea cores from neighboring waters.

Milankovitch cycles are the only known major climate forcing function identified in ocean and ice cores. Yet despite the small resulting radiation changes, large climate changes occur. Thus, there are strong nonlinear and positive feedbacks in the system. In ways not yet explained, the climate system is driven by feedbacks in CO_2, water vapor, sea ice, large-scale ocean circulation as well as albedo due to snow and ice cover changes. Ice sheet growth, once begun, may act as a driving force and become the dominant cause of climate change. Subsequent changes in CO_2 may amplify the climate signal, but the exact manner and magnitude of the effect is unknown. The availability of unique records of CO_2 and methane in ice cores allows this aspect of the climate system to be more thoroughly studied.

In summary, a wealth of information is being assembled for both polar regions with respect to long-term climate and atmospheric chemistry changes.

Collaborative efforts to reconcile and synthesize data from Arctic and Antarctic ice cores, continental lake sediments, and deep-sea sediments are needed.

The establishment of accurate chronologies in ice cores by all available means is a high priority.

SUMMARY OF DISCUSSION ON METHANE AND BIOLOGICAL CHANGES

Concurrent with an increase over the past 100–200 yrs of CO_2 in the atmosphere, methane has risen more than twofold from about 650 ppb to nearly 1700 ppb. This spectacular increase in methane is difficult to explain from current knowledge of the sources of methane and their carbon isotopic composition. For example, the overall isotope composition of methane sources appears to have changed in value from about −52‰ during the preindustrial era to a value of about −50‰ today. None of the known sources of methane have isotopic composition of the right magnitude to explain this change. It is implied that an unidentified source of heavy isotopic methane is contributing to the increase in historic times. A source of methane of about −40‰ should be looked for. Measurements of the isotopic composition of methane outgassing from tundra is essential to determine how this region might contribute to the atmospheric methane budget (McElroy, this volume).

It is possible that the increase in modern methane may relate to decreases in the efficiency of the OH radical as a sink for methane. The increasing burden of CO in the atmosphere may be responsible for a decrease in OH radicals. However, current calculations suggest that this mechanism can explain only about one-half of the observed increase in CH_4. Thus, it is clear that extensive laboratory and field measurements of ^{13}C from methane in various source regions should be undertaken.

Methane is a significant greenhouse gas. As pointed out by Blake and Rowland (1988), each additional molecule of methane is about 20 times more effective as a greenhouse gas than each additional molecule of CO_2. Thus the increase of methane in the atmosphere is a significant positive amplifier of global warming. In addition, the growth of tropospheric methane may have increased the water concentration in the stratosphere by as much as 28% since the 1940s and 45% over the past two centuries (Blake and Rowland 1988). Such an increase in water vapor could enhance the formation of polar stratospheric clouds which in turn may be contributing to the observed sharp decrease in ozone in the polar stratosphere.

Measurements of methane in ice cores provide additional insight into the complex interaction of the atmosphere and biosphere. Understanding the cause of past methane fluctuations will require improved knowledge of current sources and sinks. It is possible that in the past, changes in biological systems and tropical forests may be related to changing levels of the gas. To the extent possible, such paleoecologic information shall be assembled.

At the same time, measurements of carbon monoxide and hydrogen peroxide in ice cores, in addition to methane, would provide useful information with respect to removal rates of methane and hydrogen chemistry; however, the feasibility of such analyses is not yet established.

Methane, like most atmospheric gases other than rare ones, is primarily of biological origin. Because the major chemical cycles of O, N, C, H, and S pass through both the biosphere and atmosphere, it is therefore not surprising that the biosphere plays a significant role in affecting and responding to climate change.

An example of the complex and potentially very important biosphere-chemistry-climate interactions is the recent hypothesis by Charlson et al. (1987) that variations in oceanic photosynthesis might be important in regulating cloud reflectivity. Many kinds of planktonic algae produce the gas dimethyl sulphide (DMS), which escapes from the ocean to the atmosphere and forms a sulfate aerosol. Charlson proposes that DMS is a major source of cloud condensation nuclei over the oceans. Variations in DMS nuclei could conceivably modulate cloud formation and albedo, and consequently the radiation balance of the Earth.

Measurements are required to test this theory, and some approaches have been suggested (Pylc 1988).

SUMMARY OF DISCUSSION ON COSMOGENIC PARTICLES

The changing concentrations of ^{10}Be in ice cores may provide evidence of changes in cosmic ray intensity, solar variability, or changes in the strength of the geomagnetic field. Each of these factors can contribute to the modification of ^{10}Be in the atmosphere and its subsequent preservation in ice cores. Of the above factors, cosmic ray intensity is assumed to be generally constant over the last several hundred thousand years.

Correlation of ^{10}Be with evidence of past solar variability (e.g., Maunder minimum and direct observations of changes in solar constant) suggests that this fraction may be the most important (Raisbeck et al. 1987).

Comparisons of ^{10}Be and ^{14}C variations in Greenland ice cores and with ^{14}C variations in tree-rings show striking similarities over the past 4,000 yrs. Beyond this point ^{14}C and ^{10}Be trends tend to diverge. It is not clear what the causes of this divergence are, or how widespread the phenomenon is since the observations are only from a few ice cores.

The ^{10}Be and isotopic records in Antarctica show similar long-term trends. The ^{10}Be record is, however, characterized by major increases at two intervals: around 35 K and 60 K. It is not clear whether these ^{10}Be concentration peaks are actually due to increased ^{10}Be deposition or whether they reflect major changs in precipitation. Deposition of ^{10}Be is related to aerosol distribution and changes in processing through the atmosphere could also affect deposition rates.

It is likely that a decrease in precipitation would also result in increased concentration of other aerosol components such as dust, but this is not observed. Consequently, it is assumed that ^{10}Be does reflect increases in ^{10}Be deposition. During these long periods, which may have lasted for 1,000 yrs, the sun would have been "quiet" in much the same way as during the Maunder Minimum.

Measurement of solar effects and climate signals in the same reservoir opens the door to looking at the relationship of solar variability and climate. This is a new scientific opportunity. Yet, at present, it is not clear whether ^{10}Be is an integrated global signal or concentrated by regional effects. Some better estimate of the present distribution of ^{10}Be by latitude would be useful.

Despite the present uncertainty in understanding the production rate, modification, and ultimate cause of ^{10}Be deposition in ice cores, the ^{10}Be record is an important stratigraphic tool becuase it represents a well-mixed global signal.

SUMMARY OF DISCUSSION ON ICE SHEET RECONSTRUCTIONS

The isotopic and environmental record preserved in ice cores is a function of the source of material, processes of transference through the atmosphere, and final site of preservation on the ice cap. For example, oxygen isotopic composition derived from ice cores reflects the original source of the meteorological condition in the moisture, subsequent depletion of $\delta^{18}O$, mixing in the atmosphere, and elevation of site of deposition. Many of these problems are discussed by White et al. (this volume). Discussion in this group led to only general comments.

The $\delta^{18}O$ variations in Greenland ice cores largely reflect surface temperature changes, which are partly related to shifts in the source of precipitation and the position of the polar front. However, the extent to which a part of the $\delta^{18}O$ variations can be related to shifts in the polar front is unclear.

It is also known that isotopic variations in ice cores are related to ice thickness and motion away from the original site of deposition. Variations in total gas content of the ice are a good indicator of ice elevation and have been used in the Vostok core to adjust the isotopic values due to changes in elevation (Jouzel et al. 1987).

Reconstruction of ice sheet thickness from gas content and modeling suggests that the ice margin in northwest Greenland retreated far inland during the last interglacial and that the ice thickness there was on the order of 100 m. If so, the Greenland ice sheet was significantly smaller than today and may have contributed to the proposed 6 m sea level rise at that time.

In Antarctica, total volume of ice cannot be estimated, but analyses of existing data suggest a thinning in the central part and thickening in coastal regions of ice during the ice age.

REFERENCES

Barnola, J.M.; Raynaud, D.; Korotkevich, Y.S.; Lorius, C. 1987. Vostok ice core provides 160,000-year record of atmospheric CO_2. *Nature* **329**: 408–414.

Blake, D., and Rowland, F.S. 1988. Continuing worldwide increase in tropospheric methane 1978–1987. *Science* **239**: 1129–1131.

Charlson, R.J.; Lovelock, J.E.; Andreae, M.O.; and Warren, S.G. 1987. Phytoplankton, atmospheric sulfur, cloud albedo, and climate. *Nature* **326**: 655–661.

Dansgaard, W.; Johnsen, S.J.; Clausen, H.B.; and Gundestrup, 1973. Stable isotope glaciology. *Meddelelser on Gronland* **197**: 1–53.

Dansgaard, W.; Johnsen, S.J.; Clausen, H.B.; and Langway, Jr. C.C. 1971. Climatic record revealed by the Camp Century Ice Core. In: The Late Cenozoic Glacial Ages, ed. K.K. Turekian, pp. 43–56. New Haven: Yale Univ. Press.

Dansgaard, W.; Johnsen, S.J.; Miller, J.; and Langway, Jr. C.C. 1969. One thousand centuries of climatic record from Camp Century on the Greenland Ice Sheet. *Science* **166**: 377–381.

Hansen, B.L., and Langway, C.C., Jr. 1966. Deep core drilling in ice and core analyses at Camp Century, Greenland 1961–1966. *Antarctic Journal*, October.

Jouzel, J.; Lorius, C.; Petit, J.R.; Genthon, C.; Barkov, N.I.; Kotlyakov, V.M.; and Petrov, V.M. 1987. Vostok ice core: continuous isotope temperature record over the last climatic cycle (160,000). *Nature* **329**: 403–408.

Langway, C.C., Jr. 1967. Stratigraphic analysis of a deep ice core from Greenland. Research Report 77, 130 pp. USA Cold Regions Research and Engineering Laboratory, Hanover, N.H.

Langway, C.C., Jr.; Oeschger, H.; and Dansgaard, W., eds. 1985. Greenland Ice Core: Geophysics, Geochemistry and the Environment, Geophysical Monograph 33, p. 118. Washington, D.C.: American Geophysical Union.

Pyle, J.A., 1988. Trace Substances, Radiation Balance, and the Climate of the Earth. The Changing Atmosphere, eds. F.S. Rowland and I.S.A. Isaksen. Dahlem Konferenzen, Chichester: Wiley, in press.

Raisbeck, G.M.; Yiou, F.; Bourks, D.; Lorius, C.; Jouzel, J.; and Barkov, N.I. 1987. Evidence for two intervals of enhanced ^{10}Be deposition in antarctic ice during the last glacial period. *Nature* **326**: 273–276.

Ruddiman, W.F., and McIntyre, A. 1981. The North Atlantic Ocean during the last deglaciation. *Paleogeog. Pal. Pal.* **35**: 145–214.

List of Participants with Fields of Research

J. ANDREWS
Department of Geological Sciences
INSTAAR, Box 450
University of Colorado
Boulder, CO 80309, U.S.A.

Chronology of the Lauren tide ice sheet; paleoceanography of Baffin Bay and N. Labrador Sea; holocene climates in arctic Canada.

P. BRIMBLECOMBE
School of Environmental Sciences
University of East Anglia
Norwich NR4 7TJ, England

Chemistry and history of the atmosphere

C. BRÜHL
Max-Planck-Institut für Chemie
Abt. Chemie der Atmosphäre
Postfach 3060
6500 Mainz, F.R. Germany

Photochemical climate modeling

W.F. BUDD
Department of Meteorology
University of Melbourne
Parkville, Victoria 3052, Australia

Ice sheet and climate modeling

R.J. CHARLSON
Department of Atmospheric Sciences AK-40
University of Washington
Seattle, WA 98195, U.S.A.

Atmospheric chemistry

T. CLASS
Pesticide Chemistry and Toxicology Laboratory
Department of Entomology
Wellman Hall
University of California
Berkeley, CA 94720, U.S.A.

Analytical chemistry; chemistry of organic traces in air; photochemistry; pesticide chemistry

H.B. CLAUSEN
Geophysical Institute
Glaciological Department
Haraldsgade 6
2200 Copenhagen N, Denmark

Isotope glaciology (stable and radioactive); glaciochemistry (anions and cations)

P.J. CRUTZEN
Max-Planck-Institut für Chemie
Abt. Chemie der Atmosphäre
Postfach 3060
6500 Mainz, F.R. Germany

Atmospheric chemistry

W. DANSGAARD
Geophysical Institute
University of Copenhagen
Haraldsgade 6
2200 Copenhagen N, Denmark

Glaciology/climatology

C.I. DAVIDSON
Civil Engineering
Carnegie Mellon University
Pittsburgh, PA 15213, U.S.A.

Transport of contaminants from the atmosphere to snow, focusing on the Greenland ice sheet

R.J. DELMAS
CNRS – Laboratoire de Glaciologie et Géophysique de l'Environnement
B.P. 96
38402 St Martin d'Hères Cedex, France

Glaciochemistry

J.A. EDDY
University Corporation for Atmospheric Research, Office for Interdisciplinary Earth Studies
P.O. Box 3000
Boulder, CO 80307, U.S.A.

History of solar behavior: Earth system science

R.C. FINKEL
Lawrence Livermore National Laboratory
Box 808 (L-232)
Livermore, CA 94550, U.S.A.

The use of cosmogenic nuclides as geochronometers and as geophysical tracers

E.L. FIREMAN
Smithsonian Astrophysical Observatory, MS 13
60 Garden Street
Cambridge, MA 02138, U.S.A.

Dating polar ice by the uranium-series method; ^{14}C terrestrial ages of antarctic meteorites; instrumental neutron activation studies on polar ice

W. GRAF
Gesellschaft für Strahlen- und Umweltforschung
GmbH, Institut für Hydrologie
Ingolstädter Landstrasse 1
8042 Neuherberg, F.R. Germany

Accumulation and ice core studies on the Ronne and Ekström ice shelf

G. GRAVENHORST
Institut für Bioklimatologie
Büsgenweg 1
3400 Göttingen, F.R. Germany

Biogeochemical cycle of atmospheric substances

C.U. HAMMER
Geophysical Institute, University of Copenhagen
Haraldsgade 6
2200 Copenhagen N, Denmark

Glaciology

A.D. HECHT
National Climate Program Office/NOAA
11400 Rockville Pike
Rockville, MD 20817, U.S.A.

Paleoclimatology, global climate change: administration and research planning

T. HUGHES
Department of Geological Sciences
University of Maine
Orono, ME 04469, U.S.A.

Global ice sheet modeling: past, present, and future; field research on the dynamics of ice streams and their interaction with ice shelves

S.J. JOHNSEN
Science Institute, University of Iceland
Dunhaga 3
101 Reykjavik, Iceland

Isotope glaciology; ice dynamic modeling; geothermal heat

J. JOUZEL
LGI/DPC
CEN Saclay
B.P. 1
91191 Gif-sur-Yvette Cedex, France

Paleoclimatology (ice core studies) and associated climate modeling

M.A. LANGE
Alfred-Wegener-Institut für Polar- und Meeresforschung
Columbusstrasse
2850 Bremerhaven, F.R. Germany

Analytical investigations on ice cores; numerical models of ice shelf dynamics; evolution and internal constitutions of icy planets

C.C. LANGWAY, Jr.
Ice Core Laboratory,
Department of Geology
State University of New York at Buffalo
4240 Ridge Lea Road
Amherst, NY 14226, U.S.A.

Glaciology

C. LORIUS
CNRS – Laboratoire de Glaciologie et Géophysique de l'Environnement
B.P. 96
38402 St Martin d'Hères Cedex, France

Polar (Antarctica) ice cores

M.B. McELROY
Harvard University
29 Oxford Street
Cambridge, MA 02138, U.S.A.

Atmospheric sciences

K.O. MÜNNICH
Institut für Umweltphysik
Im Neuenheimer Feld 366
6900 Heidelberg, F.R. Germany

Environmental physics

H. OESCHGER
Physikalisches Institut der
Universität Bern
Sidlerstrasse 5
3012 Bern, Switzerland

Earth system physics; records in ice cores and other natural archives modeling

G.I. PEARMAN
CSIRO, Division of Atmospheric
Research
Private Bag No. 1
Mordialloc, Victoria 3195, Australia

Global atmospheric trace gases

D.A. PEEL
British Antarctic Survey
High Cross
Madingley Road
Cambridge CB3 0ET, England

Evaluating evidence for past climate and air pollution by analysis of antarctic ice cores

S.A. PENKETT
School of Chemical Sciences
University of East Anglia
Norwich NR4 7TJ, England

Atmospheric chemistry

K.A. RAHN
Graduate School of Oceanography
University of Rhode Island
Narragansett Bay Campus
Narragansett, RI 02882–1197
U.S.A.

Trace elements in atmospheric aerosol and deposition

G. RAISBECK
Lab. René Bernas
CNRS
Bat. 108
91406 Orsay, France

Measurement and interpretation of cosmogenic isotope concentrations in geological reservoirs

D.P. RAYNAUD
Laboratoire de Glaciologie et
Géophysique de l'Environnement
B.P. 96
38402 St Martin d'Hères Cedex
France

Ice core record of atmospheric air

N. REEH
Alfred-Wegener-Institut für Polar-
und Meeresforschung
Columbusstrasse
2850 Bremerhaven, F.R. Germany

Ice sheet dynamic modeling; ice sheet margin investigations

J. RUDOLPH
Kernforschungsanlage Jülich GmbH
Institut für Atmosphärische Chemie
Postfach 1913
5170 Jülich, F.R. Germany

The atmospheric chemistry of organic trace gases, especially hydrocarbons and related substances

P. SCHLOSSER
Institut für Umweltphysik,
Universität Heidelberg
Im Neuenheimer Feld 366
6900 Heidelberg, F.R. Germany

Tracer oceanography of polar regions; water/ice interaction

U. SCHOTTERER
Physikalisches Institut der
Universität Bern
Sidlerstrasse 5
3012 Bern, Switzerland

Analyses of alpine ice cores; isotope hydrology

J. SCHWANDER
Physikalisches Institut der
Universität Bern
Sidlerstrasse 5
3012 Bern, Switzerland

Analyses of polar ice cores

G.E. SHAW
Geophysical Institute
University of Alaska
Fairbanks, AK 99775-0800, U.S.A.

Arctic and antarctic air pollution and chemistry

H. SHOJI
Department of Earth Sciences
Faculty of Science
Toyama University
3190 Gofuku
Toyama, 930, Japan

Physical and mechanical property studies of antarctic and Greenland ice cores

U. SIEGENTHALER
Physikalisches Institut der
Universität Bern
Sidlerstrasse 5
3012 Bern, Switzerland

Carbon cycle modeling; paleoclimate research

B.R. STAUFFER
Physikalisches Institut der
Universität Bern
Sidlerstrasse 5
3012 Bern, Switzerland

Analyses of polar ice cores

D. WAGENBACH
Institut für Umweltphysik der
Universität Heidelberg
Im Neuenheimer Feld 366
6900 Heidelberg, F.R. Germany

Background aerosol composition; glaciochemistry of alpine and antarctic glaciers

J. WEERTMAN
Materials Science and Engineering
Department
Northwestern University
Evanston, IL 60208, U.S.A.

Glaciology; fatigue and fracture of metals

J.W.C. WHITE
University of Colorado at Boulder
Institute of Arctic and Alpine
Research
1560 30th Street
Boulder, CO 80309–0450, U.S.A.

Stable isotope ratios in the hydrologic cycle and in plant tissues

D.S. ZARDINI
Fraunhofer-Institut für
Atmosphärische Umweltforschung
Kreuzeckbahnstraße 19
8100 Garmisch-Partenkirchen,
F.R. Germany

Air chemistry

Subject Index

Author Index

Dahlem Konferenzen Workshop Reports

Physical, Chemical and Earth Science Research Reports (PC)

PC 6 **Resources and World Development: Energy and Minerals, Water and Land**
Editors: D.J. McLaren and B.J. Skinner
0 471 91568 8 958pp 1987

PC 7 **The Changing Atmosphere**
Editors: F.S. Rowland and I.S.A. Isaksen
0 471 92047 9 296pp 1988

Prices on application from the Publisher

WILEY